SCIENCE IN AGRICULTURE

SCIENCE IN AGRICULTURE

The Professional's Edge

by
Arden B. Andersen, Ph.D.

Acres U.S.A.
P.O. Box 9547
Kansas City, Missouri 64133

For information and ordering, contact the publisher:

Acres U.S.A.
P.O. Box 9547
Kansas City, Missouri 64133
816-737-0064

or the author:

Arden B. Andersen
2499 Cannonsville Rd. N.W.
Stanton, Michigan 48888

First Printing 1992

ISBN: 0-911311-35-1
Library of Congress: 92-071851

To the elders, senior citizens of the world who graciously share their wisdom and patiently tolerate the "question boxes" who seek them out.

A NOTE FROM THE PUBLISHER

This book contains an overview of the basic sciences necessary for the farmer to understand the language of the trade followed by their application in growing plants, regenerating soil and producing the highest quality, nutritionally sound commodities possible.

Ag-chemical enthusiasts/activists have long claimed that food in sufficient quantities to feed the world's population cannot be grown without the use of pesticides and herbicides. "We would all starve," they contend, "if we did not use these agricultural poisons."

The truth is that millions of people are starving on food raised under the petrochemical system despite having full stomachs because this food supply is so poorly mineralized. Science points out that low-to-mediocre crop production, weed, disease, and insect pressures are all symptoms of nutritional imbalances and inadequacies, *not* the result of a lack of pesticides and herbicides.

The progressive farmer who knows this science and who adds a measure of common sense will grow bountiful, appealing, tasty, disease- and pest-free commodities without the use of toxic chemicals.

The progressive turf manager who knows this science will have lush, thatch-free, child-safe turf, also without using toxic chemicals.

Today's professional consultants who are "reading" the public's pulse know better than to ignore the changes already taking place, which will reduce and/or restrict their client's use of toxic chemicals. They know that, whether by their vote at the polls or by their buying habits at the market, consumers ultimately determine farm-management practices. As consumers become more educated and challenge today's conventional thinking, consultants must be prepared to meet public demand as well as public law, and at the same time show a farm profit. Combining *Science in Agriculture* with a measure of common sense, they will be able to do just that.

ACKNOWLEDGMENTS

Every work has many individuals in the background who contributed a little or a lot to the manifestation of the work. This book is no exception. I convey my respect and greatest thanks to Nancy Berg for her drawings and art work, Charles Walters, for his continued encouragement; and to Dan Skow, Hersel Robertson, Gene Logue, Jerry Brunetti, Ron Ward and the staff at TransNational AGronomy, and Rose and Ted Baroody, as well as to the many farmers and acquaintances, for their contributions. Special thanks is given to Sue Cooley for her professional editing work.

TABLE OF CONTENTS

Note from the Publisher . . . . *vii*
Foreword . . . . *xi*
Introduction . . . . *xv*
1. Paradigms of Agriculture 1
2. Chemistry 11
3. Physics 31
4. Biology 43
5. Energy: The Basis of Life 60
6. Plant Function 72
7. Concepts in Microbiology 86
8. Microbiology in the Field 116
9. Clay Chemistry 134
10. Carey Reams' Testing and Evaluation Methods . . . 150
11. The Terminology of Carey Reams Explained 169
12. Weeds: Caretakers of the Soil—What They Tell Us . 189
13. Fertility Programming 204
14. Nutrients and Their Basic Functions 230
15. Cultural Management 243
16. Foliar Spray Programming 246
17. Management: Making it Happen 252
18. Turf, Landscape, and Ornamental Plant Care 271
19. Basics, or First Principles 279
20. Common-Sense Principles 301
21. Philip Callahan, Ph.D. 322
22. Electronic Scanner Testing 330
Afterword 336
References 345
Footnotes 347
Appendix 1 354
Appendix 2 357
Appendix 3 359
Index 363

FOREWORD

Do you know that most of us today are suffering from certain dangerous diet deficiencies [1936] which cannot be remedied until the depleted soils from which our food comes are brought into proper mineral balance.

A healthy plant grown in soil properly balanced, can and will resist most insect pests. Tomato and cucumber plants grown alongside each other, some in good mineral balanced soil alternated with plants grown in mineral devitalized soil proved the point. "He [Dr. Charles Northern] had grown tomato and cucumber plants, both in healthy [mineral-balanced soil] and diseased [typical, depleted soil], where the vines intertwined. The bugs ate up the diseased plants and refused to touch the healthy plants." This research shows that insects, in nature's balance, are intended to eat and destroy that which is not fit for human consumption. Thus, in all of our perverted wisdom, we eat that which is worthless in nutritional value, but, garnished with insecticide. It now becomes more nutritious according to medical authorities—Congressional Record, *74th Congress, 2nd Session, SENATE, June 5, 1936.*

In 1958, William Albrecht, Ph.D. Emeritus Professor of Agronomy, Department of Soils, University of Missouri reported: "Over 40 years ago, worldwide samples of soils were studied and established an average viable protein output of 12%; the minimum necessary for animal and human health being 25% viable protein. In the United States the samples averaged not 25%, not 12%, but 6%." During the late 1960s, early 1970s, the U.S. Department of Agriculture released figures showing that the average U.S. soil had a viable protein factor of 1.5% to 3%. Biomedical Critique, Biomedical Health Foundation, Inc., Ocala, Florida, September, 1984, Vol. 5. No.6.

The infamous Green Revolution touted by the Earl Butzes of the twentieth century has nearly bankrupted American agriculture—at least environmentally, if not economically. The Green Revolution implemented and accelerated the mining of our soils and declared a chemical war on nature. Over the past

50 years, half of America's topsoil has been lost. Since 1945, pesticide use has increased tenfold, yet crop loss due to insects has almost doubled—from 7% to 13%. As long ago as 1936, Charles Northern warned of depleted nutrient levels in crops sufficient to cause human health disorders.

Farmers were deluded into thinking that pesticides and chemical fertilizers would solve all their problems. Now, in the last decade of the twentieth century, farmers and American consumers are realizing that the Green Revolution was green only for the chemical manufacturing industry. However, the farmers and the public bought their sales pitch, and it is now our responsibility to rectify the situation.

Farmers must be educated in the basic sciences. In my travels, discussions, seminars, and consultations, I have found that what farmers need most is a general understanding of the basic sciences: chemistry, physics, and biology. Unfortunately, most farmers were never exposed to these sciences in school and have not been educated about them by the Cooperative Extension Services. This is a major reason farmers have been misled so readily by fertilizer and chemical salespeople, university "experts," and farm magazines. A simple discussion about the vices of muriate of potash elicits such questions as "What is muriate of potash?" or "What else is there to use?" or "It is the cheapest form of potash available, so what's the problem?"

Farmers traditionally have been trusting people, and consequently, have believed the information they received from industry and governmental sources. Unfortunately, most farmers did not have enough working knowledge of basic science to distinguish the chaff from the grain.

Growers who were keen observers and good managers began noticing their soils hardening; more weed, disease, and insect problems; more need to supplement animal rations; more animal health problems; and, overall, more input costs per unit of production. These growers began changing their approach, although they still did not know the reasons behind the results.

Much of this lack of education among farmers is a result of society's belief that one does not need much formal education to farm. The contention has been that one goes to school to get off the farm, not to stay on it. The failure of thousands of family farmers over the last few decades is proving just the opposite: One needs knowledge to stay on the farm. Unfortunately, education and knowledge are not necessarily synonymous. Agriculture is the most complex and mentally and physically demanding of all professions. It also has the greatest influence on all mankind.

This book is intended to address the issue of basic science knowledge for farmers. Its purpose is not to make an expert of the reader, but to give him enough basic knowledge to evaluate the claims and reports with which he is bombarded daily and to evaluate his farming operation from the viewpoint of a sustainable agriculture.

INTRODUCTION

Agriculture is the oldest of all industries and one of the most complex and difficult to master and to pursue with success. Farming requires more intelligence than any other industrial occupation. [The farmer's] work is with living things; and living things, whether animal or vegetable, cannot be managed by coercion. They must be humored. They must be understood. .. The farmer is of necessity keenly observant. His whole livelihood depends daily and hourly on the keenness and faithfulness of his observation of a thousand things that the townsman is utterly blind to—Charles Mercier, *A Manual of the Electro-Chemical Treatment of Seeds,* London: University of London Press, 1919, pp. 45-47.

The words Mercier wrote in 1919 apply equally today. Agriculture has been relegated to the rank of a second-class profession, something one does if he is not inclined to "higher" education. As a result, a food chain of second-class nutrition has been developed. In 1919, Mercier stated, "A poor sample of wheat weighs 60 pounds per bushel, an average sample 62 or 63 pounds per bushel, a fine sample 64 pounds per bushel, and 65 pounds per bushel is an extraordinarily fine sample" (p. 18). Today, the standard for quality wheat is 60 pounds per bushel, a poor sample by 1919 standards. In *The Farmer's Every-Day Book,* the Reverend John Blake (1851) pointed out numerous yield reports that would seem impressive to today's farmers. He reported 172 bushels of corn per acre in 1822, followed by 170 bushels per acre in 1823, and 174 bushels per acre on another farm that same year. He reported numerous 118 bushel to 154 bushel-per-acre corn yields from 1822 to 1850, all in New York State, as well as 800 bushels of potatoes per acre in Massachusetts. Blake discussed desirable fertility and soil-management practices that are almost identical to those advocated by Carey Reams, William Albrecht, and other twentieth-century pioneers.

In the early 1930s, Charles Northern, M.D., a gastro-intestinal specialist, pointed out that the diseases of man were directly correlated to soil fertility, particularly the mineral integrity of the soil. He warned that our soils had been seriously depleted of many critical mineral nutrients and impressed on farmers the importance of remineralizing their soils. Rudolf Steiner, Carey Reams, and William Albrecht all made the same recommendation. They also pointed out that nutrition was the key to yield, crop quality, and the control of weeds, pests, and diseases. Have American farmers heeded these repeated warnings and teachings? No, they have ignored them, and now America's soils, food chain, environment, and people are in a situation of great concern. George Watson pointed out that mental as well as bodily action is directly correlated to one's nutrition. Our current state of crime, social aberrations, and terminal disease verifies the degenerated state of our nutrition.

Following World War II, we experienced what was called the Green Revolution. During those years, collective crop yields doubled and tripled, corporate farming became the thrust of political policy, and petrochemical sales became the driving force in agriculture. However, what seemed to be lush and thriving on the surface was rotten to the core. After World War II, petrochemistry was chosen over remineralization in mainline agriculture because a great industry had been built to produce synthetic nitrogen and chemical weapons for war. Rather than dismantling this industry after the war, entrepreneurs sought an outlet to perpetuate this lucrative business. At that time, agriculture was in a position of vulnerability; people were moving to urban areas, and political policy sought a more collective approach to agriculture.

Throughout the next three decades, environmental poisoning, crop demineralization, and whole communities' experimentation with nuclear and electromagnetic radiation went largely unchecked. Although collective crop yields increased dramatically, at the same time the nutritional integrity of the crops greatly decreased. Finally, in the 1980s, it became generally recognized that chemical agriculture had resulted in resistant weeds, diseases, and insects; a polluted ecosystem; and a less-than-desirable food chain. The profitability of agriculture dwindled, and thousands of farmers lost their farms; a trend that continues today. Hence, America's post-World War II Green Revolution has actually been a Brown Revolution.

I believe this trend can be reversed. The original values underlying the family farm are a good place to start. Corporate agriculture does not work, either for a healthy society or for the long term. It may function in a chemical context, but this context itself is self-destructive.

I believe knowledge is the key to returning all of agriculture, not just farming, to its natural state of sustainability and integrity. Science and nature are the basis of knowledge and common sense, and science must be reinstated into agriculture.

Because science is the basis of knowledge, I have included in this book a brief introduction to the basic sciences—chemistry, physics, and biology. I do not consider these subjects electives, but required core curriculum. I then apply these basic subjects to agriculture, discussing the workings of plants and soils as nature intends, and interjecting the teachings of Carey Reams, Philip Callahan, William Albrecht, and others for correlative comparisons. My intention is to convey basic concepts, giving sample programs so each reader can grasp the principles universal to all situations which, when combined with a measure of common sense and ingenuity, reward the steward with success.

Agriculture is an art whose foundation is science. In this book I pull no punches with the pseudoscience espoused by members of the petrochemical/drug industry. We can grow 200 ton dry-weight biomass at 30% true protein and 20 brix refractometer readings per acre per year, without pesticides, used as cattle feed at less than the price of alfalfa or converted to ethanol at less than 70 cents per gallon. This cannot be done using chemical agricultural methodology. We can correct soil salinity, compaction, hardpan, and chemical toxicity with nutrition. Chemical agriculturalists cannot do so. We can deliberately increase crop refractometer readings through nutrition; they cannot. We can control weeds, diseases, and insects with nutrition; they cannot. We can improve the nutrient density, test weight, and storability of commodities with nutrition; they cannot.

In 1965, the *Farm Journal* published an article entitled "Seven Wonders of American Agriculture." It painted a picture of chemical agriculture solving all the problems resulting from insects, diseases, and weeds; producing five-bale cotton; whole farms producing 200 bushel corn; and vineyards producing golf-ball-sized grapes. The author spoke of resolving famine with prompt emergency food relief followed by Peace Corpsmen teaching people in less-developed countries to produce the necessary foodstuffs to feed themselves. By these 1965 standards, petrochemical agriculture apparently peaked in that year, for such results certainly cannot be claimed today. One needs only to reflect on history to realize that petrochemical agriculture, having toxic poisons and salt-based fertilizers as its mainstay, is futile and suicidal. Throughout history, salting the land has been the warfare tactic of conquering nations. In 146 B.C., after Rome's third war with Carthage, the Romans set out to destroy Carthage, particularly because of its competitive grape and wine industry. The Romans systematically leveled the city, killed all its males, and salted the land, terminating its agriculture and any possibility of Carthage's revival. The Romans repeatedly salted the land of enemy nations, knowing that this tactic would cripple those countries. Since Roman times, salting of land through poor management has accounted for the decline of many great civilizations, most notably those of Babylon and the Fertile Crescent.

It seems inconceivable that people, many of them claiming to be agricultural scientists, would deliberately salt the soil, when such a practice has been proven fatal throughout history and is viewed as preposterous by today's professional chemists and petroleum engineers. Fortunately, the condition can be reversed, but not through continued applications of salt.

Having been raised on a farm and educated by the petrochemical agricultural system as a vocational agriculture instructor, and having had some experience in the real world of agriculture, I believe I can accurately describe and discuss both systems. Based on this background, I believe that common sense and a keen understanding of nature are the most important attributes an agriculturalist can have. Theories, formulas, and textbooks are great for erudite discussions with people, but they mean nothing to plants, animals, or nature in general, where we have to achieve our results.

My grandfather would periodically turn his dairy cows out into the meadow so that they could eat various herbs. When he thought the animals had been restored to health, he would return them to the pasture and "green chop." None of my college textbooks or professors mentioned anything resembling this practice, yet this wondrous man, who had only a second-grade formal education, intuitively knew what to do.

As you read this book, I encourage you to keep in mind that it is based on what has occurred and is occurring, not what theoretically could happen. Use it as a handbook to guide you in your decisions until you have exhausted its message. Then turn to the next source of direction to advance to new heights. Keep in mind that when the student is ready the teacher will appear. God, or Nature, is always the omniscient, ultimate source of guidance.

—Arden Andersen, PhD.

—1—

PARADIGMS OF AGRICULTURE

WHATEVER ONE'S TRADE or hobby, he acts within a belief system that sets up the rules and regulations by which he functions. These are called paradigms. Any time there is progress, one's paradigm must and does shift or expand, allowing him to perform in a progressive manner. Over the past fifty or more years, agriculture has functioned, officially anyway, in a paradigm whose philosophy says that nature is flawed and must be controlled with man-made materials. This paradigm has placed agriculture in a state of constant war with nature, continuously battling pests and diseases.

A new paradigm is gaining acceptance due, in no small part, to solid science. To understand the unraveling of the old paradigm, one must understand the fundamental aspects of both the old and the new. One might say the old and new are analogous to singing monotone versus making music.

In this chapter, I will describe the two fundamental aspects of both real-world and conventional agriculture: the model and the logic. The real-world and conventional-agricultural model and logic will be compared to linear and nonlinear physics. I will give field examples and show how science demands the practice of biological agriculture, regardless of what methodology the farmer employs.

No matter what we do in life, there is a model of what we think things should be, by which we judge our position, and there is a logic by which we solve problems and execute our actions. The model is essentially our standards, and the logic is essentially our science.

In conventional agriculture, the model has the following elements:

- Food and fiber production constitutes a war.
- Nature is the adversary.
- Insect, disease, and weed pest are "normal" and evidence the wrath of God on mankind.
- Soil is inanimate.
- Nature is random, unintelligent, and flawed.
- Man knows a better way.

The logic of conventional agriculture, more properly described as the dogma of the conventional agriculture "church," has the following elements:

- Reductionistic—The whole equals the sum of its parts and nothing more.
- Linear—Based on straight-line, in-vitro observation and principle; what you get out is only equal to or less than what you put in—purely entropic.
- If all else fails, get a bigger hammer.

On the other hand, the model of real-world agriculture has the following elements:

- Food and fiber production is a part of nature, where peaceful coexistence is the rule.
- Nature is the guide and guardian.
- Insect and disease pests are nature's garbage collectors; weeds are nature's caretakers.
- Soil is living and dynamic, analogous to the ruminant

digestive system.
- Nature is ordered, intelligent, and perfect.
- Nature is the example to follow; she possesses the ideal plant, soil, and animal characteristics.

The logic of real-world agriculture contains the following elements:
- Wholistic—The whole is greater than the sum of the parts.
- Nonlinear—Keyed to tuning, based on harmonics, in-vivo observation, and principle.
- Energetics is the fundamental basis of all physiology, animate or inanimate.

Summing up the two systems, one might say that the conventional system is linear, functions with single variables, and adheres strictly to the theory of relativity, has no harmonics, and is a driven system. In essence, it is a messenger (symptom)-oriented system. The real-world system, on the other hand, is nonlinear, functions with many variables, makes use of harmonics, and is a functioning system. In essence, it is a message (cause)-oriented system.

The conventional system, being the old paradigm, is solidly instituted in the annals of modern society. It therefore needs no further elaboration. The real-world system, being the new paradigm, warrants further elaboration, to establish its right as successor. To do this, I will begin by pointing out three landmark studies confirming Lakhovsky's and Abrams' contentions that biological systems are energetic and nonlinear.

The first such study was done by Philip Callahan in 1956. He showed that ants would home in on a candle flame when there was no barrier between the candle and the ant, and even when he placed a plastic sheet between the candle and the ant. However, when he placed a glass sheet between the candle and the ant, the ant did not home in on the candle.[1] Vlail Kaznacheyev demonstrated a similar phenomenon thousands of times in the 1970s with cultures in petri dishes, showing that responses between the cultures were observed when quartz sheets separated them but were not observed when glass sheets separated them.[2] Both of these studies demonstrated that ultraviolet and infrared radiation played a part in organism function because these radiations would pass through quartz and plastic but not glass.

Then in the mid-and late 1980s, Fritz-Albert Popp demonstrated that photons, particularly ultraviolet photons, are the medium of biological communication. He also showed that these photons precede chemical and physical action and that any alteration or response observed in chemical or physical matter is preceded by a conversion of the biophotons associated with the matter.[3] Popp's work was a magnificent capstone to the growing mountain of proof that biological systems are fundamentally energetic.

A simple side note to this premise is the "polar solar phenomenon." Northeastern University scientists found that polar bears, through their hair, are about 90% efficient in using ultraviolet radiation to maintain body heat.[4] The quark for the uninformed reader in this finding lies in what is called the ultraviolet catastrophe. This is the fact that very little heat energy is derived from ultraviolet radiation, certainly not enough to maintain directly the polar bear's body heat. Linear conversion simply does not account for the benefit gained by the polar bear. Only when nonlinear considerations are taken into account can the polar-bear phenomenon be explained. Biological cells are energy accumulators and converters.[5] But they are nonlinear in these capacities. As such, they evolve and are subject to harmonics. The collection of ultraviolet radiation allows for a kindling effect of all the harmonics associated with the ultraviolet collection. The polar bear's hair is a very good ultraviolet wave guide.

Warren Hamerman's biophysical studies in 1989 lend further credibility to harmonic phenomena in living organisms. He laid out the musical-scale harmonics correlated to biological processes. Hamerman found that DNA (material that forms the gene) is the "tuning fork" for the rest of the biological systems, e.g., photosynthesis, mitosis, vision, respiration, protein synthesis, and neurotransmission. Most interesting is that the frequency of the DNA signal at 1.128×10^{15} Hz (cycles

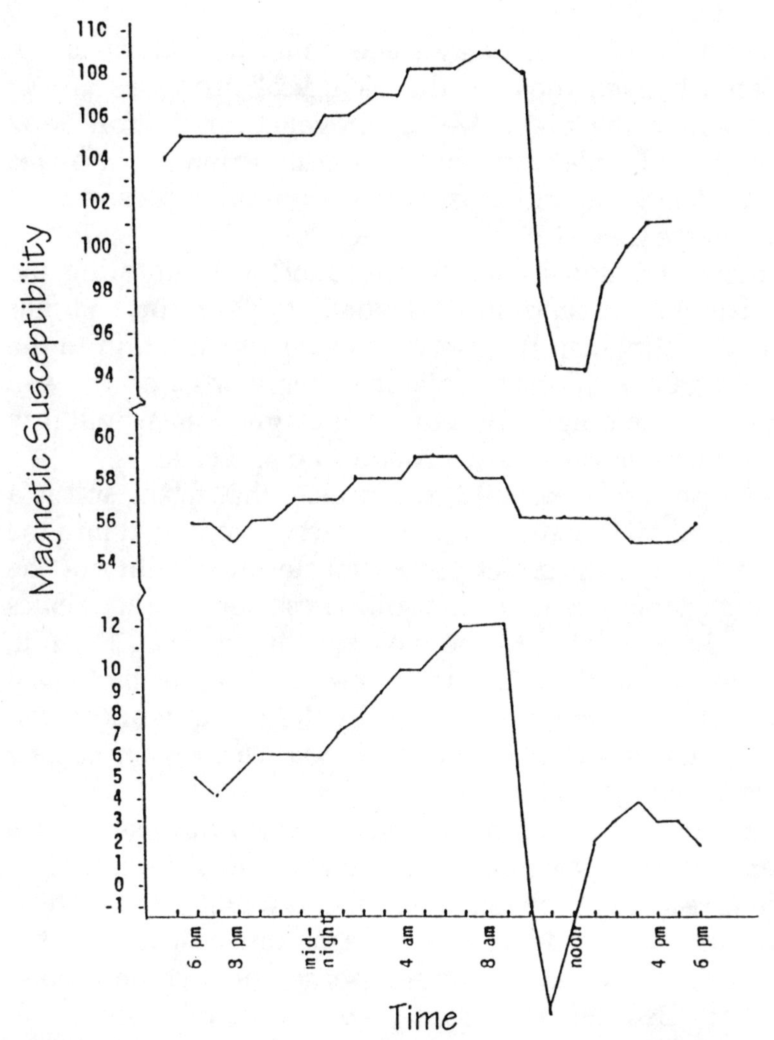

The above graph shows 24 hour magnetic susceptibility readings for a good California soil (top), a good Indiana soil (middle), and a poor Indiana soil (bottom). Notice that all three dip during the hottest part of the day when solar radiation is greatest. The California soil, though it has the highest magnetic susceptibility, still dips considerably. This is because it has the mineral spectrum for initial magnetic susceptibility but lacks the humus system to utilize the collected solar energy and to maintain its stability. The poor Indiana soil even goes negative (diamagnetic) during greatest solar radiation. It neither has the initial mineral nor the humus system. The good Indiana soil, however, has a fair initial mineral base for a fair magnetic susceptibility and as well has a fair humus system to maintain some stability during greatest solar radiation. This means that the Indiana soil is best equipped to utilize solar radiation while the other two soils are not and the poor Indiana soil even becomes a repeller of solar radiation making it unable to utilize this energy. Mineral is important to establish the magnetic suseptibility characteristics and humus is important in allowing the soil to utilize what it is now collecting. Both components are necessary for optimum fertility.

per second) is exactly 42 octaves above middle C at 256 Hz on the piano. In fact, most of the biological processes are 42 octaves above the piano keys. Mitogenetic radiation is 42 octaves above F at 341 Hz, photosynthesis action is 42 octaves above C at 128 Hz, and biosphere maximum radiation is 42 octaves above C at 64 Hz.

Harmonic relationships also are found in monitoring the soil's magnetic susceptibility. The daily cycles fit into a lunar cycle and distinguish themselves in direct proportion to the biological integrity of the soil. In other words, as the soil becomes more biologically active, the magnetic susceptibility cycles seem to become more pronounced and clear.[6]

From a physics viewpoint, we can say that fertile soil is a living biological system. Plants are antennas plugged into the soil and function in direct proportion to the stability of the energetic characteristics of the soil. These soil characteristics are directly correlated to the biological integrity of the soil. Therefore, soil fertility is directly proportional to biological activity. This is further confirmed by the Soviet work in soil microbiology reported in the book *Soil Microorganisms and Higher Plants* by Krasil'nikov.

According to the old paradigm of agriculture, soil is inanimate and therefore linear. The new paradigm demanded by science says that soil is biological and therefore nonlinear. Nonlinearity accounts for some of the interesting activities that take place, particularly pumped-phase conjugation, something Tom Bearden discusses regularly in relationship to military weapon applications.

In the field we can compare conventional fertilization to real-world fertilization using a physics model. In the conventional set-up, we have a strictly linear system and approach. Calcium will be used as the example. We take a conventional soil test and arrive at a soil pH. If the soil pH is below 6.5, the conventional recommendation is to add lime (the conventional source of calcium and/or magnesium) to raise the pH to between 6.5 and 6.8. The amount required is arrived at by determining the soil cation exchange capacity and then calculating the number of pounds, usually tons, of lime, based on the lime's neutralizing capacity compared to calcium carbonate equivalent (CCE), that are needed to raise the pH the desired amount. This looks workable on paper and can be programmed into a computer and reported in textbooks as something of real significance. It is significant for paper work, but plants and microbes do not read books or computer printouts.

With this linear approach we have a base calcium level (pounds per acre) in the soil, but that is considered secondary to the pH. Because of the pH, several tons of lime are added to the soil. The result is usually that pH rises, but the pounds per acre of calcium available to the plant is only slightly different or often remains unchanged. People neglect even the basic chemistry principle—that nutrient calcium is made soluble in acid solutions. Calcium hydroxide is that which becomes more soluble in alkaline solutions. The linear approach here completely misses the fact that pH is the result, not the cause, of nutrient interaction in the soil. Therefore, the linear approach to fertilization may change the pH but not necessarily the calcium available for plant and microbial growth. (See diagram A.)

Conventional (linear) evaluation and solution is analogous to evaluating the attunement of a symphony using a decibel meter—the job is broader than the tool—and then attempting to alter this attunement by increasing the volume of the symphony.

Approaching the same situation from a nonlinear perspective, we test the soil and notice the pounds of calcium per acre that are available. We then select some material that will pump the unavailable calcium, giving us a greater calcium availability in the end. This can be done simply, as with liquid calcium, or it can be done with more sophistication, as with sugar and vitamin B_{12}. (See diagrams B and C.)

The real-world system, being nonlinear, is multivariable, requiring measuring tools and methodology of an equal variability. The key to attunement in this system is one's proper perspective, that being nonlinear, one of a naturologist.

Conventional Fertilization (A)

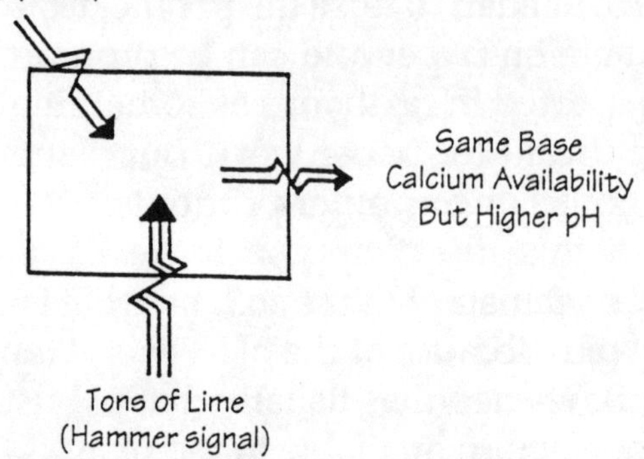

Real World Fertilization (B)
(Single Input)
Non-Linear Input Stimulates Chain of Events

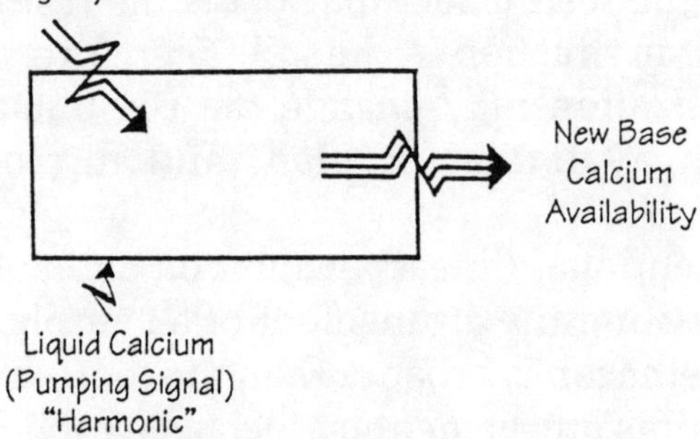

Real World Fertilization (C)
(Multiple Inputs)

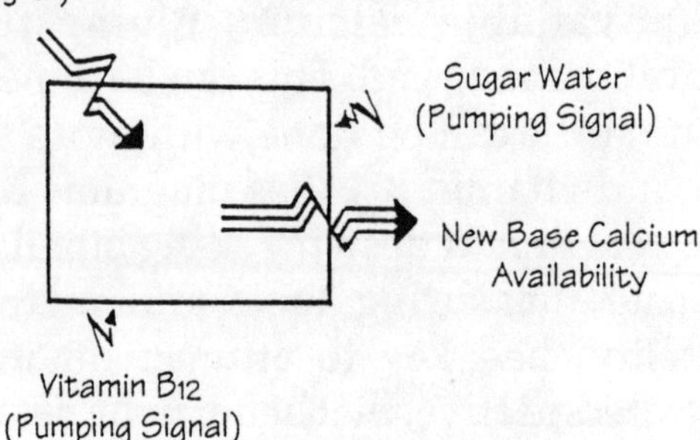

We must learn to fertilize our nonlinear system nonlinearly. All fertilizers, from compost to pure acid/caustic, are potentially nonlinear. The manner in which they are used determines whether one is dealing in a linear or a nonlinear context.

One of the most effective ways that soils, plants, and fertilizer materials can be evaluated with the new paradigm in mind is by using radionics. Unfortunately, radionics can be and often is applied linearly. This methodology follows entropy and its logical conclusion. Correctly used, radionics would be applied nonlinearly, both in evaluation and treatment. This methodology follows neg-entropy and its logical conclusion. In a nutshell, one must look for causes, understand the basic principles of the life processes, and then find solutions that aid, not circumvent, nature in the manifestation of the desired outcome. Killing weeds, insects, and disease organisms is a linear approach, whether using radionics or toxic chemicals. In some cases, these seem to be appropriate practices so that one can at least harvest a crop to sell to the market, pay the bills, and farm again next season. However, if the nutritional cause of these symptoms is not addressed, one is working strictly within the old paradigm, and, regardless of all the treatment and sophisticated gadgetry, entropy will prevail because one is working in the entropic side of physics. War is inherently entropic, and war between man and nature is the foundation of the old paradigm—a war that will never be won.

There is an old story that seems to be the epitome of linear perception, the old paradigm of conventional agriculture. Six blind men each attempted to describe an elephant after touching it. Each one, however, touched a different part of the elephant and, consequently, each described the elephant differently. One man contacted the elephant's leg and described the elephant as a tree; another felt the elephant's trunk and proclaimed it was a large snake; another touched the elephant's ear and described the animal as a great fan; another man felt the tail and described the elephant as a hanging rope; another touched the elephant's side and contended it was a great wall; and the sixth man contacted the elephant's tusk and described the animal as a spear.[7] Each description was accurate from each man's viewpoint, but none described the real elephant, showing that the whole is much greater than the sum of its parts.

The moral of the story for purposes of this discussion is that the old paradigm, although accurate from its individual perspectives, has missed the true portrait of agriculture and all living systems. It has a flawed model and logic, rooted in entropy (perpetual degeneration), and therefore will inevitably decline. The new paradigm allows us to get at the causes of disharmony by selecting and using the proper tools, materials, and methodology to attune nature's symphony.

In our studies of science and nature Heisenberg once said, "What we observe is not nature itself, but nature exposed to our method of questioning."[8]

—2—

CHEMISTRY

CHEMISTRY IS THE STUDY of the composition, structure, and properties of matter and the changes that matter undergoes. It is the science in which the composition and interaction of compounds and elements, as well as the synthesis of natural and artificial compounds, are investigated. Chemistry is divided into several branches of study. Inorganic chemistry is the study of compounds that do not contain carbon, whereas organic chemistry deals with carbon compounds. Biochemistry is the study of chemical processes in living systems, and stereochemistry is the study of the structure and form or shape of chemical compounds.

Every agriculturalist should be familiar with several terms and definitions that are important in discussions of chemistry. Matter is anything that occupies space and has mass. Mass is the quantity of matter an object contains. Energy refers to the capacity to do work; this does not indicate what energy is, but rather, what it does. A solid is rigid; under most conditions, it has a definite shape and constant volume. A liquid flows and assumes the shape of its container. For practical purposes, it has constant volume. A gas takes the shape and volume of its container. It is compressible and capable of infinite expansion.

Chemical properties are the characteristics of a substance relative to its chemical reaction with other substances, such as being an acid, a base, a reducing or oxidizing agent, stable or unstable, and so on. Chemical changes take place whenever elements unite to form a compound. For example, a chemical change occurs when charcoal is burned. The carbon in the charcoal is combined with oxygen, giving off heat and carbon dioxide. Milk sours when the milk sugar is converted to acetic acid. This is also a chemical change. A physical property is a characteristic of a substance that does not change its composition. Physical properties include color, hardness, crystalline form, ductility, malleability, melting point, boiling point, density, electrical and thermal conductivity, and specific heat (the amount of heat per unit of mass needed to change the temperature one degree).

Elements are pure substances that cannot be decomposed by a chemical change. Eleven elements constitute about 99% of the earth's crust and atmosphere. Oxygen makes up nearly one-half and silicon nearly one-fourth. Only about one-fourth of the elements ever occur in nature in their free state (purely alone); the rest occur in chemical combinations with other elements. Compounds are substances that are composed of two or more different elements and can be decomposed through chemical changes.

The smallest particle of an element is an atom. For a compound in a stable form, the smallest component is called a molecule. If one could lay one-hundred million molecules side by side, they would form a row about one inch long. Molecules are two or more atoms combined to form a stable, electrically neutral unit, e.g., H_2O, a water molecule. Molecules are the smallest form in which a compound can exist as a stable, independent unit. An atom is the smallest particle of an element that can enter into a chemical combination. Two atoms of hydrogen combined with one atom of oxygen make one molecule of water (H_2O). An atom is said to resemble the solar system in miniature. It has a nucleus (analogous to the sun), with electrons (analogous to the planets) spinning around the nucleus. The nucleus is said to be composed of protons (positively charged particles) and neutrons (neutral particles).

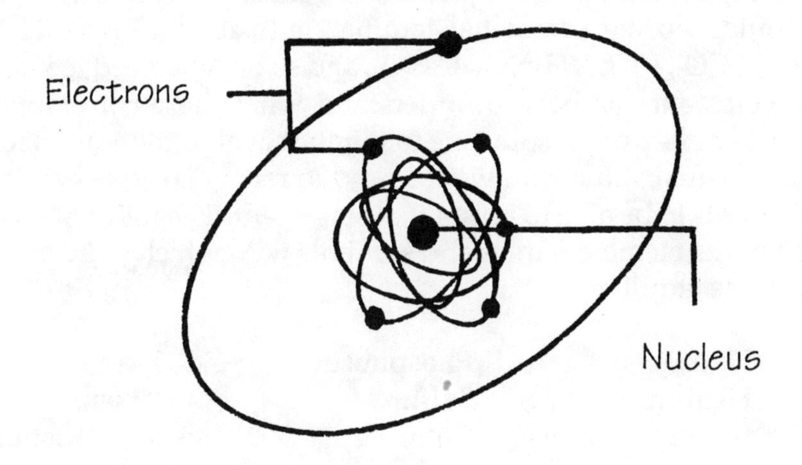

Thomas Bearden discussed the atom in a somewhat different way, explaining that the nucleus is in constant flux, charging and discharging in its interaction with the ether in which it exists. Thus, it can be seen that classical explanations of scientific terms are not absolute or necessarily totally correct. Rather, they serve as a convenient basis for discussion.

In current classical terms, the subatomic particles, namely the proton, neutron, and electron, are the units that constitute an atom. As mentioned, the proton is a positively charged particle, and a neutron is a neutral particle. If one were to analyze a neutron, one would find that it is the union of an electron (a negatively charged particle) and a proton. Electrons are the orbiting particles of the atom and are usually associated with electric current flow and with the ionization/deionization of compounds. The electron is the main basis of energy exchange between different chemical materials, and in living systems. An electron is much smaller than either a proton or a neutron; it is about five ten-thousandths of the mass of a proton. One theory is that an atom is actually a spherical energy system in a precarious balance.

This leads into the discussion of the periodic table of the elements, a systematic listing of 109 currently known elements, 90 of which are found in nature. The periodic table is to chemistry what the multiplication table is to mathematics. You should become somewhat familiar with it. If you see H_2O, H_2O_2, CO_2, KCl, NH_4NO_3, $CaSO_4$, and so on, you need to know the element symbols to understand what these compounds are. For example, if someone told you that all nitrogen sources are of equal value, you would need to recognize the chemical elements to be able to refute that statement. The following is a list of the element names and symbols with which you should become familiar:

H	Hydrogen	P	Phosphorus	Zn	Zinc
He	Helium	S	Sulfur	As	Arsenic
B	Boron	Cl	Chlorine	Mo	Molybdenum
N	Nitrogen	Ca	Calcium	Ag	Silver
O	Oxygen	Cr	Chromium	Sn	Tin
F	Fluorine	Mn	Manganese	I	Iodine
Na	Sodium	Fe	Iron	Ba	Barium
Mg	Magnesium	Co	Cobalt	Au	Gold
Al	Aluminum	Ni	Nickel	Hg	Mercury
Si	Silicon	Cu	Copper	Pb	Lead
Rn	Radon	U	Uranium	Pu	Plutonium

These are the elements most commonly mentioned in agriculture and daily life. Associated with each element is an atomic number and an atomic weight. Technically, the atomic number refers to the number of protons in the nucleus, but simply think of it as the size ranking of each element in the periodic table, with hydrogen being 1 and plutonium being 94. If I were to ask for the sixth element in the table, you would tell me carbon.

Atomic weight is fairly self-explanatory; it is the weight of each element, but relative to carbon. Carbon has an atomic

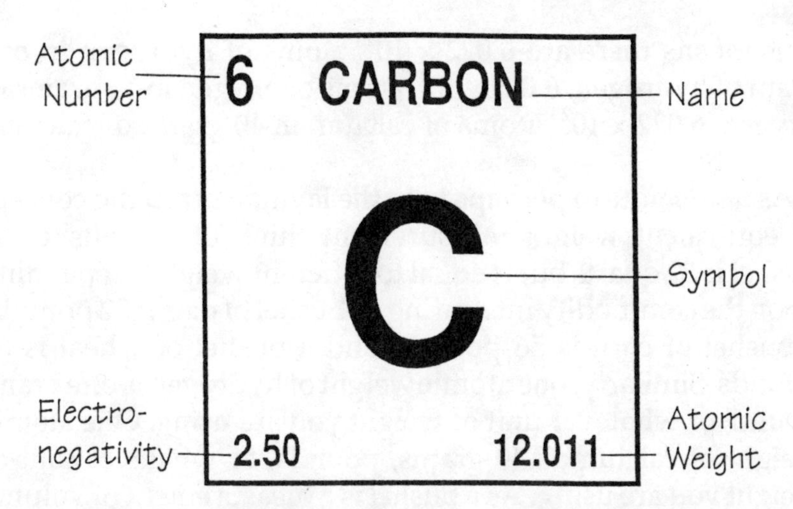

weight of 12. Hydrogen is 1⁄12—the weight of carbon—so its atomic weight is 1. This number is dimensionless. There are no units like grams or ounces because atomic weight is only relative to carbon—12. Relative weight is used because the weights of atoms and molecules in grams are quite cumbersome. It has been determined experimentally that one atom of hydrogen weighs 1.67×10^{-24} grams (0.000 000 000 000 000 000 000 001 67 grams). A carbon atom would weigh 12 times 1.67×10^{-24} grams or 2.004×10^{-23} grams. These minute weights are simply too small to comprehend, thus relative weights are used.

In measuring out a given quantity of a material, equivalent weights are used—either the atomic weight or the molecular weight in grams. This information is used to calculate the quantities of elements or compounds that are needed to make the desired product. This also explains why phosphoric acid is heavier than water and why a brick of gold is heavier than a brick of aluminum. Phosphoric acid has a molecular weight of 98, whereas water has a molecular weight of 18. Gold has an atomic weight of 197, whereas aluminum has an atomic weight of 27.

To illustrate further how small atoms are, it has been shown that there are 6.022×10^{23} atoms in one gram atomic weight of an element—that is 602,200,000,000,000,000,000,000 atoms.

This means there are 6.022×10^{23} atoms of hydrogen in one gram of hydrogen, 6.022×10^{23} atoms of oxygen in 16 grams of oxygen, 6.022×10^{23} atoms of calcium in 40 grams of calcium, and so on.

As a side note to perhaps help the layman grasp the concept of equivalent weight measurement think of the bushel. A bushel is 32 quarts but is equal to different weights depending upon the commodity measuring. A bushel of oats is 32 pounds, a bushel of corn is 56 pounds, and a bushel of wheat is 60 pounds. Similarly, one atomic weight of hydrogen is one gram, pound, or whatever unit of weight you are using. One atomic weight of calcium is 40 grams, pounds, or whatever unit of weight you are using. As a bushel is a measurement of volume having different weights depending upon the commodity measured, think of atomic weight as a measurement of volume—a given number of atoms as just mentioned—having different weights depending upon the element measured. Once we know the symbols for the elements, we can write and interpret chemical formulas, at least basic ones. H_2O is the chemical symbol for water; it means that there are two hydrogen atoms combined with one oxygen atom. H_2O is termed a molecule, and technically we would refer to its molecular weight rather than its atomic weight. Molecular weight is the sum of the individual atomic weights of each of the atoms comprising the molecule. The molecular weight of water is calculated by taking two hydrogen atoms at an atomic weight of 1 each, and adding one oxygen atom at an atomic weight of 16, for a total molecular weight of 18. CO_2 is the chemical symbol for carbon dioxide. It means there is one carbon and two oxygen atoms. The molecular weight of carbon dioxide is 44. N_2 (meaning there are two atoms of nitrogen) is the symbol for nitrogen gas, like that found in the atmosphere. The molecular weight of nitrogen gas is 28. O_2 is the chemical symbol for oxygen gas, like that found in the atmosphere. It means there are two atoms of oxygen combined. The molecular weight of oxygen gas is 32. Cl_2 is the symbol for chlorine gas, meaning there are two atoms of chlorine. Its molecular weight is 70.

Now, when someone tells you that the chlorine from muriate of potash just evaporates into the air, you will know better because the molecular weight of chlorine gas (Cl_2) is 70, compared to the lighter weights of H_2O (18), CO_2 (44), N_2 (28), and O_2 (32), which are the major components of air. Thus, because chlorine gas is heavier than air, it will remain close to the ground.

Hydrogen (H_2) originally was used in the zeppelin airships because it is the lightest element in the atmosphere, with a molecular weight of 2. Helium is now used instead of hydrogen because it is nonexplosive.

The combinations of atoms described above are called molecules, which were defined earlier in this chapter. Sometimes you will see symbols written like NH_4^+ and NO_3^- or Na^+ and Cl^-. These are called ions—the ammonium ion, nitrate ion, sodium ion, and chlorine ion, respectively—not molecules, because they are charged either negatively or positively. Being charged, they will conduct electricity and do not hold together as a complete molecule when dissolved in water. They are considered salts, according to the chemical definition of a salt, i.e., a compound made up of positive and negative ions. In water, salts ionize or split into their separate positive and negative ions. For example, sodium chloride (table salt—NaCl) ionizes into separate ions of Na^+ and Cl^-. Ammonium nitrate (NH_4NO_3) ionizes in water into separate ions of NH_4^+ and NO_3^-. These compounds remain unified only in the dry form. They separate once they are dissolved in water.

In contrast, other types of molecules, like nitrogen gas (N_2), remain intact when dissolved in water. They do not ionize into N^+ and N^- ions. As a result of ionization, salts are readily reactive in solution—in water. They will react to and exchange energy with other compounds or ions in that solution. An example of this is the reaction for making hot-mix fertilizers. The starting components for making a hot mix of 9-18-9, for instance, would be potassium hydroxide (KOH), phosphoric acid (H_3PO_4), and usually urea (H_4N_2CO). The KOH ionizes in solution into K^+ and OH^- ions, and the H_3PO_4 ionizes into $3H^+$ and PO_4^{---}. The ions react to form the salt potassium phosphate (K_3PO_4) and water. The H^+'s combine with the OH^-'s to form the water. This reaction is called neutralization, which gives off heat, hence the name "hot mix" fertilizer. The urea does not ionize; it simply dissolves and remains in solution, thus increasing the nitrogen value of the mix. The result is a mix of 9-18-9 or whatever other hot mix number that has been formulated containing urea, potassium phosphate, and water.

The positive or negative charges on the various ions result from the ions gaining an extra electron, which gives a negative charge on the ion (called an anion), or losing an electron, which gives a positive charge on the ion (called a cation). The charges on a compound are important because components with opposite charges attract to stabilize each other.

It is often said that the soil is full of nutrient ions. In general, the interaction of nutrients with soil or clay particles occurs due to the nutrients ionizing to anions and cations. pH is discussed, using anions and cations. It is said that adding lime having calcium or magnesium cations (Ca^{++} or Mg^{++}) will "sweeten" the soil because the calcium and magnesium cations will replace the hydrogen cation (H^+) on the negatively charged clay particle (colloid). This raises the soil pH in some cases, provided there is not too much sodium (Na^+) to compete with the calcium and magnesium or too many anions like Cl^-, NO_3^-, and SO_4^{--} to compete for the calcium and magnesium. Adding high-calcium lime, one in which the calcium carbonate component in extremely dominant to a high-magnesium soil might actually lower the pH. This can also happen in high-sodium soils.

Ammonium can pick up the hydrogen ions (H^+), forming ammonium ions (NH_4^+), and then replace the other hydrogen ions (H^+) on the soil particles, thus raising the pH. Over the long term, the pH might decline in the soil solution due to the NH_4^+, knocking the H^+ into solution and calcium being used up, thus lowering the pH. The colloid pH with the NH_4^+ attached merits a pH increase.

pH is a logarithmic measure of acidity or basicity. Pure water has a pH of 7. Water is H_2O or H-O-H. The H-O-H molecule will break into two parts, H^+ and OH^- ions, giving a mixture of H-O-H, H^+, OH^- in solution. When there are equal numbers of each ion, the pH is 7. If there are more H^+ ions than OH^- ions, the pH is less than 7 in multiples of 10. A pH of 6 has 10 times as many H^+ ions as a pH of 7. pH is actually the negative log of the hydrogen ion concentration. In other words, a pH of 7 means the concentration of H^+ ions is 10 to the minus 7 or 0.0000001 molecular weights per liter or 0.0000001 gram of H^+ ions per liter of water. A pH of 6 means 10 to the minus 6 or 0.000001 molecular weights of H^+ ions per liter of water. A pH of 1 means 10 to the minus 1 or 0.1 molecular weights (0.1 gram of H^+ ions) per liter of water. A pH of 1 also means that there are 10 to the minus 13 or 0.0000000000001 molecular weights of OH ions per liter of water. Whatever the pH, the sum of the H^+ concentration plus the OH^- concentration must equal 10 to the minus 14, or the negative log of this must equal 14. The key thing to understand is that pH is a convenient reference relating the concentrations of hydrogen (H^+) and hydroxyl (OH) ions to each other.

Let us briefly review logarithms now. The log, short for logarithm, is a shorthand method for writing very large or very small numbers. We will stick to base ten for now. Ten raised to the second power (10^2) means 10 squared, which equals 100. In log terminology, we would write log 100 = 2. This means: What number is 10 raised to, to equal 100? The answer is 2. Another example is: 10 raised to the minus 6 (10^{-6}) is $\frac{1}{1,000,000}$ or 0.000001 or log 0.000001 = -6. What power is 10 raised to, to equal 0.000001? The answer is -6. A pH of 6 is actually $-\log [H^+] = 6$ or $\log [H^+] = -6$ or $[H^+] = 10^{-6}$.

I recommend that you review the use of logarithms from your high school mathematics because it will help you understand what pH really means and, most important, that pH is not a measuring stick for calcium. You will realize that anything that alters the H^+ and OH^- ion ratios will alter pH. A classical example is distilled water. Immediately after distillation, distilled water has a pH of 7 and a pOH of 7. How much calcium is available in distilled water? None! If this distilled water is exposed to air, the carbon dioxide (CO_2) in the air mixes with the water, causing the pH to drop to about 6.4. How much calcium has been removed or used up? Again, none! There was none in the distilled water at the beginning. If we were to add some household ammonia to this same water, the pH would jump to 9 or 10. Did we add any lime? No! Farmers have been led to believe that pH changes only when lime is removed or added. This is simply not the case. pH merely indicates the relative concentration of hydrogen.

Carey Reams, as an ag consultant, used pH in a different way. He looked at pH as a measurement of the resistance in the soil. He observed that the higher the pH, the greater the resistance there was and the more difficult it was to get energy to flow, particularly if the pH was somewhat alkaline, in the 8 or 9 range, resulting in nutrient imbalances. On the other hand, he observed that if the pH was moderately low, below 6, there was not enough resistance. This exchange allowed the energy to flow too readily, making it difficult to contain it, again resulting in apparent nutrient imbalances. This seems to be a practical and workable use of pH, for it addresses the reality of how plants grow through energy exchange. In essence, pH is the result of the nutrient interaction, not the cause. When the nutrient ratios are balanced, the pH will stabilize automatically in the correct range.

Getting back to terms, understand that several of the terms that are encountered in chemistry are frequently used in agriculture. In particular the term anhydrous is almost a household word to many farmers, but few know what it means. Anhydrous literally means, "no water." Hence, anhydrous ammonia means pure ammonia (NH_3) with no water. It is labeled 82-0-0, meaning 82% nitrogen, 0% phosphate (P_2O_5), and 0% potash (K_2O). Take the atomic weight of nitrogen (14) and divide this into the sum of the atomic weights of ammonia, nitrogen (14) and 3 hydrogen (1), multiply by 100, and you get 82% [$14/(14+1+1+1) = .82 \times 100 = 82\%$]. Aqua ammonia results from bubbling anhydrous ammonia (NH_3) into water. It actually forms ammonium hydroxide (NH_4OH) in water (H_2O) and is generally about 28% nitrogen (N). Anhydrous ammonia is very aggressive in its quest to attract water; it is a strong base and seeks to stabilize. When anhydrous ammonia contacts the soil, it attracts the most available water it can, which generally dehydrates the local area. In doing so, microorganisms are inactivated or killed. The process dries out the medium. If you get anhydrous ammonia on your skin, it burns. This burning sensation is actually rapid dehydration of the flesh, resulting in the denaturing (destruction) of the skin protein. If the NH_3 contacts organic matter, it decomposes the organic matter, making it soluble in water and thus leachable.

Recall the discussion of anions and cations (e.g., Na^+, Cl^-, NH_4^+, NO_3^-). Ions are water soluble, and when the NH_3 reacts with the organic matter it forms an NH_4^+– humus complex. When NH_3 pulls water and forms NH_4OH or NH_4^+, it will replace calcium (Ca^{++}) ions on the soil colloid, releasing them into solution, or it will make the tied-up calcium hydroxide soluble, releasing it into solution. This gives a good response to the crop as long as a reserve of humus and calcium is present. This approach inactivates or kills the microorganisms in the process, so that the humus will not be replaced nearly as fast as it is being burned off.

In addition, since farmers are told to apply calcium according to pH, they will not replace the calcium because the pH does not drop in proportion to the calcium depletion. The ammonia also causes denaturing of soil-protein complexes, which reverses the sequestering of salt ions, namely sodium (Na^+), thus allowing them to salinate the soil; this contributes to higher pH yet calcium is being depleted. The soil pH will drop in proportion to the amount of conversion of the NH_3 to either NO_3^- or N_2 resulting in H^+ ions being added to the soil solution compared to the amount of Na^+, particularly, and other alkalizing salts which are de-chelated or de-sequestered, so to speak, in the soil as a result of the anhydrous ammonia use.

Using anhydrous ammonia is something like putting racing fuel in the family car. It will run like a scared cat for awhile, but eventually the repairs or replacement will cost more than the short-lived thrill. More than 50% of America's topsoil has been lost over the past fifty years. This is why anhydrous ammonia should not be used directly on the soil. Instead, it should be mixed with water to form aqua ammonia and a carbohydrate like sugar or molasses to help retain it in the soil, and some humic acid to help chelate it for better use rather than reducing further the soil's already depleted humic acids.

Picking up with new terms again we recognize that many compounds like NH_3 are hydrophilic, meaning they love water. Oils, however, are compounds that are hydrophobic, meaning they repel water. If you get a substance to accept water, you have hydrated it. The term "hydrated lime" means that calcium oxide (CaO) has had water added to it to get $Ca(OH)_2$. Its proper name is calcium hydroxide. Dehydrated lime, burnt or calcined lime has had the water removed and is termed calcium oxide (CaO).

Most farmers use carbonated calcium (calcium carbonate, $CaCO_3$) and/or magnesium ($MgCO_3$, $CaCO_3$-$MgCO_3$, termed dolomite). These are just CaO or MgO plus carbon dioxide (CO_2). All biological systems are organic, therefore it is preferable to use nutrients in organic rather than inorganic forms. Calcium carbonate ($CaCO_3$), though not technically considered an organic chemical, is preferable to dehydrated lime (calcium oxide, CaO), hydrated lime (calcium hydroxide, $Ca(OH)_2$), or even gypsum (calcium sulfate, $CaSO_4$), if one is seeking the nutrient calcium. That is true for any nutrient except phosphorus. Phosphorus (P), actually as phosphate (PO_4^{---}), is used in both its inorganic and organic forms in the energy-storage and transfer mechanisms of living systems. However, when phosphorous is supplied and absorbed in the inorganic form, it reacts prematurely, particularly with calcium, causing undesirable calcium and phosphate loss. Carbon often acts as a buffer to take the edge off certain compounds. Ultimately, phosphate would be in the phospho-carbonate complex. Many dangerous compounds contain carbon, but relative to fertilization and plant feeding, carbon is desirable.

The presence of carbon in a mix is probably the best indicator of biological compatibility. It eliminates the following from the list of usable products: anhydrous ammonia, straight N-P-K's, high concentrations of strong acids such as phosphoric or sulfuric, and strong bases like potassium hydroxide. They simply cannot be used as harsh materials and must have carbon mixed with them. Note that I said *high concentrations of acids and bases.* When diluted sufficiently in water, acids and bases may be tolerated, although they are not always preferred. A carbohydrate still needs to be added.

Why not use concentrated acids and bases? Have you ever spilled battery acid on your skin? It burns, that is, it neutralizes and thus denatures the skin proteins. This is similar to what happens to plant and microbial proteins in the soil when such materials contact them. Strong bases like NH_3 have already been discussed and are similarly destructive to living proteins. The soil must be viewed as a living system, susceptible to the same hazardous chemicals as people.

It is quite simple to get carbon into a fertilizer mix. Carbon is in carbohydrates. Common carbohydrates are sugar, molasses, humic acid, humates, fish meal, seaweeds, algae, yeasts, enzymes, biological brews, whey, and so on. The purpose of fertilization is to feed the soil and microorganisms, and to energize plants. Therefore, it is important to provide nutrients in such a way as to minimize the amount of energy required for assimilation and to maximize net energy. In every case, nutrients eventually combine with an organic compound to become useful to the living system. By providing a carbon source in the fertilizer, the microorganisms or plants need to find carbon is reduced. This is most important with nitrogen. Nitrogen is a key component of amino acids and, subsequently, proteins. But nitrogen must be combined with carbon, hydrogen, and oxygen in order to form an amino acid. Amino acids are then combined, as discussed in the chapter on biology, to form proteins. This is rarely discussed in connection with fertilization. The most simple amino acid is glycine (NH_2-CH_2-COOH). Next are L-alanine (NH_2-CH_3-CH-COOH) and lysine ($NH_2(CH_2)_4CH(NH_2)COOH$ [$C_6H_{14}N_2O_2$]). The word amino comes from the word *amine.* NH_2^- is an amine group. The word *acid* comes from carboxylic acid, which is the other end of the amino acid molecule (-COOH). Insulin, hemoglobin, oxytocin, steroids, and so on, are proteins. Meat or muscle is also protein. Proteins can have molecular weights (remember, H_2O has a molecular weight of 18) into the 100,000s. Remember, the major chemical components of an amino acid are C, H, O, and N. Nitrogen alone does not constitute protein or guarantee its eventual participation in a protein molecule. The other building blocks, the carbohydrates, plus the phosphate catalyst, must be present with the nitrogen, hence the addition of carbohydrate to fertilizer.

Keep in mind that the purpose of this discussion is simply to acquaint you with the terminology that is regularly used at farm meetings where real world agriculture is discussed. The intention is for you to recognize the connection between the fertilizer you apply and the product nature is attempting to make with that fertilizer. Your good judgment should then be able to guide you in your decisions.

Carboxylic acids are organic acids. Vinegar is a dilute solution of a carboxylic acid called acetic acid, with a molecular formula of CH_3COOH. The formula for citric acid is $C_6H_8O_7$, described as HOOC-CH_2-COHCOOH-CH_2-COOH. Organic acids are important in dissolving and holding soil nutrients for subsequent use by microorganisms and plants. Some organic acids, like ascorbic acid, are used directly. Organic acids are obtained directly from microorganism metabolism of sugars or from humus as humic acid. The latter, however, also depends on microorganisms for its manufacture.

This brings us to the subject of sugars. The most common sugar is glucose ($C_6H_{12}O_6$). In nature, sugars form five or six sided polygons, pentagons, or hexagons. A square is a four-sided polygon. Glucose forms a six-sided polygon, and is called a hexose sugar, and looks like this:

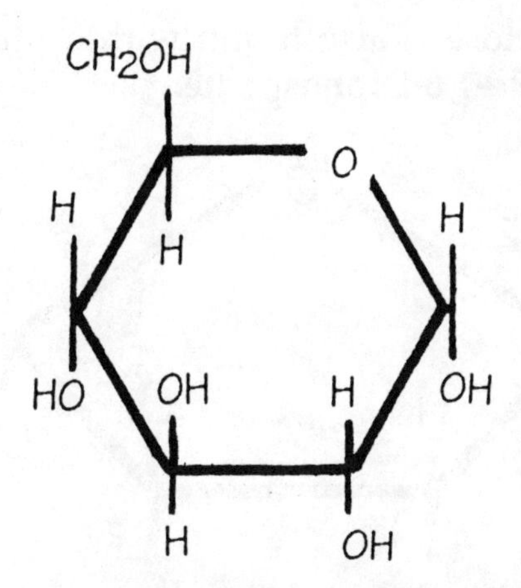

If a phosphate ion were added to this sugar, we would have alpha-glucose-1-phosphate. This sugared phosphate or phosphated sugar is the immediate precursor of starch in plants and of glycogen in animals; it is also the first product of the breakdown and use of these products. This compound is a major reason the refractometer reading of a crop is correlated to the phosphate/mineral level of the plant and also why it is recommended that sugar be added with all acid phosphate fertilizers. I am not suggesting that the simple combination of sugar and phosphate in the spray tank results in the formation of alpha-glucose-1-phosphate, but rather that the necessary building blocks are being provided for such a compound to be formed.

Fructose ($C_6H_{12}O_6$), also a hexose sugar, forms a pentagon and looks like this:

O
HOCH2 CH2OH
H OH
H OH
OH H

In nature, fructose is also found with a phosphate ion attached, as fructose-1,6-Diphosphate:

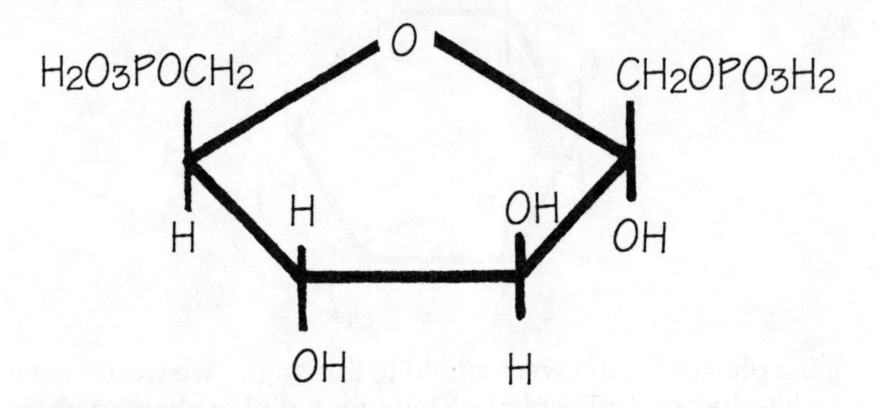

and as fructose-6-phosphate:

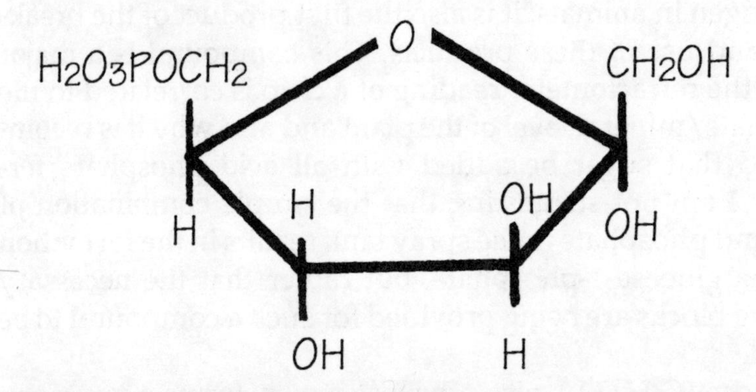

Sucrose is the combination of glucose and fructose. In living systems, sugars are used for two purposes: as energy stores and as structural components. Providing energy/food is a major reason sugar is used in fertilization. Microorganisms need food to survive and multiply. Most of the soils in the United States have been voided of their sugars due to the use of harsh chemicals, fertilizers, and bad management practices. To regenerate the microorganism populations rapidly, they must be fed. Then and only then can they digest crop residues and produce organic acids, humus, and nutrients. Very few crops have adequate sugar contents, as attested by their low refractometer values; hence, crop residues do not contain sufficient sugar for the microbes to use for optimum efficiency. Proteins provide very little energy for this purpose, and straight nitrogen sources provide none.

To maximize fertilizer efficiency, some type of sugar or at least an organic acid should always be applied with nitrogen to provide energy to the microorganisms. For example, molasses or table sugar can be mixed with liquid 28% nitrogen or ammonium thio sulfate and sprayed on corn stubble or straw to aid in its decomposition. The microbes' efficiency can then be maximized. Urea (CH_4N_2O, described structurally as $(NH_2)_2CO$), although technically an organic compound, does not provide a carbohydrate source. Urea is a waste product from metabolic systems; a sugar or organic acid needs to be applied with it to maximize efficiency.

The other use of sugars in living systems is as structural components. Sugars are one of three components of nucleic acids (DNA and RNA), which are the templates or blueprints of the entire organism. Structurally, nucleic acids resemble spiral staircases. Most farmers have heard about dominant and recessive genes in plant or animal breeding. The information for the traits or characteristics you want is contained in the gene, which is actually a piece of DNA. Nucleic acids are made of purine or pyrimidine bases, e.g., cytosine, arginine, thymine, guanine, and uracil, which are analogous to the steps of the staircase, with alternating phosphate-sugar links, which are analogous to the supporting side wall of the staircase. The sequence in which the bases (stair steps) are placed determines the information contained in each gene. Nucleic acids are responsible for the eventual manufacture of complex proteins. This is the RNA's job. It is somewhat like the reader, messenger, and executor of the DNA's (gene's) orders. Protein synthesis is crucial to the survival of the organism. The first step in protein synthesis is the reaction of an amino acid with ATP (a high-energy phosphate). This complex then binds with transfer RNA. This complex moves to the ribosome, where the amino acid is combined with another amino acid, eventually forming a protein. The point to remember is that protein manufacture requires sugar and phosphate in addition to nitrogen.[1]

An important structural use of sugar is in the formation of cellulose. Cellulose or fiber is the material that gives plants their rigidity, from the oat stem to the tree trunk. Cellulose is a glucose chain, as shown in the following diagram:

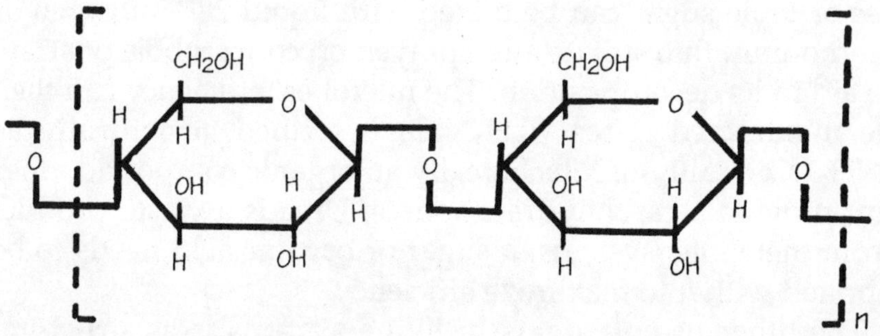

If cellulose is nitrated it forms nitrocellulose, which is used in the manufacture of explosives, collodion, and lacquers. Add excess potash to alfalfa, displacing calcium, and you will have "gunpowder hay" by the formation of potassium nitrate and nitrocellulose, which form when phosphate is insufficient to catalyze the proper formation of protein and other metabolites.

Animals, bacteria, and fungi all form glycogen, branched chains of glucose, for storage of reserve energy. To manufacture glycogen, however, adenosine triphosphate (ATP) or uridine triphosphate (UTP) is required. If one is short on glucose to begin with, not enough triphosphate compounds will be made through what is called the Krebs cycle (the primary metabolic cycle of all living things). ATP, in particular, is the main energy "wagon" for all biological reactions. Plants need ATP to make glucose during photosynthesis, and ATP is then manufactured when glucose is metabolized (used, broken down, or consumed, such as in making proteins, enzymes, and cell components) by the plant, by microorganisms, and even by humans. Glucose is also the starting material for synthesizing lipids (glycerol and fatty acids), which are then phosphated to form biological membranes.

In reviewing what has just been said about glucose, it becomes evident that sugar is an important nutrient for healthy soils and crops. One other fact should be apparent: phosphate is a close associate of sugar. In fact, without phosphates, no sugars will be manufactured or metabolized regardless of the sunshine, CO_2, and H_2O present, nor will there be enough for all the uses or needs of manufacturing. There might be enough phosphate so that some sugar can be made through photosynthesis, but there will not be enough to build sufficient cellulose for cell-wall rigidity or phospholipids for the cell membrane, or to produce all the nucleic acids, amino acids, proteins, enzymes, vitamins, and glycogen necessary for the plant to be healthy. "Healthy" means a condition in which insects and diseases do not see the plant as garbage and subsequently infest it. Insect infestation is not due to pesticide deficiencies.

There is much more to chemistry than what has been covered in this chapter. However, this discussion will provide the foundation necessary to evaluate fertilizers, programs, and statements made by industrial and educational representatives. Keep in mind that there is much more to providing a plant's nutritional needs than simply applying some nitrogen and potash, especially if one is seeking high quality, sustainable crops. Learn the basic elements and terms commonly used in agriculture, and understand the basic components of sugars, amino acids, proteins, and genetic material—nucleic acids. Be aware of the interaction of sugar and phosphate and the formation of living matter.

Concluding this section is a list of many common fertilizers and products and their chemical formulas. Become familiar with them so you can make an informed decision about what to purchase.

Anhydrous ammonia, NH_3
Aqua ammonia, $N_4OH \cdot H_2O$
Ammonium sulfate, NH_4SO_4
Ammonium nitrate, NH_4NO_3
Monoammonium phosphate, (MAP), $NH_4H_2PO_4$
Diammonium phosphate, (DAP), $(NH_4)_2HPO_4$

Ammonium thiosulfate, $NH_4S_2O_3$
Ammonium thiocyanate, (found in some pesticides), NH_4SCN; cyanide is CN^-
Urea $(NH_2)_2CO$
UAN 28% or 32% urea ammonium nitrate blend
Calcium nitrate, $Ca(NO_3)_2$
Potassium sulfate, K_2SO_4
Potassium nitrate KNO_3
Potassium chloride, KCl
Potassium hydroxide, KOH
Sodium nitrate, $NaNO_3$
Nitric acid, HNO_3
Sulfuric acid, (battery acid) H_2SO_4
Hydrochloric acid, (also muriatic acid, stomach acid) HCl
Phosphoric acid, H_3PO_4
Ascorbic acid, (vitamin C) $C_6H_8O_6$
Calcium ascorbate, $Ca(C_6H_7O_6)_2$
Calcium carbonate, (aragonite or high-calcium lime) $CaCO_3$
Calcium chloride, (road brine, wood preservative) $CaCl_2$
Calcium cyanide, (fumigant, rodenticide) $Ca(CN)_2$
Calcium cyanamide, (defoliant, herbicide, pesticide) $N{\equiv}C{-}N{=}Ca$
Calcium hydroxide, (hydrated lime, slaked lime) $Ca(OH)_2$
Calcium oxide, (burnt, dehydrated, or quick lime) CaO
Tricalcium phosphate, (rock phosphate) $Ca_3(PO_4)_2$
Super phosphate 30% $Ca(H_2PO_4)_2$, 48% $CaSO_4$, 9%$Ca_2(HPO_4)_2$
Triple super phosphate—super phosphate treated with H_3PO_4 (phosphoric acid)
Calcium phosphate, (animal feed) $CaHPO_4$
Calcium sulfate, (gypsum) $CaSO_4$
Cobaltous sulfate, $CoSO_4$
Cupric sulfate, (copper sulfate, fungicide, trace mineral) $CuSO_4$
Sodium borate, (borax) $Na_2B_4O_7$
Pyrophosphoric acid, (reverts to ortho form in hot water) $H_4P_2O_7$
Sul-po-mag: sulfur-potassium-magnesium, (Dynamate)

—3—

PHYSICS

PHYSICS IS THE FIELD OF SCIENCE that deals with how our physical environment behaves—the principles, properties, and interactions of matter, motion, and energy. Physicists and chemists often see themselves as rivals, but in reality there can be no separation between the two fields. Because all chemical matter is made up of energy, the principles of energy encompass the theories of chemistry. This fact constitutes perhaps the greatest impasse between physicists and chemists, particularly biochemists. Biochemists like to formulate theories about chemical attraction, molecular-fit mechanisms, and chemical-information transfer in living systems. Yet very few of them acknowledge that the principles of physics already explain these and almost all other biochemical phenomena.

Biochemists, insect morphologists, and entomologists contend that insects detect their target, whether it be food or a mate, by scent molecules hitting the antennas and, if they are the proper shape, fitting into holes in the antennas. This is the fit theory. If the fit theory is correct, why are not all insect antennas the same shape and size, differing only in the shapes of their holes? If the theory is correct, why can insect behavior be altered by electromagnetic signals when no scent molecules are present, or when a concentration of one molecule of scent per milliliter of air—the equivalent of a golfball in the Houston Astrodome—is sufficient to alter an insect's behavior? What do you think is the probability of a molecule colliding with an antenna in just the right way so as to fit into a specifically shaped opening, one that only this molecule will fit? That would be somewhat like drilling a hole the size of a golfball in a two by four and then walking along with the board and catching a flying golfball in the hole. A baseball, football, softball, or basketball would not fit, only the golfball would fit. Would you care to wager your next meal on whether you could catch the flying golfball in the golfball size hole in the two by four without deliberately attempting to line up the two? Common sense would also indicate that the scent molecule fit theory is not very plausible since insect behavior can be altered by devices downwind.

More than twenty-five years ago, Phil Callahan demonstrated that insect communication was electromagnetic and that insects could be controlled by electromagnetic devices. I recommend that you read his book *Tuning In To Nature*!

Everything in nature functions electromagnetically. This means an electrical signal causes every action. The simple task of reading the words on this page is electromagnetic. Light reflects off the page and into your eye. The pattern of words is formed on the retina. This pattern stimulates the millions of nerve fibers in the retina, which in turn send specific electrical signals to the brain via the optic nerve. The brain processes these electrical messages in such a way as to make sense out of them.

If a sensitive instrument that could detect minute electrical currents were hooked up to your head, it would show that your brain is very active electrically. An electroretinogram (ERG) records the differences in electrical potential across the retina of the eye when stimulated by light. An electroencephalogram (EEG) records the potential differences in the brain characterizing brain behavior. These various potential differences in the human body are in the range of 30 to 500 microvolts (μV), or .000030 to .000500 volts.

Most farmers are familiar with basic electrical wiring for lights, welders, grain driers, and so on. The same principles that govern these electrical systems also govern nature. Let us examine a basic light-wiring diagram, shown below:

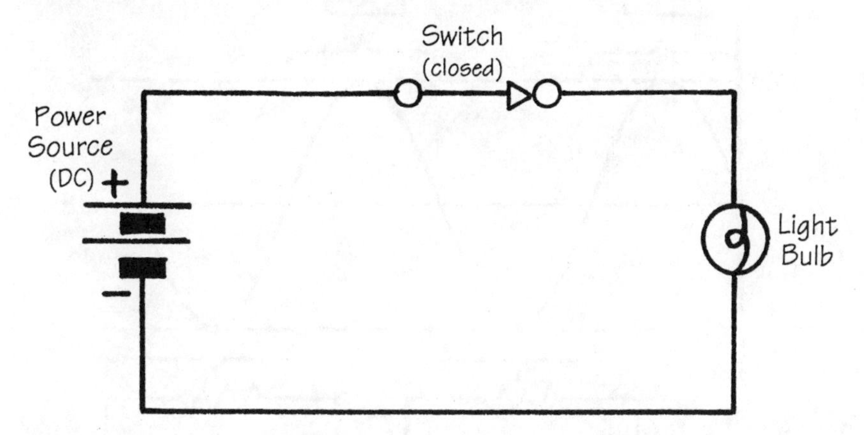

We have a power source, a conductor (wire), a resistor (light bulb), and a switch. If this were a direct current (DC) system, electricity would flow only in one direction and, therefore, the positive and negative would always be the same. If, however, this were an alternating current (AC) system, electricity would flow in one direction and then the other, alternating sixty times per second (fifty times per second in Europe). Electricity usually is discussed in terms of wave motion, i.e., flowing like the waves on a lake. Every wave has a crest and a valley. The following diagram shows both the cross-sectional view of a wave and the polarity shift of an AC power system.

This is called a sine wave function. The crest is given a positive value, the valley a negative value. DC is only the positive crest. The negative valley is blocked by a diode. AC is both the positive crest and the negative valley. Older model tractors have generators that produce DC. Newer model tractors have alternators producing AC, but diodes are installed

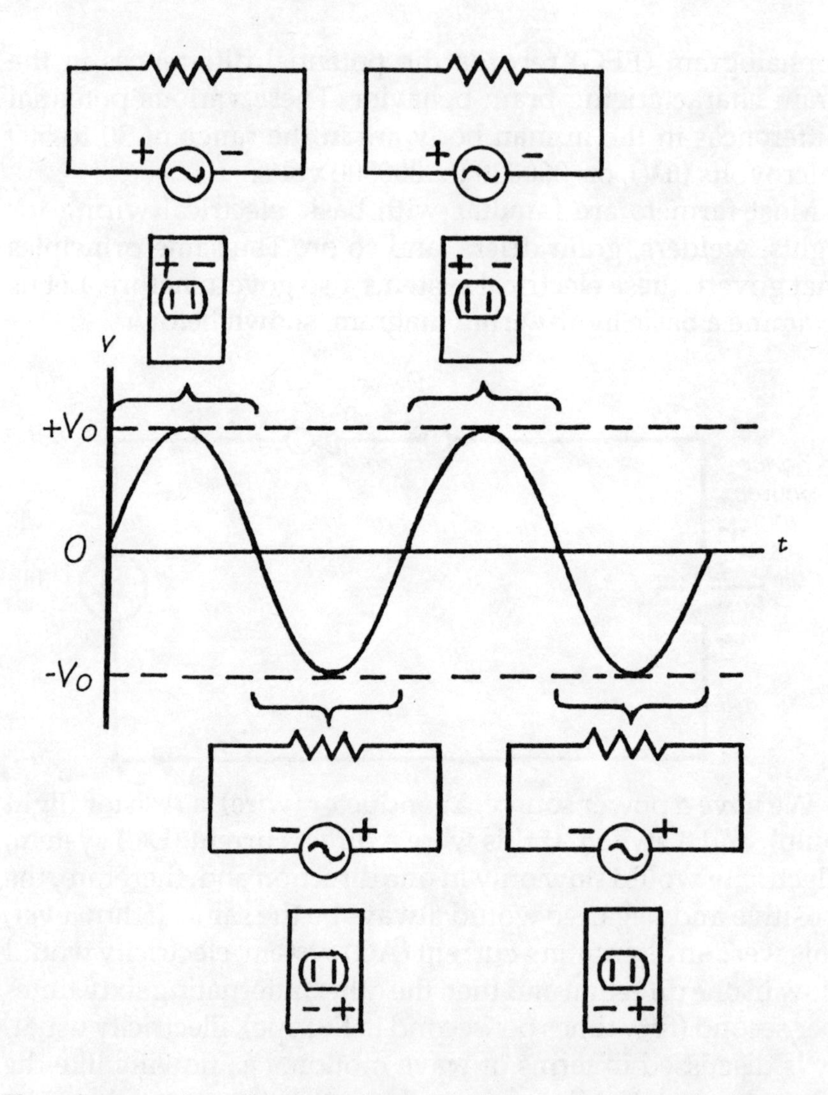

The voltage produced between the terminals of an AC generator fluctuates from moment to moment. A graph of the voltage, *V*, versus time, *t*, is sinusoidal in the most common case. The circuits and wall sockets indicate the relative polarity of the generator terminals during the positive and negative parts of the sinusoidal graph.

in the alternator, allowing the current to flow in only one direction.

The length along the horizontal (t) axis is the length of the wave. The distance up or down the vertical axis is the amplitude (height or depth) of the wave. On a lake, the distance from one crest to another varies, depending on the weather, as does the distance from the top of a crest to the bottom of a valley.

With electromagnetism, the depth or height of a wave (amplitude) correlates to the voltage. The higher the voltage, the higher the wave. The distance along the horizontal axis from the beginning of a crest to the end of a valley is termed the wavelength, thus:

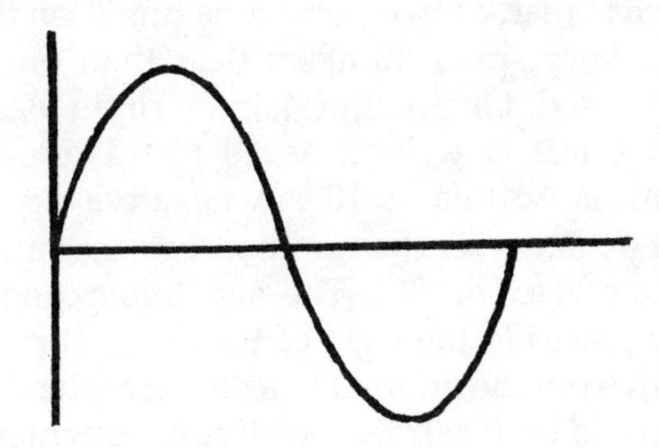

The frequency is then the number of crests and valleys per second, giving cycles per second, commonly called hertz (Hz).

The amount of charge flowing per second is called amperage (I). In drawing an analogy between the flow of a charge and water moving in a pipe, voltage is analogous to pounds per square inch (psi), whereas amperage is analogous to gallons per minute (gpm). Resistance (R) is the impedance to the flow of current in the circuit and is a function of both amperage and voltage ($R = V/I$). This equation can be rearranged to find any one component if the other two are known: $V = IR$, $I = V/R$.

Going back to wavelengths, most people have heard of various electromagnetic waves, such as X-rays, microwaves, radio waves, and infrared waves. Technically speaking, all are light waves, but most people think of light as only that which they can see, which is called the visual spectrum. Actually, the visual spectrum is only a small fraction of the entire light or electromagnetic spectrum.

Before proceeding, we need to review some mathematics. Scientific notation often is used to write very large or very small numbers. This is like a numerical shorthand. It is also called exponential notation, which means taking a number and raising it to a certain power. One can be written as 10^0 (ten to the zero power). Ten can be written as 10^1. One hundred can be written as 10^2. One thousand can be written as 10^3, and one billion can be written as 10^9 or $10\times10\times10\times10\times10\times10\times10\times10\times10$. To write 29,000,000 in scientific notation, one would write 2.9×10^7, which is read as two point nine times ten to the seventh power. For very small numbers (less than one), a negative exponent is used. One-tenth ($1/10$) is written as 10^{-1}. One-one hundredth ($1/100$) is written as 10^{-2}, and one-one billionth ($1/1{,}000{,}000{,}000$) is written as 10^{-9}. A negative exponent can be thought of as one over the number. For example, 10^{-3} is $1/10^3$, which equals $1/1{,}000$ or .001. The negative exponent indicates how many places to the right of the decimal point you must end up with from the original placement of the decimal. Thus, 7.5×10^{-5} would be .000075. A positive exponent indicates how many zeros to add or the total number of places to the right of the number's original decimal point at which the final decimal point will be placed. Thus, 7.5×10^5 would be 750,000.

Scientific notation often is used when giving frequencies or wavelengths of various electromagnetic waves. Also, the metric system is commonly used when giving units to various values. For purposes of the following discussion, remember that there are about 39 inches in a meter.

The average person can see only the light wavelengths of the rainbow. These range from 7.5×10^{-7} meters long in the red color to 3.8×10^{-7} meters long in the violet color. These are quite short. In inches, they are .000029 inches (29 one-millionths of an inch) to .000015 inches (15 one-millionths of an inch). Because electromagnetic wavelengths range from more than 10^{10} meters in length to less than 10^{-14} meters in length, one can see that the visual spectrum is indeed a very small fraction of the total electromagnetic spectrum.

In the relationship of wavelength equals velocity divided by

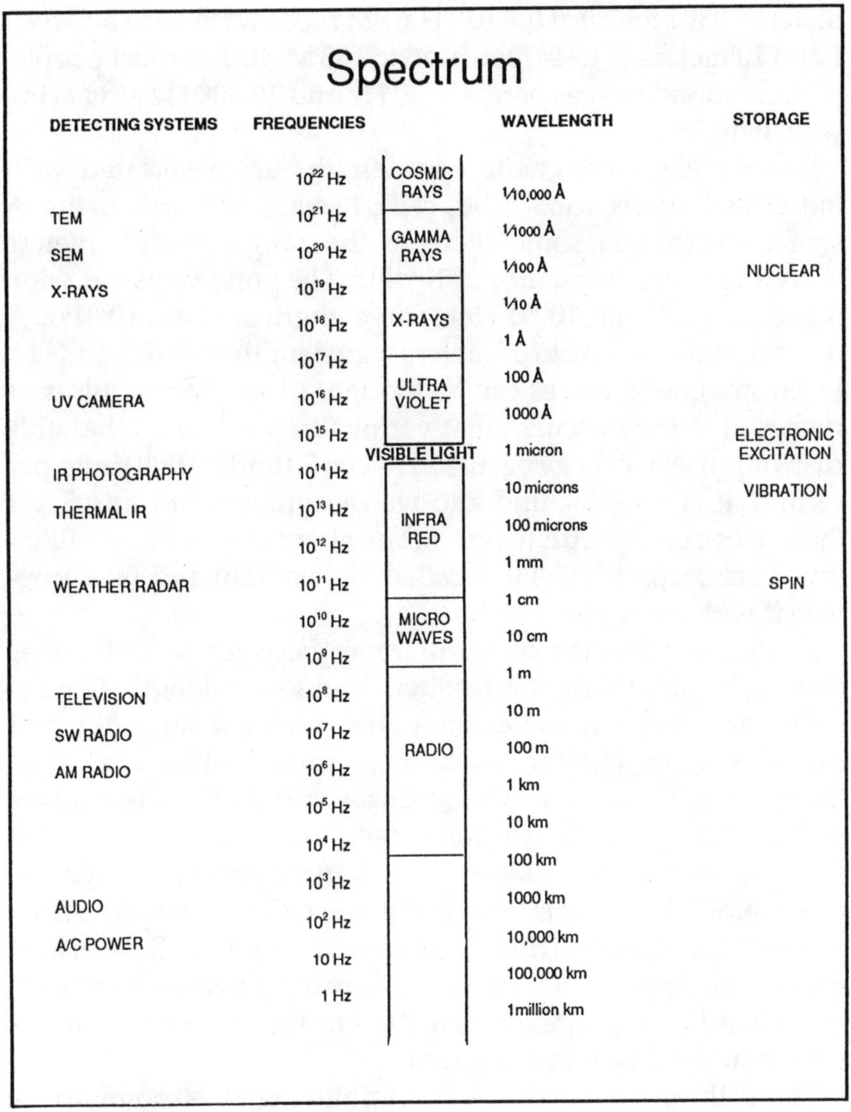

Spectrum

DETECTING SYSTEMS	FREQUENCIES		WAVELENGTH	STORAGE
	10^{22} Hz	COSMIC RAYS	1/10,000 Å	
TEM	10^{21} Hz		1/1000 Å	
SEM	10^{20} Hz	GAMMA RAYS	1/100 Å	NUCLEAR
X-RAYS	10^{19} Hz		1/10 Å	
	10^{18} Hz	X-RAYS	1 Å	
	10^{17} Hz		100 Å	
UV CAMERA	10^{16} Hz	ULTRA VIOLET	1000 Å	
	10^{15} Hz	VISIBLE LIGHT	1 micron	ELECTRONIC EXCITATION
IR PHOTOGRAPHY	10^{14} Hz		10 microns	VIBRATION
THERMAL IR	10^{13} Hz	INFRA RED	100 microns	
	10^{12} Hz		1 mm	
WEATHER RADAR	10^{11} Hz		1 cm	SPIN
	10^{10} Hz	MICRO WAVES	10 cm	
	10^{9} Hz		1 m	
TELEVISION	10^{8} Hz		10 m	
SW RADIO	10^{7} Hz	RADIO	100 m	
AM RADIO	10^{6} Hz		1 km	
	10^{5} Hz		10 km	
	10^{4} Hz		100 km	
	10^{3} Hz		1000 km	
AUDIO	10^{2} Hz		10,000 km	
A/C POWER	10 Hz		100,000 km	
	1 Hz		1 million km	

frequency, the frequency changes as the wavelength changes. Frequency equals velocity divided by wavelength. The wavelength of 7.5×10^{-7} meters corresponds to a frequency of 4.0×10^{14} Hz (cycles per second); 3.8×10^{-7} meters corresponds to 7.9×10^{14} Hz. As a comparison, infrared waves range from 7.5×10^{-7} meters (.000029 inches) at 4.0×10^{14} Hz to 1×10^{-3}

meters (.039 inches) at 5×10^{11} Hz. FM radio waves are between 1 and 10 meters (3 to 40 feet) at 88 to 108 MHz. For most people, audible sound ranges between 20 Hz and 20,000 Hz at varying wavelengths.

Specific electromagnetic wavelengths are associated with individual atoms, molecules, cells, tissues, and organisms. A single electron is somewhere in the range of 10^{-12} meters wavelength at a frequency of 10^{23} Hz. The polio virus emission is between 10^{-5} and 10^{-6} meters wavelength at about 10^{16} Hz. A red blood cell is about 10^{-3} meters wavelength at about 10^{14} Hz. Electromagnetic waves can be thought of simply as pulses or rushes of energy occurring at various frequencies. When this rush or pulse of energy occurs between 20 and 20,000 times per second, it is called sound and we recognize it in our ears. If these electromagnetic waves occur about one million billion times per second (10^{15}) it is called visible light and we recognize it with our eyes.

Fundamentally, the only difference between sound waves and visible light waves is frequency and wavelength, like the difference between waves on a lake during a calm day and during a storm. They are made of the same "substance." Also, the smaller waves on a lake occur faster than the large waves in the ocean, yet both are still water waves.

Any time electricity is considered, magnetism must also be considered because it is inherent in every electrical field. When an electrical current (I) is generated, a magnetic field also is generated. The magnetic field (B), in turn, generates a force (F) in a third plane perpendicular (90°) to the magnetic field, as shown in the following diagram.

This is the principle that, when applied, makes motor armatures turn and audio speakers function. This force equals the current (I) the length of the wire (L) the strength of the magnetic field (B) the sine of the angle (sin Θ). $F = I \times L \times B \times (\sin \Theta)$.

Let us examine briefly how this principle is used in the recording industry to make audio tapes. As shown in the diagram, the sound is converted to an electrical signal, which creates a magnetic field. This magnetic field changes as the

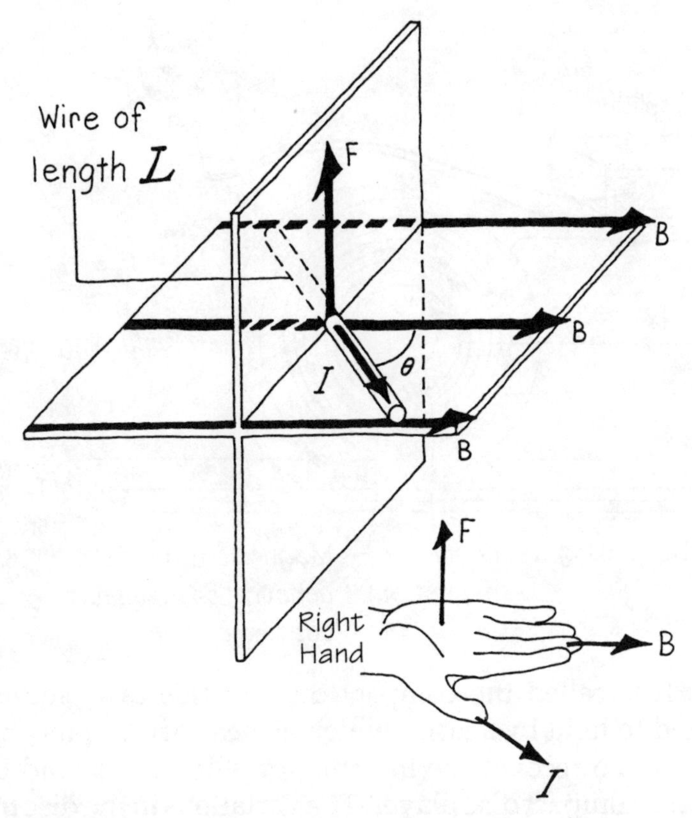

electrical signal varies, which varies as the sound changes. A blank magnetic tape is passed by the magnetic field and is imprinted with the variations of the magnetic field as time passes. This leaves a magnetic "braille," which can then be read by the tape head of a tape player, converted back into an electrical signal, and then converted into sound. The next diagram of audio conversion illustrates the point.

Marcel Vogel invented this method for storing computer information when he worked for International Business Machines. His later research proved that a crystal of any type, synthetic or natural, has the capability of storing electronic information to be recovered later. Vogel even proved that thought can be stored on such crystals, measured on the laboratory bench, and recovered fully. A somewhat different means of storing and retrieving sound has recently been in-

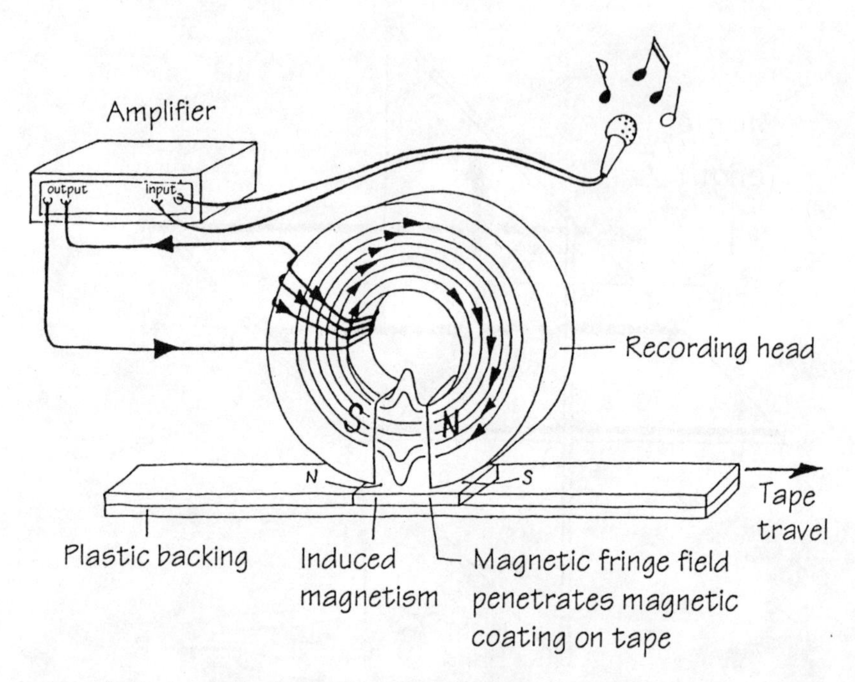

vented. It is called the compact disc. In this case, sound is converted to light by a laser, which etches various pits onto a blank disc. To recover the information, the disc is read by a laser in the compact disc player. The variations in the disc track cause variations in the laser light's reflections off of them, giving a mirror image of the original laser-light etching. This light is converted to an electrical signal, which is converted to sound. The laser is kept on track by two tracking lasers on either side of the reading laser.

The compact disc mechanism is important for the study of biological systems because it bears out the fact that complex information can be transferred via light. A particle of light is a photon, and, as Callahan and Popp repeatedly have shown, the photon is the medium (carrier) of cellular communication in living organisms. The compact disc system, electronic tapes, and crystal chips, are simply crude versions of nature's communication system.

Continuing the discussion of magnetic fields, it is readily accepted that a magnetic field flows over the earth's surface.

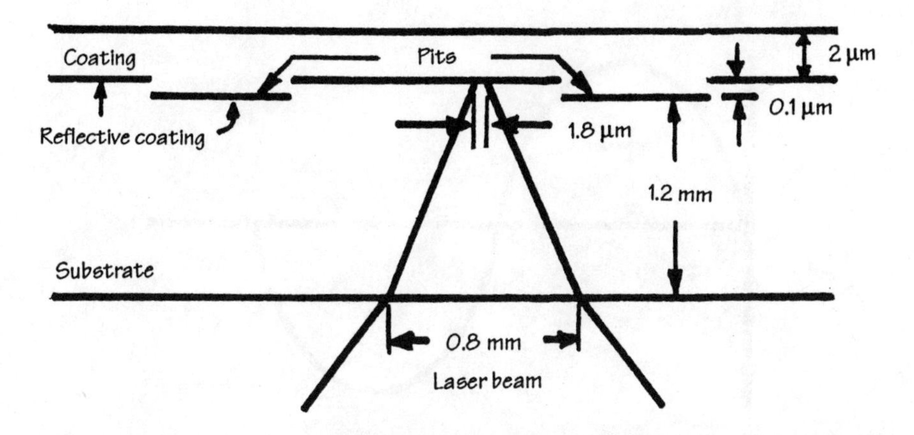

Cross sectional of a compact disc. Information is placed onto a plastic disc in the form of pits and a reflective coating is applied. When a laser beam hits a flat area it is totally reflected. When it hits a pit, which now appears as a bump, the optical path of the reflected light is shorter by twice the pit depth so that the reflected light arrives at the detector out of phase and interferes destructively. Thus as the disc rotates past the beam, the pits appear as no reflected light and are characterized as '1' whereas the flat areas appear as reflected light '0,' providing the binary information to the digital circuitry.

A simple compass verifies this. Because there is a magnetic field, there must be an appropriate electrical field at 90° to this field—that is, perpendicular to the earth's surface. This fact is rarely considered or even mentioned, yet it is very important for the proper function of plants because they are the antennas/conductors of this electrical field. The electrical field is actually an alternating one because the magnetic field is a wave function, which produces a wave function electrical field. A wave function, as discussed earlier, has a positive and a negative component, as shown in the following diagram.

An alternating field is vital to plant function. Plants are the conductors or, as Callahan showed in his tachyon experiment, superconductors of this natural electrical field, which is what drives the system. Like any electrical system, a given current flow is needed, in order to have proper function. If the current is too low, a brown-out or complete dysfunction occurs. If the current is too high, the system will overheat, blow a fuse, or burn out. So it is with the plant, a living electrical system. Too little nutrient energy results in poor growth, malfunction, and

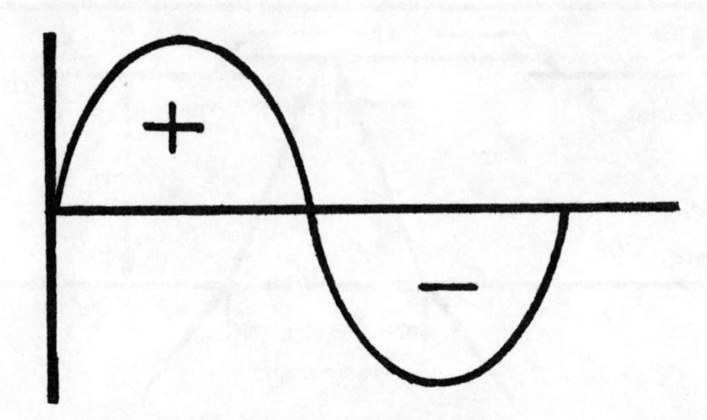

illness. Too much nutrient energy leads to leaf and/or root burn.

It is hoped that, by reading this chapter, the reader has gained an appreciation for the application of physics to plant growth and a realization that nature's communication and operating systems encompass and surpass all the sophistication of man's communication systems. With this overview, students of plant growth should be able to expand their perspective for more effective problem solving.

—4—

BIOLOGY

BIOLOGY IS THE STUDY OF LIVING ORGANISMS and systems. It is the investigation of life cycles and the structures and functions of organisms. Biological scientists often set themselves apart from physicists and chemists, which is unfortunate because, in nature, the sciences cannot be separated. They all apply simultaneously. This section contains definitions of some common biological terms: a look at the structures and functions of plants, microorganisms, and animals; a discussion of some basics of metabolism and reproduction; and an examination of mutual symbiosis—sustainable coexistence—as life is intended to be.

THE CELL

The cell is the basic unit of life. Every cell is a complete living system. We often talk of the microorganisms in the soil. The majority of these microbes are single celled. They are the algae, fungi, which include yeasts, and bacteria. Some algae and fungi are also multi-celled. Cells are divided into two basic categories—procaryotes and eucaryotes. Procaryotic cells are bacteria, whereas eucaryotes include plants and animals. The procaryotes are more primitive than eucaryotic cells. The major distinguishing characteristic between the two is that procaryotic cells have no nucleus, whereas eucaryotic cells do. The nucleus contains the genetic material (DNA) of the cell. Procaryotic cells have a nucleoidal area, which is less sophisticated than a nucleus.

Other differences between these cells are important to this discussion. The cell walls of procaryotes contain a substance called peptidoglycan. This is an amino acid layer, sandwiched between two carbohydrate layers. Bacteria with thick layers of peptidoglycan are gram-positive. Some familiar gram-positive (+) bacteria are *Actinomyces, Bacillus, Clostridium, Staphylococcus,* and *Streptococcus.* Bacteria with a layer of peptidoglycan

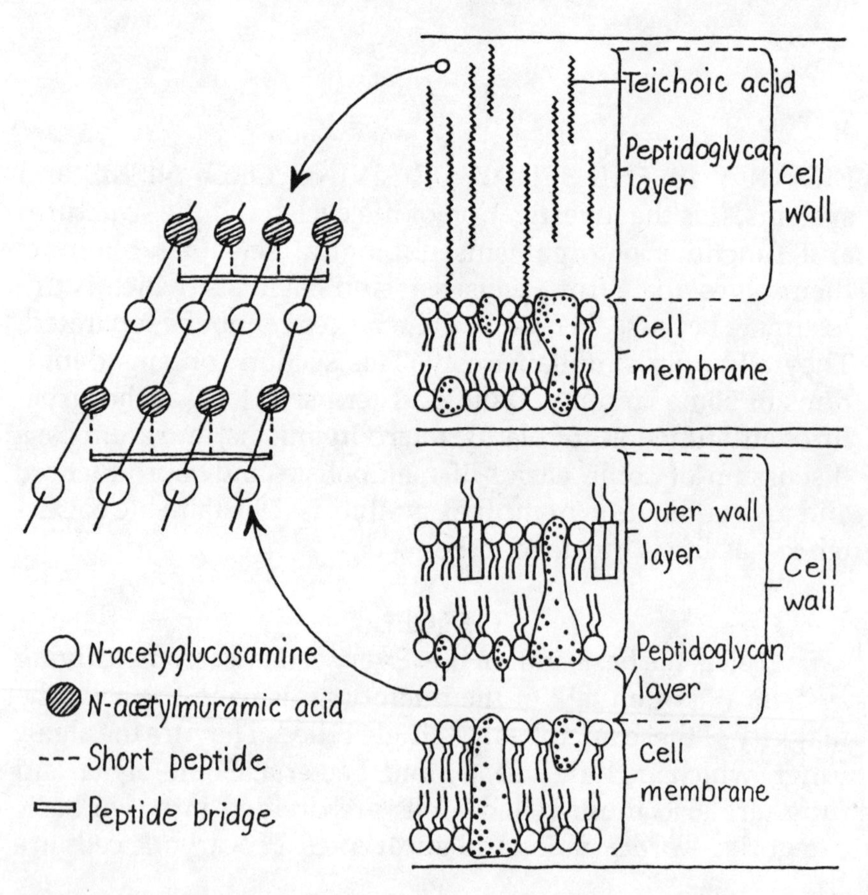

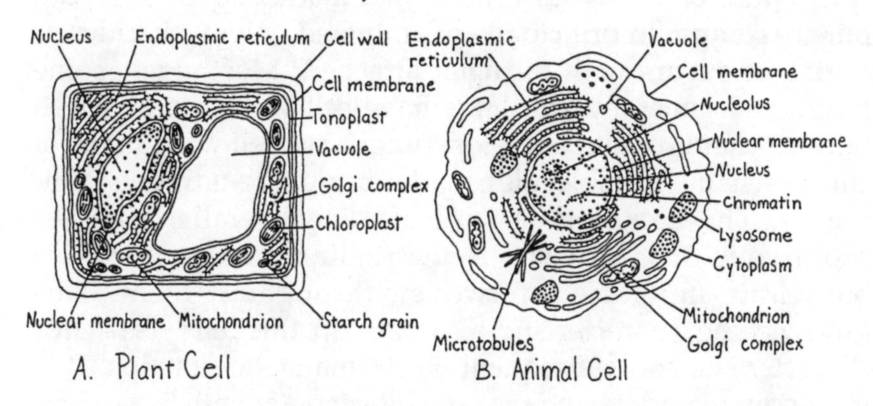

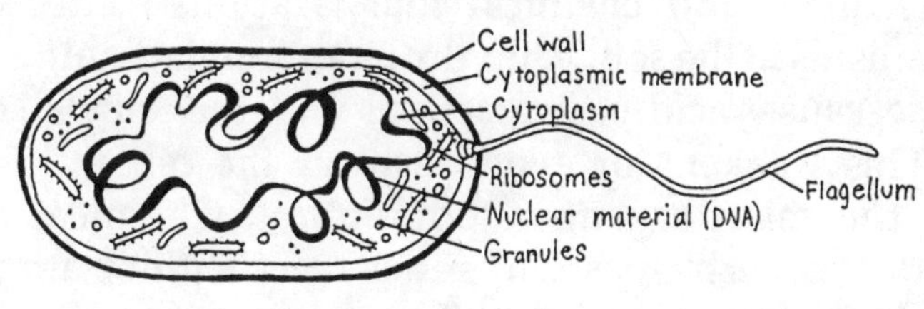

This diagram shows the cell walls of gram-positive (top) and gram-negative (bottom) bacteria. Though there is a difference between the two, both have a layer of peptidoglycan, shown enlarged to the left. This layer consists of a polysaccharide (sugar) made up of alternating N-acetylglucosamine and N-acetylmuramic acid and short peptides attached to the N-acetylmuramic acid. The point for the farmer to understand is that bacteria cannot reproduce without these nutrient complexes nor can they survive if harsh chemicals that will denature these cellular complexes are used on the soil. Pay particular attention to the fact that sugar is a very important component of bacterial cell wall and membrane complexes. Thus, our use of sugars in many fertilizer formulations. Salt based fertilizer programs do not provide nor leave intact these needed sugars.

inside a phospholipid layer are gram-negative (-). These include *Salmonella, Hemophilus, Pseudomonas, Brucella,* and *Agrobacterium.* In medicine, bacteria are controlled by antibiotics. Penicillin, for example, interrupts the formation of peptidoglycan on gram-positive bacteria. They are thus left without a cell wall and subsequently die.

The walls of eucaryotic cells are not made of peptidoglycan but are composed principally of aminated polysaccharides or chains of sugars with nitrogen attached. Most algae, some fungi, and generally all plants have cell walls made of cellulose—chains of glucose. Most fungi have cell walls made of chitin—chains of molecules made of glucose-nitrogen complexes. The common feature of all cell walls, whether procaryotic or eucaryotic, is that amino acid/protein/carbohydrate units are interwoven throughout them. This sandwiching characteristic also sets up the cells' capacitor characteristic and subsequent electromagnetic function. It is important to understand that any interference with these compounds causes the inhibition or death of the cell.

Every fertilizer and chemical that is applied affects the microorganisms in the soil. Anhydrous ammonia, being a very strong base, causes acid-base reactions with the cell wall compounds. This weakens or even destroys the cell wall, thus reducing the microorganism population. Chemical compounds such as herbicides will either react with or attach to cell wall compounds, thereby inhibiting proper function. Herbicides especially are detrimental to soil algae, which are simply small plants. Excess chlorine from potassium chloride, forming hypochlorous acid ($HClO$), oxidizes—breaks chemical bonds in cell wall compounds. This weakens or destroys the cell walls. In addition, excess chloride forms numerous chloride salts. Salts create a hypertonic, dehydrating environment, which draws water out of the cells, thereby stopping the microbial growth or even killing the microbe.

In many areas, fertilization practices have brought about the formation of aldehydes rather than sugars, proteins, and humus. An area may have much organic matter but very little actual humus because humus formation requires plenty of oxygen and energy for the correct microorganisms to work properly. If these conditions are not met, the crop residue, manure, and other organic materials are simply converted to ashes, alcohols, aldehydes, or other nonhumus compounds. Aldehydes inactivate proteins, which are important components of cell walls, as previously mentioned, usually to the point of cellular death.

Notice in the figure that the cell walls are much more than nitrogen and potash, the two most emphasized materials in chemical agriculture. In addition, because cell walls are made of layered compounds with a layered cell membrane inside the wall, an electrical potential is set up. One side carries a positive charge, and the other a negative charge, very similar to a capacitor. Strong acids and bases, heavy metals, chemicals, and electrical fields disrupt this gradient, thus inhibiting cell growth or killing the cell.

Humans' skin is primarily protein. The outer layer is a protein called keratin. Have you ever noticed what happens if you spill acid or caustic (anhydrous ammonia, potassium hydroxide) on your skin? The acid eats through the skin. How does your nose feel if you get a whiff of formaldehyde or ammonia fumes? It burns. Our cells are eucaryotic and therefore are made of compounds similar to those of microorganisms. Any fertilizer or chemical that injures the skin and mucous membranes, or reduces their function, does similar things to the living organisms in the soil. Likewise, any chemical that is readily absorbed through the skin most likely is readily absorbed into the bacterial cell, as well. Some people think this is ridiculous. They say soil is just dirt and that the microbes are a minor consideration in crop production. Such an attitude has killed our living soil. However, a little common sense can go a long way toward regenerating the soil. Sometimes, all we have to do is dilute the fertilizer with water to make it safe; a good example is phosphoric acid diluted sufficiently in water.

Although there are billions of microorganisms in the soil, the important point is what varieties of microorganisms are present. If the soil does not include enough desirable microbes like mycolytic bacteria, which kill fusarium, verticillium, and rhizoctonia fungi, or hyphomycetes fungi, which kill undesirable nematodes, or perhaps actinomycetes, which digest fodder and synthesize vitamin B_{12} for plant uptake, the soil is not regenerating or producing to its potential. If the soil contains disease organisms, it is deficient in beneficial organisms, including those just mentioned. These desirable microorganisms also increase plants' nutrient uptake from two to eight times, converting these nutrients to bio-organic forms; microbes also increase the metabolism of amino acids in the roots by converting inorganic nitrogen to organic nitrogen compounds.

Increasing nutrient uptake tremendously changes the efficacy of chemical soil tests because two soils with the same chemical-analysis nutrient levels would have marked differences in nutrient use if one was a typical sterile American soil and the other a biologically active soil. This also is why we seek fertilizers that enhance microbial growth or directly feed microorganisms. Neither traditional soil tests nor fertilization practices consider the microbial aspect of the soil. As a result, fertilization is inefficient and crops are inundated with pests.

GENETICS

Most farmers are somewhat familiar with genetics because they have been involved in animal and/or plant breeding, at least to the extent of selecting animal blood lines or crop varieties with desirable genetic traits. A large industry has been built around the hybridization of both animals and plants. Traits or characteristics are coded for, or imprinted, in genes. Genes are combined into clusters, forming chromosomes. Every species of plant or animal has a constant number of chromosomes. Corn has 20 chromosomes, pine trees 24, houseflies 12, fruit flies 8, and human beings 46. Species with the same chromosome number can be crossed. Genetic-engineering specialists seeking to alter genetic information first isolate the chromosome holding the gene in question and then isolate the specific gene on that chromosome.

Chromosomes can be thought of as gene "packages." Chromosomes are like computer programs or software packages, and genes are like the individual commands and data within the program or package. Mutations are spontaneous alterations of gene coding. They can occur as a result of viruses, chemicals, electromagnetic fields, faulty gene expression, or random occurrences in nature.

Nutrition plays a key role in gene expression. If there is not sufficient nutrient to form what the gene is coded for, it goes unexpressed. For example, bacteria that ferment glucose have suppressed the gene that allows them to ferment lactose. If glucose is replaced by lactose, after a lag phase to alter gene expression, the bacteria will begin to ferment lactose. This can be likened to a computer program that has the command for giving depreciation schedules, but just gives zeros if no depreciable item values are entered into the computer.

Genetics is the basis of modern hybrid seed selection. Varieties have been selected and bred that produce the greatest volumes of crops using chemical fertilization, primarily nitrogen and potash. Little or no consideration has been given to the resultant feed value of imbalanced fertilization. Natural quality indicators such as insects, weeds, and disease have been concealed with toxic pesticides, thus creating a facade of healthy crops. As such, most open-pollinated native varieties have been discarded because their genetic expression requires a fertilization program that contains all nutritive elements.

Under natural selection, plants that have genes expressing a predominant affinity for nitrogen and potash would be eliminated either directly by insects and diseases or indirectly through selective avoidance by the consumer. The brix reading of these plants would be lower and, therefore, these plants would be less desirable to animals and more susceptible to storage rot. Man, however, has specifically selected varieties with such affinity in order to perpetuate the sales of nitrogen and potash and the subsequent sale of pesticides and herbicides to cover up the mistakes. To get an indication of what high nitrogen fertilization achieves or does, run a normal Reams soil test. Then run the same soil audit again using household ammonia as the extracting agent. You will find that nitrogen and potash stand out in the latter test. Calcium and phosphate drop almost completely out of the available range.

This indicates that high nitrogen fertilization, particularly using anhydrous ammonia, creates a nutrient availability condition in the soil that is almost exclusively nitrogen and potash occupied.

What this means for variety selection and hybridization is that selection—based upon plant growth volume—will take only those varieties that grow discriminately on simplistic nitrogen and potash fertilization, consequently eliminating those varieties—particularly the open-pollinated varieties—that require a full banquet table of nutrition to grow satisfactorily. Under natural conditions these discriminating hybrid varieties would be eliminated by nature via insects and diseases, but man kills the clean-up crew and perpetuates the simplistic anhydrous ammonia and potassium chloride fertilization to his own demise. Selecting varieties with the traits to grow best under "chemical" fertilization is, in principle, the same as selecting varieties for height, color, standability, and so on. It is interesting that a majority of the seed companies are now owned by major chemical companies, but the reason for this becomes clear when one realizes that the latest hybridization and genetic engineering trend is toward varieties that are resistant to pesticides.

Selective breeding of plants and animals is simply a matter of selecting gene expression. Every cell contains all the same information, but certain cells express the information to be a leaf or a stem, or roots, or grain. Certain genes express traits like color throughout the organism. In animals, certain genes express small calves, longer legs, greater weight gain under feedlot conditions, and so on, all dependent upon the nutrition and the subsequent allocation of growth factors. The selection of crops to grow best under high nitrogen/potash fertilization is no different, in principle, from the selection of beef cattle for maximum weight gain on feedlot rations versus maximum weight gain on western rangeland. This principle correlates to matching a seed to a specific soil or field.

SEED AND PLANT STRUCTURES

In the ensuing discussion of seed and plant structures, keep in mind the perfection in nature and the connection of all things. Have you ever cut a seed in half and closely examined the parts? The figure on the following page shows cross-sectional views of bean and corn seeds. Notice that the first leaves and first roots are already formed in the seed. The endosperm is stored food for the young seedling; it is largely starch, which is made of glucose chains. If sugar production in the plant is limited, so will be the endosperm, resulting in poor sprout and seedling vigor, as well as susceptibility to pests. The cotyledons are food-storage organs that make use of the food in the endosperm. After germination, cotyledons may also function as leaves, as they do in squash plants. The epicotyl

Cross Sectional View

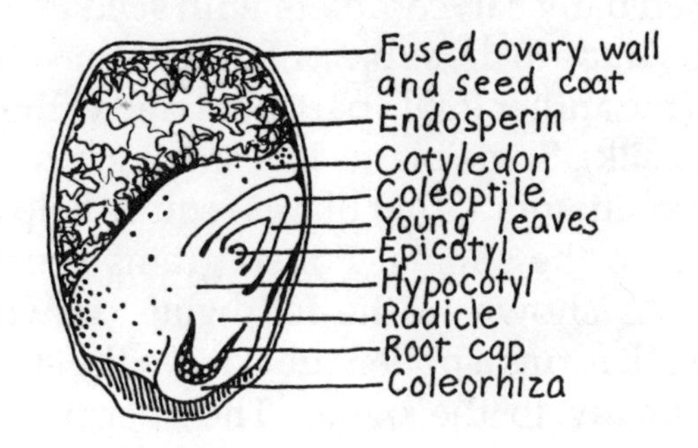

Corn Seed

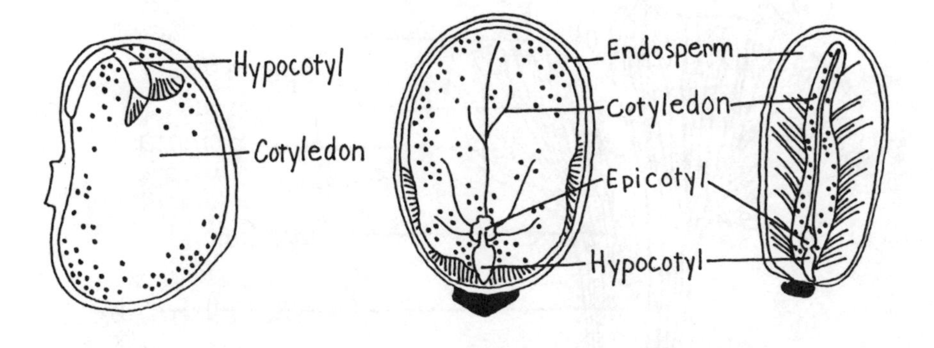

Kidney Bean

Castor Bean

becomes the shoot or stalk system of the new plant. The hypocotyl also becomes a part of the stem, and the radicle becomes the root system. The coleorhiza protects the root tip, and the coleoptile encloses the young primary shoot.

It is important to realize that the primary parts of the mature plant are all present in the seed because it indicates that the seed greatly determines the potential of the mature crop. The better the nutrition that was provided during seed formation, the healthier the seedling. The healthier the seedling, the greater its ability to assimilate nutrients and grow. This is very important because the number of seeds set is determined early in the growth cycle. Corn kernel set is determined between the twentieth and twenty-eighth day after emergence. Cut open a corn stalk that is about waist high, and you will see the fully formed baby ear, complete with seed embryos and silk.

This leads to the subject of pollination—the transfer of pollen from the anther (male part, or tassel) to the stigma (female part, or ear silk). Some plants like small grains and corn are termed self-pollinators. Other plants require help from bees to transfer pollen to the stigma. Pollen grains contain sperm and tube cells. As shown in the following drawing, as the tube cell grows, it forms a pollen tube down the style or silk to the ovary and finally to the ovule. The sperm migrates through the

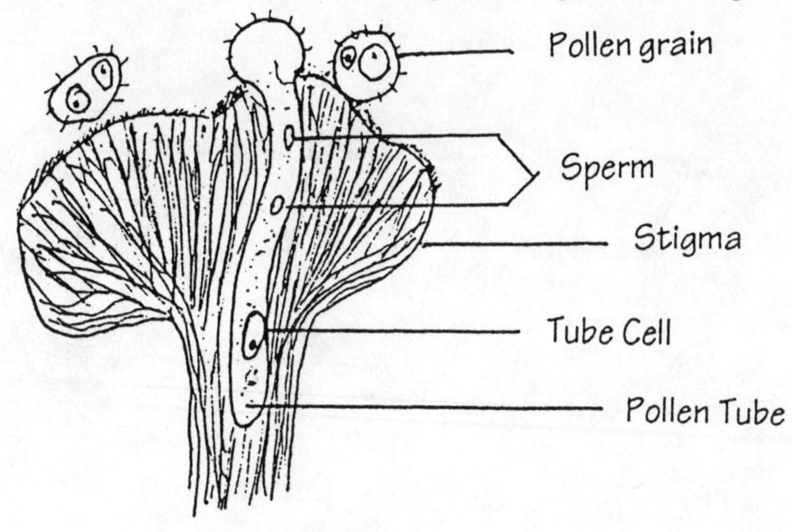

pollen tube to the ovule and fertilizes the egg. As can be seen, a considerable amount of stored energy is needed in a grain of pollen for successful fertilization to occur. The quality of the pollen is only as good as the nutrition given to the plant.

Pollination is a critical stage in crop production. In corn, extremely hot days often impair pollination; adult corn root worm beetles hinder pollination by chewing off the ear silks. Nutritional levels seem to be the most limiting factor, however. If the refractometer reading in the corn stalk opposite the ear shank is 8 brix or higher, root worm beetles will not cut the silks; below 8 brix, they will. In addition, the higher the refractometer reading, the less heat stress the plants encounter. The more biologically active a soil is, the more stable the field temperature and the cooler the field will be on extremely hot days. Considerable field experience on this has been logged by David Larson of Larson Farm Management in Princeton, Illinois, and Dan Skow of International Ag Labs, Fairmont, Minnesota. Floyd Buchelew, of Ripon, Wisconsin, has assembled the records of his considerable field experience with the brix-nectar connection. Bees will work in flowers whose nectar has refractometer readings of 7 brix or higher. If the nectar is below 7 brix, the bee expends more energy than it can possibly recover from the nectar. Even at 7 brix, the nectar is of questionable value.

In terms of nutrition, it has been found that manganese is important in pollination and seed production. This is closely linked to calcium, phosphate, copper, iron, and vitamin C.

LEAF AND STEM STRUCTURES

Farmers need to become familiar with leaf and stem structures so they can better evaluate fertilization, tillage, and cultural practices. The figure on the following page shows a cross-sectional view of a bean leaf—a typical dicotyledonous leaf. Monocotyledons like grasses and corn are a little different, but their major anatomy is the same as that of dicotyledons. The major part is the leaf "skin," which has an outer cuticle layer and an upper epidermis. This "skin" is usually covered with hairs called trichomes. Typically, these are said to be present to ward off certain insects, but in reality they function as antennas for the collection of energy and nutrients. This skin has a wax covering in proportion to the refractometer reading of the plant. The higher the brix, the higher the wax density, which gives a greater dielectric property to the leaf, increasing the antenna capacity of the leaf and the trichomes.

Next is the mesophyll, which consists of layers of cells containing chloroplasts. The number of layers varies, principally due to nutrition. More layers mean a thicker leaf, more photosynthesis, and more crop. The mesophyll also consists of the "veins" of the leaf—the xylem, which transports water and minerals up the plant, and the phloem, which transports sugars and food compounds from place to place throughout the plant. Biophysically, the veins function as waveguides for intercellular communication and energy flow, much like a fiber-optic system and the vein and artery system of animals.

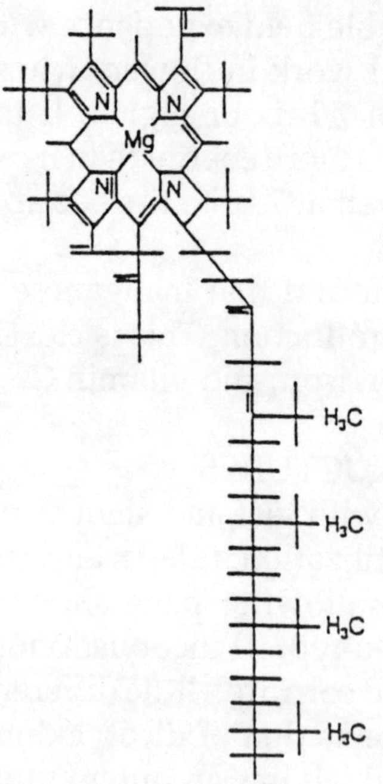

Chlorophyll-a

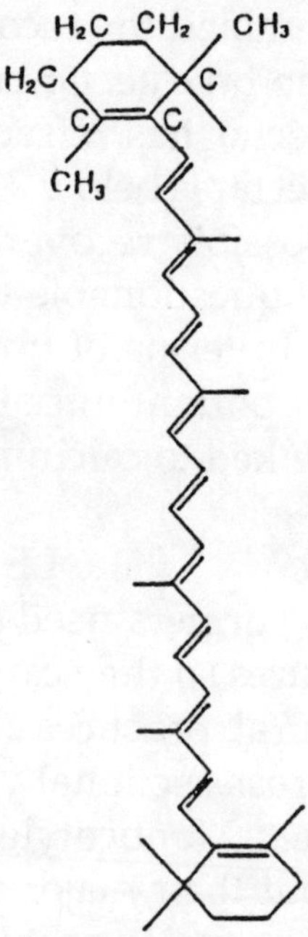

Beta-carotene

Chloroplasts are fascinating structures; they contain the chlorophyll. Chloroplasts resemble the mitochondrion in animal cells, which are called the power house of the cell. Both structures are biological lenses, meaning they focus electromagnetic radiation. Chlorophyll is closely associated with beta-carotene (provitamin A); both are tuning antennas. According to Hamerman (*21st Century Science & Technology*, March-April 1989), beta-carotene absorbs and funnels biophoton energy (light energy) to the chloroplasts, which in turn prepare and transmit this energy to the photosynthetic mechanism for the manufacture of sugar. Hamerman has actually correlated this process to the musical scale. Susumu Ohno did similar work at the Beckman Research Institute of the City of Hope in Duarte, California. Chlorophyll receives the energy specifically in two regions, 42 octaves above E—158 Hz (one octave below middle C—256 Hz) and 42 octaves above A-flat—102.25 Hz. The ratio between these two gives us green, C—128 Hz.

Thus, what exists in a plant is a series of tuning antennas or mechanisms, all being controlled by the "central data processor," called the DNA genetic material. Leaf hairs initially capture and concentrate environmental energy in the ultraviolet, visible, and infrared regions of the electromagnetic spectrum. Like lasers, the chloroplasts focus this energy for the beta-carotene to separate discriminately and pass on to the chlorophyll. The chlorophyll then refines this energy for use in manufacturing carbohydrates, i.e., photosynthesis. This ingenious bioenergetic system is the foundation of life on earth. This may sound complex, yet its premise is basic biophysics common to all living systems, as Phil Callahan has taught.

Looking at the anatomy of the leaf, you will notice small openings called stoma. The size of these openings is controlled by guard cells. The stoma are important for the exchange of moisture and nutrients. Proper opening and closing of the stoma is correlated to the nutrient potassium. Because the stoma are so minute (a magnifying glass is needed to see them), mist blowers or foggers work best for foliar feeding since these

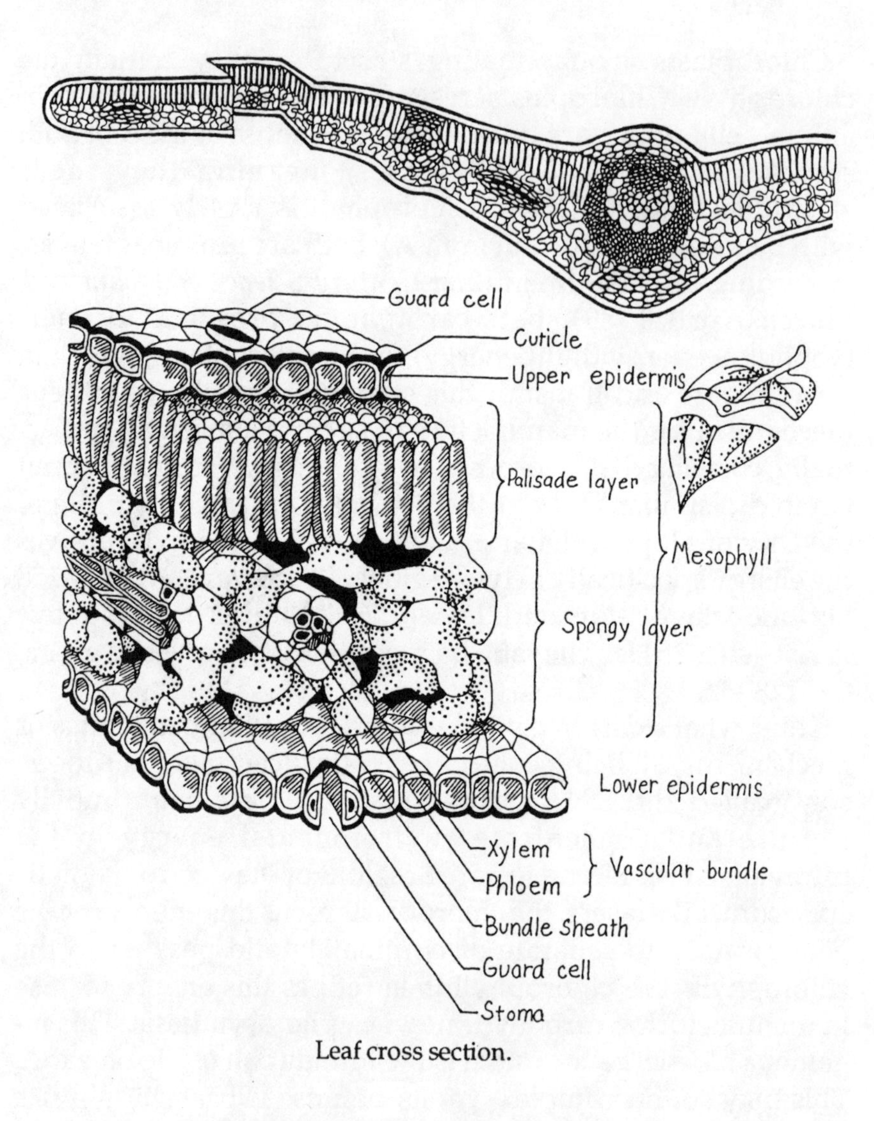

Leaf cross section.

machines produce a tiny droplet of mist. Properly charging the nutrient solution facilitates nutrient absorption because the solution will be more attracted to the leaf antennas.

PLANT STEMS AND ROOTS

The following figure shows the major parts of both monocotyledon and dicotyledon stems. The functions of phloem and xylem in conducting materials have been discussed. If you cut a corn stalk perpendicularly, giving a cross- sectional view of the stalk, it would look similar to diagram (A). The cells between the vascular bundles, "veins," should be pearly white, and the stalk should be perfectly round. A dull white or even a brownish color indicates deficiencies in nutrients, especially calcium, and a condition of congestion termed "gummosis." This condition can be partially alleviated by applying Sul-po-mag between July 15 and September 15 in the Northern Hemisphere. Out-of-roundness of the stalk indicates phosphate, potash, and molybdenum deficiencies.

If you cut the corn stalk lengthwise from the first leaf down through the root mass, you can observe several interesting characteristics. The area from the bottom tip up to the top brace root is discolored or brown and very hard. Also, the initial taproot has died off, and most of the roots radiating from this discolored area are brown and have many fewer root hairs. The area close to and enclosing the root mass is called the rhizosphere. This is where the exchange of nutrients between plant and microbes takes place.

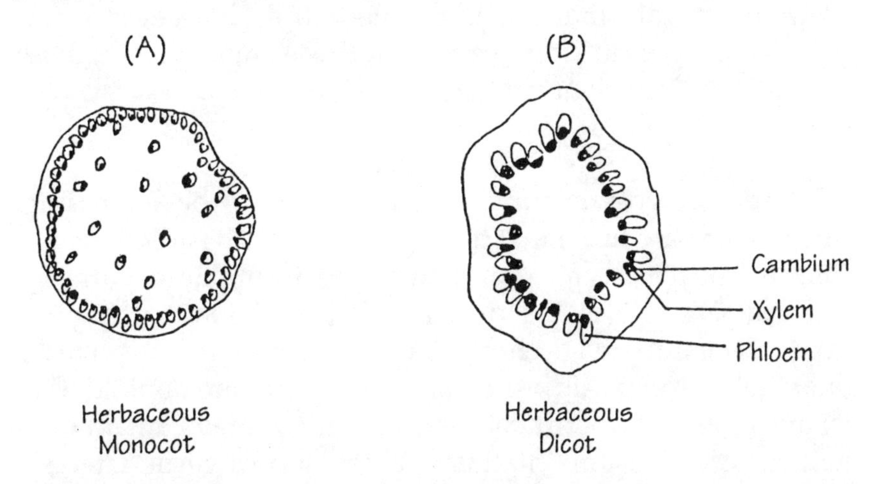

The corn plant just described is typical of those found throughout the United States today. Is this normal? Yes, if normal means commonplace. No, if normal means perfect health. Most farmers have been taught that corn has brace roots to prevent the plant from falling over. Actually, brace roots are the plant's emergency response in order to exchange nutrients and prevent starvation and death. The browning of the stalk interior is the result of congestion in the vascular system. The plant's plumbing is plugged, shutting off the movement of nutrients. The plant then sends out brace adventitious roots above the plugged area to make up for the reduction in flow from the primary roots. This is similar to a heart bypass operation. Each successive growth of brace roots indicates increased vascular plugging below. It is a rescue operation by nature. The plugging is caused by many things—chemical toxicity such as herbicides, putrefaction products of an anaerobic soil, excess nitrogen, and premature death of vascular tissue—all related to lack of nutritional integrity. Proper farming practices can eventually correct these problems, making brace roots unnecessary. Considerable field experience on this has been logged by David Larson of Larson Farm Management in Princeton, Illinois, and Dan Skow of International Ag Labs in Fairmont, Minnesota.

Alfalfa and small grains commonly have hollow stems. Farmers are told that this is a genetic trait. However, a few years of proper nutrition can fill in those stems, raising both the yield and nutrient content of the crop.

MUSHROOMS

Interest in growing mushrooms seems to be increasing. Mushrooms are fungi that belong to the class Basidiomycetes. They are not plants *per se* and do not perform photosynthesis. Their major role is the primary decomposition of organic matter. Mushrooms do not grow from seed but from spawn, or actually from basidiospores. These spores are haploid; that is, like eggs and sperm before fertilization, they contain only half the chromosome material. These spores germinate and grow into hyphae. A mass of hyphae is called a mycelium; it looks like a mass of fuzz growing throughout the medium. This is the feeding body of the fungus. It is underground and makes up much more volume than the above-ground mushroom. When two hyphae fuse, they form a basidiocarp or fruiting body, that grows into the structure we see and know as a mushroom. The underside of the mushroom cap is lined with gills, where the spores form and mature. Upon maturation, the spores are released to begin the cycle again.

SUMMARY

As can be seen, all of the sciences must be integrated; they cannot be viewed as disparate fields. A discussion of biology necessarily includes physics and chemistry. It is hoped that this brief overview of chemistry, physics, and biology will help you evaluate your soils, crops, farming practices, and the information you receive from diverse sources. There is no substitute for common sense, but informed common sense works even better!

—5—

ENERGY: THE BASIS OF LIFE

THE TERM ENERGY is a commonly used term in our society yet very few people actually know what it means. Most people can describe what it does and generally associate it with such things as fossil fuel, nuclear power, electricity, and so on. These associations do not, however, define what energy actually is. Thomas Bearden perhaps defines energy in the most encompassing manner of all. He defines energy as the *amount of ordering* which exists at any given point in time and/or space. This definition fits agriculture very well, for in the discussion of plant health, weed and pest infestation, and disease presence, plant health correlates to *order*, whereas weeds, diseases, and insect pests correlate to soil/plant *disorder*.

We and all living things live off the energy derived from our food and environment, not the physical substance, *per se*. This seems to be a difficult concept for many people to grasp, but a review of some well documented studies may help to clarify this concept.

One of the first was accounted for by Francis M. Pottenger Jr., M.D., and made a matter of record in *Pottenger's Cats*. This research involved nine-hundred cats and stretched out over a ten year period, starting in 1932. Pottenger had developed an adrenal extract that proved important in the treatment of tuberculosis. Since there was no procedure for standardizing biological extracts, animals were used to determine potency.

Cats in residence at Pottenger's Monrovia clinic were fed liver, sweet breads, brains, hearts and muscles. The potency of the adrenal extract was ascertained by testing the cats after their adrenal glands were removed. Pragmatic observations led to the design of a study that has never been equaled. The metes and bounds of that study are set out on this page.

THE MEAT STUDY

Two diets were used in this study:

Adequate Diet A:	⅓ raw milk and cod-liver oil ⅔ RAW MEAT.
Deficient Diet B:	⅓ raw milk and cod-liver oil ⅔ COOKED MEAT.

THE MILK STUDY

Five diets were used in this study:

Adequate Diet A:	⅓ raw meat and cod-liver oil ⅔ PATEURIZED MILK.
Deficient Diet B:	⅓ raw meat and cod-liver oil ⅔ EVAPORATED MILK.
Deficient Diet C:	⅓ raw meat and cod-liver oil ⅔ SWEETENED CONDENSED MILK.
Deficient Diet D:	⅓ raw meat and cod-liver oil ⅔ RAW MILK.
Deficient Diet E:	RAW METABOLIZED VIT. D MILK. 1. from cows on dry feed 2. from cows on green feed.

Acres U.S.A. summarized the study.[1] Generally speaking, the cats fed a healthy diet were gregarious, friendly and predictable in behavior. Cats on the deficient diets became quarrelsome, irritable and sometimes dangerous to handle. The raw food cats reproduced one homogeneous generation after another with the average weight of kittens being 119 grams at birth. Miscarriages were rare and litters averaged five kittens which the mother cat nursed with no difficulty. Two other factors of great interest and importance emerged from this research: deficient kittens from the pasteurized, evaporated, sweetened-condensed and cooked food groups that were placed on the total raw food diet improved immensely in health, but it took four generations to bring a degenerated second generation cat back to normal.

This study was unique as there is no record of a similar experiment in the medical literature. Accurate records meeting the most rigorous scientific standards of the day were carefuly kept with six-hundred of the nine-hundred cats. The pathological and chemical findings were done in conjunction with Alarin G. Foord, professor of pathology at the University of Southern California and pathologist at the Huntington Memorial Hospital in Pasadena.

Another unexpected result of the study showed up after the study was closed: the growth of weeds in the pens after the cats were gone. The weeds grew luxuriantly in the pens of the healthy adequate diet cats, and were sparse and spindly in pens that had been occupied by the cats on deficient diets.

More recently, a group of cats who were fed microwave cooked food not only failed to thrive, but all died in a few months—well before the age set by Pottenger "cooked meat" diet cats.[2]

In 1939 an experiment was designed in which pigs were fed natural vitamin B_1 vs. synthetic B_1. The pigs fed the synthetic B_1 became sterile. This experiment was repeated in 1940 in Scandinavia and again in 1949 in America. It is interesting to note that a University of Florida study revealed that in 1929 the sperm count of the average American male was 100 million/ml semen. In 1973 it had dropped to 60 million and in 1980 had dropped to 20 million/ml semen. This becomes even more interesting when one learns that the mandatory use of synthetic vitamin B_1 to fortify white bleached flour went into law in the United States in 1939. Further evidence of the energetic difference between the synthetic and the natural vitamin was revealed by Phil Callahan while working at the Garvey Center in Wichita, Kansas. Callahan found that there is a profound difference between the infrared pattern of "fortified" white flour and "unfortified" flour. The unfortified has a much more organized and complete pattern.

Finally, a Swiss study by Kouchahoff evaluated the response of human leukocytes (white blood cells) when raw, cooked, and processed food are ingested. A known toxin was ingested as the control for immune cell response. When a known toxin was ingested the leukocytes increased in numbers as expected to fight the toxin. When raw food was ingested no leukocyte increase was observed indicating that the food was not "viewed" as a threat to the body. When cooked food was ingested, the leukocytes increased as they did when a toxin was ingested, indicating that the cooked food was viewed as a threat to the body. When processed food was ingested not only did the leukocyte count increase, but the ratios of the various types of leukocytes changed significantly, indicating that not only was the processed food a threat to the body but also it was diverse in its threat.

In all the studies mentioned the differences in the foods lie in the life sustaining *energy* of the food because the chemistry and physical nutritional aspects of the food are not significantly different. This understanding is, of course, very difficult for the microwave and irradiation advocates who wish to squelch such information in order to perpetuate their profits. Irradiation of "food" only covers up the mistakes of growing nutritionally inferior crops and perpetuates the *myth* that crop diseases, insect pests, and cosmetic flaws are due to pesticide and radiation deficiencies.

Since most people are entrenched in the physical chemistry

aspects of living systems at the exclusion of the energetic aspects, they find it near difficult to impossible to perceive of measuring such aspects. Such measurement and evaluation, fortunately, has been extensively researched and practiced, forming the conclusion that the root of all life is truly the energy component. Let's look at a few of the various systems that have been developed.

Starting with the basic electrical circuit we see that the way we evaluate it is via various meters (volt, amp, ohm) and alter it by changing the components we place in the circuit.

Moving to the radio and television system, we measure its quality by the sound and/or picture we receive. We generally alter this quality, by changing the components of the system, but we can also alter either by simply placing local bodies near the system. Anyone who has used a radio and/or television knows that his or her own body positioned near the radio and/or TV can greatly alter the reception quality yet no physical contact is made with the system. Physical alteration of the system need not take place in order for its function to be altered.

This phenomenon holds true especially for living systems. The simple muscle test using applied kinesiology on a person proves the point. Simply by placing a substance near a person's body the strength of any given muscle can be tested. If the substance in question promotes greater order (if its energy field facilitates greater order in the body's energy field) the muscles will be stronger. If the substance in question promotes greater disorder, the muscles will be weaker. This can all be determined without having to actually ingest or introduce the substance physically into the body. This technique is employed by progressive holistic health practitioners saving patients the often timely and expensive regimen of the normal trial and error treatment of taking experimental products.

Further substantiation of energetic evaluation of biological systems is given by many progressive scientists. Phil Callahan, working at the Garvey Center in Wichita, Kansas, showed that the infrared spectrometer pattern of the breath of a healthy person was significantly different from the breath pattern of a person with cancer.

The Voll instrument places the client in the circuit of an instrument. It utilizes the body's electrical meridian system—proved to exist by Doctors Darras and Devernejoul, both M.D.'s in France—to isolate specific imbalances and disorders, any and every place in the body. It can then be used to determine what substances(s) will work to balance them.

Kirliean photography, developed by a Soviet scientist, utilizes a technique to photograph the energy field of living and non-living things by placing the object in a high frequency, high voltage field and photographing it. An advancement in this technique has been developed by Dieter Knapp of Germany, who utilizes different colored photographic plates for more definitive evaluation data.

Further evidence of energetic characteristics of living systems is revealed in the Reich Blood Test, developed by the late Wilhelm Reich, M.D. This blood test reveals under 2500 to 5000× magnification that blood cells will disintegrate into energy vesicles—called "bions" by Reich—as the blood breaks down. A healthy person's blood devolves to one form and a sick person's blood devolves to another form.

The Electromagnetic Blood Test developed by Dieter Aschoff, M.D. of Germany shows that healthy peoples' blood exemplifies a strong magnetic field as well as a strong electrical field, whereas the blood from cancer patients has a very weak magnetic field and a dominant electrical field.

Finally, one of the pioneers of energetic analysis of living things, particularly people, was Albert Abrams, M.D. at the Cooper Medical Clinic, now Stanford Medical School. He is listed in *Who Was Who in America*, authored several medical text books, and was revered as a world authority in medical pathology and the perfector of diagnostic percussion; percussing various parts of the body to locate and evaluate organs from the tonal qualities heard. Abrams is now recognized as the "father" of radionics.

There are other significant methods of evaluating living

systems energetically and there are additional pioneers not mentioned. The question we can ask about it all is, *What have we learned about energy, particularly, its application and implications in nature?*

From the Russian cytopathic experiments replicated 5,000 times and duplicated by Wilbanks at the University of Chicago, we learned that cell cultures can and are affected by what happens to "sister" cell cultures whether that affect be stimulatory or fatal, even though there is no physical contact between the two cultures. This was determined by growing duplicate cell cultures and contaminating, irradiating, or stimulating one of the cultures. The duplicate culture, though not contaminated, irradiated, or stimulated developed all the characteristics of the culture that was contaminated, irradiated, or stimulated.

We have learned from the Australians, who placed beakers of sodium chloride (table salt) in series with a DC electrical circuit, that when protein is added to one beaker, thus altering crystal formation of the sodium chloride, the crystal formation in the second beaker is equally affected as in the first beaker though no protein is added to the second beaker. Crystal formation "information" is transferred via the DC circuit from one beaker to the other.

We have learned from Robert Becker's work that tissue and limb regeneration is obtainable via energetic stimulation. From his work we also learned that the external, man-made electrical pollution can be as "toxic" as the most toxic pesticides and industrial chemicals.

From Noordenstrom's work in Sweden we have learned that cancerous tumors can be shrunk by weak electrical currents imposed across them.

Further, the Russian microwave work has taught us many things. First, cell growth can be altered for three-hundred generations by simply irradiating the first generation with a specific frequency and wavelength microwave, yet at the same time causing no genetic mutations. Second, enzyme synthesis by microorganisms can be altered at will by irradiating the microbes with the proper resonant frequency or harmonic. Third, the pH of a solution containing microorganisms can be changed simply by irradiating the microbes with a frequency that changes their cell membrane dielectric constant. Forth, the detrimental effects, particularly on lymphocytes and bone marrow cells, of ionizing radiation such as X-ray can be totally eliminated by the simultaneous irradiation of the living organism with a specific frequency of microwave radiation. The Soviets have verified that interactions in nonliving and living nature are electromagnetic and non-linear, meaning their effectiveness depends upon "tuning," not "power."[3]

Finally, the work of Cleve Backster in monitoring plants with a polygraph, now combined with his work with human cells, has conclusively taught us that energy is the universal link between living organisms.

With all of these research data and experimental proof why do so many people still have such difficulty in grasping and believing biological energy concepts? Very simple. Biological energy systems are like "whispers" in a society of mechanical "shouts." The gaudy, high energy instruments, employed by conventional scientists, in and of themselves drown out the subtle, low energy biological impulses that drive all living systems. The majority of researchers and non-researchers alike are not "quiet" or attentive enough to detect these subtle energies, so they arrogantly claim they do not exist. They are so used to listening to "shouts" that they totally ignore the "whispers."

Let's take a moment to preview two methods employed in "listening" to nature's whispers. As the Bible suggests, "Be still and know. . ."

Perhaps the most intriguing discoveries of biological energies has been made fairly recently by Philip S. Callahan. Using a Tektronic 222 oscilloscope operated by battery pack and a suitable biological detector (a leaf, his body, or a Photonic Ionic Cloth Radio Amplifier Maser) he found that the ancient structures in Ireland were built with a diamagnetic-paramagnetic sandwich design and as such amplified the 8 Hz. (alpha) earth frequency and the 2000 Hz. electrical anaesthesia radio wave frequencies. Further, he found that the megalithic chambers additionally synchronize the 20,000 Hz. signal to the beat of a healthy human heart. The power of these signals Callahan found to be in the range of only 5 to 60 millivolts. Callahan was able to draw several conclusions from this research, three of which will be mentioned.

One, the 60 Hz. signal from man made power systems definitely distorts the 8 Hz. and 2,000 Hz. natural frequencies. Two, "There is absolutely no doubt at all that these ELF radio waves are involved in enhancing the human immune system and seed germination and plant growth." Three, "In absolutely no case did the structures demonstrate amplification if they were not a paramagnetic-diamagnetic sandwich."[4]

These three conclusions are particularly significant to our discussion because they further explain the energetic aspect of farming. The fact that 60 Hz. AC distorts natural radio waves is significant, because it means that our evaluation instrumentation and technique must be void of this distortion. One system that fits these criteria is radionics, that will be touched upon here, but discussed further later. We know this system is minimally a radio wave system that allows the technician to evaluate the energetic interaction (potential ordering or disordering) between fertilizers and soils or fertilizers and plants, that ultimately determines the material response seen in the field. This technique is similar to "muscle testing" the field and crop.

With radionics one can reveal the causes of problems and with this knowledge become personally responsible for solving or causing one's own problems. Many people would rather shrug this responsibility off onto their Extension agent, chemical or fertilizer salesman, the government and power brokers, or the weather. Many people are simply afraid of the truth, for the truth may demand a change in management or a change in products used.

Another problem is the lack of knowledge. "My people perish for the lack of knowledge." Many people do not understand and do not want to understand. Carey Reams repeatedly stated, "When the student is ready the teacher will appear."

Perhaps the biggest problem people have with radionics, and even with Callahan's work, is with philosophy. Some people feel it is inappropriate to take a photograph (liquid crystal antenna) and drive insect pests out of a crop by broadcasting with a radionic instrument, but fully appropriate to spray a poison on this crop, that will be eaten by one's fellow man. Or people feel it is inappropriate to formulate a fertilizer prescription with the aid of a radionics instrument resulting in a healthier crop, that naturally resists disease and insect pests.

Unfortunately the mindset opposing bio-energy concepts is based upon fear and fear filled people are restless listeners. Many people believe that aphid infestation, for example, is due to a lack of pesticide or a lack of predator lady bugs. In actuality aphid infestation is due to nutritional disorder, particularly non-protein nitrogen. Management of energy nutrition is the key to farm "peace." If the farmer manages for disorder, then God sends in the sanitation engineers, insects and diseases to clean up the mess the farmer created. Caretaker plants called weeds restore order. If we as farmers heed this teaching we can manage our farms for *order*.

Since energy is the amount of order present, how do we build a soil and cropping energy system as encouraged by Carey Reams and Dan Skow? We do it very much like building a home.

First we must lay the foundation. This includes the calcium, necessary to establish the capacitor characteristic, and the base minerals, necessary to initiate the magnetic susceptibility characteristic (antenna), particularly the paramagnetism Phil Callahan noted in his research. This foundation initiates ordering and determines the amount of energy that can be collected in this soil.

Next, we frame the home. This includes providing the needed trace elements but most importantly includes establishing the fundamental nutrient ratios measured with the "Reams Soil Test." These include the 7 to 1 calcium to magnesium and 2 to 1 phosphate to potash ratios, particularly. This phase of building allows us to tap or plug into the electromagnetic grid we laid as the foundation. This is like our "plug-in receptacles" for our appliances, the plants.

Finally, we can finish our home. This includes most importantly establishing the humus complex, that allows us to regulate, store, and transform the energy into which we have tapped. This provides us with the paramagnetic-diamagnetic sandwich noted in Callahan's research necessary for optimum seed germination and plant growth.

The architect of our building endeavor is God. We, the farmers, are the contractors, and the microbes and plants are the carpenters. In this divine plan we are building order or actually setting up the circumstances for order to be built. The epitome of this building of order is photosynthesis, the ordering of collected components into matter, i.e. sugar.

Somewhere in the building process we need a method of monitoring our progress. Weeds, diseases, and insect pests are nature's litmus tests for our progress, yet it is often helpful to have an intermediate monitoring tool or tools to satisfy man's instrumental desires. There are actually two such tools that very effectively serve this purpose. The first is the refractometer, that measures sugar, the product of photosynthesis, and dissolved solids, the overall mineral concentration of the plant sap. The higher the refractometer reading, the greater the order/energy present in our cropping system, whether it be a conventionally fertilized system or an "organically" fertilized system. There are people in both camps who vehemently bad-mouth the use of the refractometer for monitoring crop quality and progress. Upon investigation, I find that their criticism is rooted in their inability and lack of understanding in getting crop refractometer readings to increase consistently and remain stable. They arrogantly contend that since *they* cannot do it, the refractometer is an invalid tool for such work.

Part of their failure lies in their lack of understanding and disuse of our second tool, the conductivity meter. This meter measures the current flow of ions in the soil at 540 cycles per second. The reading is an indicator of the sufficiency of "ordered power" available to drive the "appliances/machinery," the crop and microbes. It does not measure every aspect of soil energy but, regardless, if your soil does not read at least 200 microsiemen (net) of energy on the conductivity meter, it does not have sufficient "power" to either increase or maintain crop refractometer readings above 12. In this case of too low power you experience a "brown out." If the conductivity reading gets too high you experience "burn out." The true artist-farmer learns how to maintain a satisfactory soil conductivity reading with the desirable crop refractometer reading, that can be done only in combination with each other.

Thus, our goal is to build a formidable foundation with calcium and base minerals; a suitable frame with desirable trace elements and nutrient ratios; and a well crafted finish of sufficient humus and microbes. As farmers our goal is to build order, that relates to the amount of energy present. To do this we must recognize that every fertilizer is different, energetically, and so must be evaluated as such *a la* Callahan and radionics. Our progress should be monitored regularly with the refractometer and conductivity meter to gauge our management.

Some may contend that, if this were all so true, why doesn't everyone know it and why don't our lords of wisdom in the agricultural colleges teach it. H.T. Urt in his book, *Masterful Persuasion* perhaps answers this question best. "Daily doses of bits of nonsense repeated for only ten years wipes out one hundred years of truth." We have had a least fifty years of petrochemical warfare nonsense in order to perpetuate petrochemical agriculture. Truth is not determined by majority opinion. Though nonsense can obscure and confuse, as it has for decades, it cannot change the truth.

Every farmer can build a house. He must simply use this same common sense to build an energy system in the soil. He must establish that same "silent harmony" in his soil as that which he exhibits in a community barn raising.

—6—

PLANT FUNCTION

NUMEROUS BOOKS HAVE BEEN WRITTEN on the function or physiology of living organisms. These publications, however, have been mechanistic in their approach, asserting that every activity of life is mechanical or solely physio-chemical. Nowhere in traditional texts on plant or animal physiology has it been suggested that the function of living organisms is energetic or electromagnetic. Thanks to pioneering scientists like Philip Callahan, Fritz-Albert Popp, Carey Reams, and Robert Becker, we know that the function of living organisms is fundamentally energetic, not mechanical or physio-chemical. In this chapter, the function of plants is discussed, and a possible scenario for nutrient-soil-plant interaction is presented.

From reading *The Secret Life of Plants,* we know that plants actively affect electrical instruments. From the basics of chemistry and physics, we know that all matter is fundamentally a compilation of positive and negative charges. From this we also know that all chemical reactions, inorganic and organic, biological and nonbiological, are fundamentally the exchange or rearrangement of positive and negative charges. A simple acid-base reaction in the manufacture of calcium nitrate shows this: $2\ HNO_3 + CaCO_3 \rightarrow Ca(NO_3)_2 + H_2CO_3$ (goes to $H_2O + CO_2$) + heat. The hydrogen on the nitric acid molecule ionizes to a hydrogen ion that is a proton, a positively charged particle, leaving the negatively charged nitrate ion free to react with the calcium (Ca^{++}) ion, forming $Ca(NO_3)_2$. The hydrogen ion reacts with the carbonate ion (CO_3^{--}), forming carbonic acid (H_2CO_3), which immediately decomposes to water (H_2O) and carbon dioxide (CO_2). Acid-base reactions are termed neutralizations; they produce considerable heat.

Electrical charge is what holds molecules together. The point here is that both physical/observable phenomena and non-physical/unobservable phenomena are fundamentally electromagnetic or energetic. Lund and Waller have demonstrated and determined definite potential gradients in plants—a constantly fluctuating potential difference between the aboveground part of the plant, the root, and the soil in which the plant is growing. Their investigations show a definite bioelectric current in the plant and therefore lend support to Breazeale's theory of plant nutrition as an electrical phenomenon.[1] Thus it is appropriate to discuss plant function in terms of electromagnetics. Why discuss plant function? The better one understands plant function, the more insight he has regarding fertilization, plant feeding, and getting plants to perform as they should.

Most farmers are familiar with electric motors, starters, alternators, and generators, or at least magnetos. How does a motor run? How is it that an armature can be made to rotate by running electrical current through it? In 1820, Hans Christian Oersted discovered that a current-carrying wire produces a magnetic field. A magnetic field exerts a force perpendicular to the magnetic field; this force causes the armature to move. Now, we know that a magnetic field flows over the earth's surface and that there is a force perpendicular to that. We know there is a magnetic field because compasses and geomagnetometers indicate there is one. We know there is a force because the laws of physics say there is and because the earth is turning like the armature of a motor. By deduction, then, we know there must be a current flow perpendicular to the earth. The incoming current source is sunlight and solar light energy. The outgoing current is somewhat more elusive, but it also is in the form of light.

Light possesses the characteristics of both particles and waves. As a wave function, it inherently has crests and valleys like the waves on a body of water. Have you ever seen waves on a lake where there were no depressions or valleys between the crests? Of course not. In nature, waves always have valleys and crests. These valleys and crests model alternating current, that automatically creates an alternating magnetic field. Because there is no diode to convert from AC to DC, current flows both into and out of the earth. This is where the connection is made between electromagnetism and plant function.

Plants receive and use light from the sun; this is current flow in one direction. Soviet researchers have proven that energy does, in fact, flow in this direction when the system is operating as it should. Through observation, we know that energy (nutrient) flows in the opposite direction as nutrients are extracted from the soil. The Soviet work showed that 40% to 80% of a plant's weight in the form of usable nutrients (amino acids, proteins, vitamins, and enzymes excreted by the roots and metabolized by the plant) can and should be deposited into the soil. Thus we have current flow in both directions, manifested as physical nutrient synthesis movement in both directions.

A discussion of plant function must ultimately be related to the farmer's fertilization practices and his gaining a better understanding of what he is accomplishing with those methods. It is hoped that the farmer will translate this understanding into innovative management practices in the field that achieve the desired outcome—a healthful food supply, a healthy planet, and a successful farm operation. The discussion of plant function in terms of biophysics allows us to appreciate how all disciplines fit together. Most farmers are familiar with basic wiring and general electrical maintenance

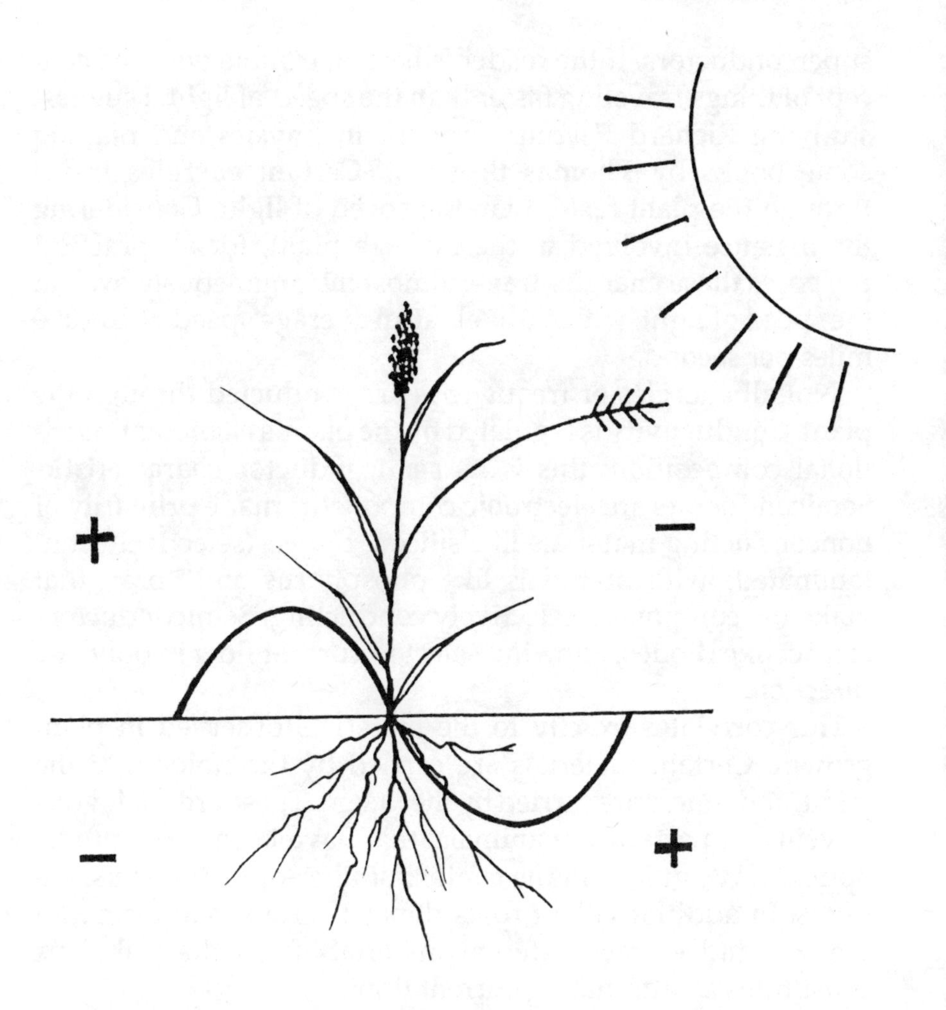

Plant Growth Is Electromagnetic Non-Linear

on the farm. This is an important foundation to have, in order to grasp the way plants grow.

Carey Reams emphasized that plant growth is independent of time; it is regulated entirely by energy. It is possible, he said, to grow plants from seed to harvest in just a few hours. From a mechanistic viewpoint, this seems miraculous, but from a biophysics viewpoint it becomes logical. Ideally, plants function as both semiconductors and superconductors. In his Tachyon experiments, Callahan proved that plants function as superconductors. If the reader is having trouble with the concept of things traveling faster than the speed of light, I suggest studying Richard Feynman's work in physics and reading some books by Thomas Bearden.[2] Certain energies travel through the plant faster than the speed of light. Considering the distance involved in the average plant, for all practical purposes these energies travel almost instantaneously even at the speed of light, which travels at an average speed of 186,000 miles per second.

Not all energies or frequencies are conducted through the plant. Conductivity is regulated by the plant's elemental-nutritional composition; this is its semiconductor characteristic. Semiconductors are electronic components made primarily of nonconducting materials like silicon, doped (selectively contaminated) with materials like phosphorus and boron that make the component selectively conducting. Semiconductors can act like diodes, allowing selected current flow in only one direction.

This correlates exactly to the materials observed in plant growth. Certain materials are carried by the phloem of the plant, and others are carried by the xylem. These are analogous to veins and arteries in animals, that have been shown to be optical wave guides in the context of fiber-optic transmission fibers. In addition, plant roots deposit certain materials into the soil and extract different materials from the soil. This constitutes an alternating current flow.

As with any electronic equipment, the quality of the equipment's output is correlated directly to the quality of its components. A faulty semiconductor can shut down the entire component system or reduce its efficiency considerably. The key to the function of the plant system is its design, determined by the genes, and its construction, determined by nutrition. Major nutrients are the foundation or core of the plant cell. The most important of these major nutrients is calcium.

As Hans Nieper stated, calcium gives the animal or plant cell its capacitor characteristic. Having a capacitor characteristic, the cell is able to store and subsequently discharge—in a word, pulse—electrical charges, which are necessary for the cell to operate efficiently. The greater the cell's capacitance, the crisper its electromagnetic function. In classical electronics, a capacitor operates as a capacitor in only an AC system. It acts as a spark gap or circuit interrupter in a DC system.

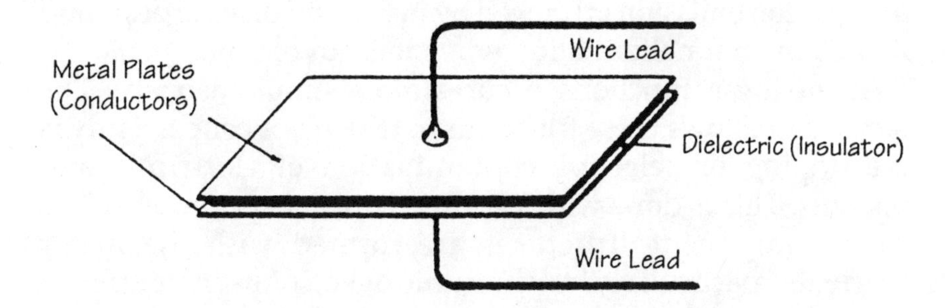

Basic capacitor construction

In the AC mechanism of the cell, the capacitor characteristic functions to maintain the electrical potential within the cell that is necessary for cell function. Also, as in classical electronics, the capacitor characteristic acts as a frequency tuner. In the DC mode, the spark-gap characteristic functions as a pulse generator. A charge is built up on one side of the cell membrane to the point at which it arcs across the membrane, creating a pulse that can then be read and interpreted by surrounding cells if it is expelled from the cell or read by the cell's own "'brain,'"—the mitochondria and nucleus—if it is absorbed by the cell.

Cell membrane nutrition determines the overall electrical capacitance of the cell. This is where one can easily acknowledge the link between physio-chemical and electrical energy. The two cannot be separated. As the membrane's composition and resulting integrity variations resulting in cell capacitance change, so does its frequency tuning, e.g., the communication signals it transmits and receives. Of particular importance in this communication process is the "reading" and "interpretation" of the DNA, or genetic information. The reading and interpretation of the DNA determine the cell's execution of the information that determines the integrity of new cells, tissues, and organisms. This integrity then determines the subsequent integrity of communication signals, the reading and interpretation of genetic information, and the continuing cycle of new cell formation. This composite communication emission is tracked by insect and disease pests and homed in on for elimination when it is out of tune.

All of these functions occur almost simultaneously. The factor that allows these functions to take place concurrently is the doping or selective contamination of electrical components. This is done with various elements or nutrients. The basic component structures are formed with the major nutrients, particularly calcium, nitrogen, phosphate, potassium, carbon, hydrogen, and oxygen, categorized and fine-tuned by trace elements such as boron, silicon, copper, and so on. Depending on the design conferred by the genes, each frequency and wavelength of the electromagnetic spectrum, including scaler waves[3] and photons[4], will participate in the life of the cell. Each is guided, blocked, pulsed, chopped, or altered according to the specific cell component it contacts. The key to this system's functioning is its quality of construction. Sloppy construction of sophisticated electronics equipment using inferior or second-rate components results in disappointing performance, regardless of the original design.

In nature, component quality correlates to nutritional integrity. In the genes, nature provides man with the design; it is his job to supervise the construction. Nature even provides the inspectors—insects and diseases—as well as helpers—weeds—to ensure success. Man has ignored these gifts, as well as the quality control signs on the plants themselves. One important sign is the glossy sheen on the plant leaf, which indicates a satisfactory wax or dielectric coating on the leaf surface. As Callahan has pointed out, this dielectric coating allows the antenna—the plant, in this case—to function most effectively. The less integrity this wax or dielectric coating has, the more static there is and the weaker the signal reception.

This is because the wax or dielectric coating on the leaf determines the impedance match between the atmosphere and the leaf. As this match changes, so does the integrity of the signal. In a nutshell, the poorer the plant nutrition, the poorer the dielectric coating on the leaf surface. This causes a less-than-optimum impedance match, resulting in greater static in the signals emitted by the plant as well as in those collected by the plant and used for its life processes. This static in the emitted signals is what nature's garbage collectors respond to and home in on. One problem encountered in detecting the subtle signals of living systems, such as plants, using mechanical instruments is matching the instrument/probe's impedance with the plant's impedance. If these impedances are not matched satisfactorily, no signal will be observed. This accounts for reductionist scientists' repeated contention that biological systems are nonenergetic. However, scientists like Phil Callahan, who have been successful in detecting biological signals, use biological detectors like doped linen or the human body to match the impedances between subject and instrument. Some people contend that the human being or any other living system is not an energy detector. One simply needs to hear a radio program get louder when a person stands in a certain position near the radio to know that biological systems are at least collectors of radio waves.

For all practical purposes, an antenna is simply a wave guide collector and a concentration amplifier. It is a collector of the energy it is designed to accumulate. A rain gauge collects rain, but it does not generally collect birds, insects, or animals. The first hearing aids were large turned horns, that constituted traveling wave guides. They simply collected a large volume of sound waves and focused or collectively guided them to the ear. Thus, the hearing-impaired person received a greater concentration of sound and was enabled to hear better. If one coated the inside of a hearing horn with foam padding, the padding would absorb the sound and prevent the sound waves from entering the ear. If one coated the inside of this hearing horn with beeswax or polished the interior of the horn, this would decrease the absorption of sound into the horn material, thus increasing the horn's efficiency as a hearing aid.

Electronic and biological antennas are of various designs and shapes, appropriate to the types and shapes of energy they collect. If the antenna surface gets even a little dirty, corroded, or absorptive, the antenna's efficiency declines. The sheen observed on highly mineralized plants is a result of improved surfacing of the leaves, which results in a more "polished" antenna that collects energy more efficiently. The less of this coating there is, the greater the plant-energy and respiration loss. These are signals to the quality-control inspectors to take a closer look. In many cases, the soil in which these plants are growing is spewing free ammonia into the atmosphere, either from ammonia fertilization or anaerobic soil digestion. This further pumps up the plant signal—turns the volume up, as one can do with modern hearing aids—notifying the quality-control inspectors to reject this production run due to inferior construction.

If the efficiency of the antenna, the plant, in this case, is less than optimum, even if the soil is in excellent condition and provides a satisfactory electrical ground, that can happen when a poor seed is planted in good soil, there will not be completely satisfactory collection of solar and cosmic energy by the plant. This results in a deficiency in the energy inventory, that results in a deficiency in the necessary components the plant needs to manufacture all of its products. Here the plant will skip certain manufacturing functions and cope the best it can with the materials it has. As fewer and fewer components are available, less and less of the plant's total function is performed, energy leakage occurs, the circulation system gets clogged, cells age prematurely, short or faulty circuits develop, and finally the plant becomes diseased and dies. This faulty circuit is what we observe in most of crop production today relative to energy flow from the plant into the soil. Energy flows in only one direction, on a limited basis at that, from the soil into the plant. This results in the mining of our soils.

As the current flows in one direction and then the other, as it is truly meant to happen, nutrients are moved in the corresponding directions. The fact that plants actually extract and deposit nutrients is well documented in *Soil Microorganisms and Higher Plants.*[5]

> Plants . . . excrete various volatile and non-volatile substances—with the aid of special glands or by guttation. Vigorov in 1954 observed that the intensity of guttation depends on illumination, soil humidity, on the presence of nitrogen and other nutrient elements.[6]

Plant nutrition is the key. Introduction of ammonium salts into the soil elevates the excretion of nitrogen compounds in the guttation drop. If these ammonium salts accompany a balanced nutritional profile, particularly carbohydrate, phosphate, and calcium, then the excreted nitrogen compounds will be functional organic units. One of the tests for the presence of organic substances in the fluids is the growth of microorganisms in them. Fungi grow abundantly in guttation drops. These are beneficial fungi if the guttation drops contain functional biological metabolites. They are pathogens if these guttation drops contain mostly non-protein nitrogen that results from poor nutritional balances. Plant nutrition is the key to what the plant excretes and consequently the species of microbe that thrives. Pathogens are present because of what the plant excretes—garbage relative to the needs of beneficial microbes.

The efficiency of this alternating circuit system of the plant depends on the quality of the soil, the plant (especially the initial seed), and the atmosphere. The quality of the soil is related to its mineral balance, humus content, moisture, and particularly its microorganism activity. The mineral provides the structural components, resulting in a substantial paramagnetic value. The humus acts as the insulator, stabilizer or choke, variable resistor, and partly the capacitor of the system. The active microorganisms act as regulators, variable capacitors (tuners), circuit breakers, and programmers for the system. It is difficult to assign each component of the soil a distinct role because the components are interdependent.

The quality of the plant starts with the seed. The plant will be no better than the potential of the seed. Traditional hybrids selected for and grown primarily with nitrogen and potash, anhydrous ammonia and potassium chloride, must have a similar fertilizer program in order to grow as expected. The plant needs a soil that is tuned or designed for it, like matching radio antennas to radios, and television antennas to televisions. If you put this narrow-spectrum, hybrid type of seed on a balanced living soil, neither the soil nor the plant would fare well.

Another analogy is that of putting inferior speakers onto a high-quality stereo system. The system is only as good as its weakest component. In this case, the soil will be able to give up the narrow spectrum of nutrients in one direction, but the plant will not be able to conduct the broad spectrum of nutrients in the opposite direction. The plant or soil can conduct energy corresponding to only the quality the plant or soil possesses. Any energy pattern passing through a system that surpasses the system in sophistication and quality will result in energy loss, debris accumulation, and premature aging of the system; in the plant this is evidenced by clogging of the vascular system, disease and insect infestation, and early death.

To summarize the analogy between the soil/plant system and a stereo system, one could say that the soil-mineral component relates to the stereo antenna and receiving mechanism, whereas the biological component relates to the multiple-component tuner and speaker mechanism. The stereo system as a unit emits music at some degree of quality, at some level of yield or power (watts). By analogy, our natural system produces a crop at some degree of quality, at some level of yield or volume. Both stereo and soil/plant systems are energy converters. One converts radio waves to sound, and the other converts solar energy to matter. The design or blueprint for the biological system is held by the DNA, which is the master "information chip" for designing the system. It then acts as the feedback mechanism to keep the system on track.

Now, a poor seed will not produce good seed on poor soil, but it will produce the quantity of poor seed it was bred to produce. A poor seed on good soil results in an impedance to the flow of energy back into the soil. A good seed on a poor soil causes an impedance to the flow out of the soil into the plant. Therefore, seed matching is very important. The analogy can be made to two people talking to each other on their CB radios. If both CB's are tuned to the same frequency, communication is successful. If one or the other is out of tune and can either transmit or receive but cannot do both, communication is unsuccessful. I have experienced seed matching on many acres, and, without exception, those farmers employing anhydrous ammonia, potassium chloride, must use certain hybrids to obtain the desired volume of yield. The feed value is very poor, but that is of little concern to these farmers because they are selling the crop. Farmers who have well-balanced soils on biological mineralization programs will fail using these same hybrids. They must use seed grown on similar programs in order to achieve maximum efficiency.

Carey Reams repeatedly asserted that plants absorb much nutrition from the air. But they can do this only if the plant is a good conductor and if the soil acts as a good electrical ground. To be a good ground, the soil must be highly paramagnetic; this means that the soil is a good antenna for solar and cosmic radiation. Paramagnetism does not guarantee fertile soil, but it is a prerequisite for fertile soil. Sterile soils are diamagnetic; they are poor antennas. Paramagnetism is achieved by both the mineral composition and the physical structure or form of the materials in the soil. The microorganisms are responsible for creating structured materials in the soil that are paramagnetic.

When the direction of nutrient flow from the soil to the plant is reversed, the atmosphere acts as the electrical ground. The quality of the atmosphere is determined by the sun, solar and cosmic radiation, pollution, and atmospheric gases. Superior soil and plant quality can partially compensate for mediocre atmospheric conditions. Always view the soil/plant/atmosphere system as an electrical apparatus. If you apply something to the soil that interrupts the mineral balance, organic stability, or microorganism activity, you de-tune the circuit, reducing its efficiency. At first, nature attempts to re-tune the circuit with additional antennas and conductors in the form of weeds. If the circuit is de-tuned badly enough, nature calls in the garbage collectors, the insects, to clean it up. Disease organisms are simply an indication that the circuit is way out of tune and is building up debris as a result. Have you ever noticed how an overloaded electrical circuit will start to erode or break down, or how an engine that is out of tune will build up debris?

All of the chemical analyses, CEC tests, and tissue analyses in the world will not indicate how the circuit as a whole is functioning. In trouble-shooting an electrical circuit, one might find that, when the components are removed and tested individually, each is intact and functions well. However, the circuit itself does not function at 100% efficiency. Finally, one discovers that, under load, in the circuit, one of the components is faulty. As Fritz-Albert Popp, Vlail Kaznakeyev, Robert Becker, Philip Callahan, Carey Reams, and Rudolf Steiner have proved repeatedly, the functional foundation of nature is electromagnetic. The chemistry of nature is secondary.

Before concluding this chapter, I shall discuss one other electromagnetic function of plants, especially trees. Trees serve as highly directional capacitors, altering the electrical tension in the atmosphere and thus the weather. Wilhelm Reich found that he could simulate this function with metal tubes grounded to water. He demonstrated that clouds, atmospheric pressure and storm fronts, and ultimately the weather could be altered with such devices. Many scientists have replicated Reich's experiments and confirmed his results. Trees are nature's weather regulators. It is well known that trees improve the atmosphere of an area and that massive clear-cutting alters the rainfall of that same region. This is simply nature at work, applying the laws of physics that man has ignored, denied, or exploited.

There are no accidents in nature. Every cause has its effect. Scientists like Callahan and Reams have realized this and been able to see the perfect divine order in all things. Plant function, like the function of every living system, is a marvel. It is much more sophisticated than any man-made system, yet it is no less straightforward in its principles of operation and design. The principles of nature are valid throughout; they apply equally to a living cell and to a semiconductor or a microchip used by NASA.

Understanding these principles and laws does not in itself raise good crops or regenerate soils. It does, however, give you a blueprint by which to grow good crops and regenerate poverty soils. You must read the blueprint and then assemble the necessary materials to carry out the plan.

—7—

CONCEPTS IN MICROBIOLOGY

MICROBIOLOGY IS THE STUDY of organisms—both plants and animals—that are too small for one to see without magnification. These include bacteria, algae, fungi, yeasts, actinomycetes, nematodes, protozoa, viruses, and a few others. Mainstream academic discussion of microorganisms in reference to agriculture focuses almost entirely on microorganisms associated with disease and their control with toxic chemistry. However, an increasing number of people are discussing a new approach to pathogenic microorganisms—that of biocontrol, the use of natural predators and antagonists rather than toxic chemistry to control undesirable microorganisms. Some universities even have begun soliciting funds for research on the possibilities of biocontrol.

Numerous university and field study reports are available on the integral function of microorganisms in agriculture and the subsequent biocontrol of undesirable micro-pests. The answers to problems of crop diseases, soil infertility, fertilization, and commodity integrity are all available to read about, consider, and implement immediately. Biocontrol is not a recent revelation, although many people are claiming to have made the discovery. Obtaining information about biocontrol does require some personal initiative and research because it is not reported on the nightly news, in the *Farm Journal*, or by local Cooperative Extension agents.

One of the most informative publications available on agricultural microbiology is *Soil Microorganisms and Higher Plants* by N. A. Krasil'nikov of the Institute of Microbiology, Academy of Science of the U.S.S.R.[1] The book contains more than 400 pages of case-study information and more than 640 bibliographical entries. I regularly suggest to farmers that they read this book, but few have actually done so. Consequently, this chapter is devoted to a thorough review of the work. Many passages from *Soil Microorganisms and Higher Plants* are quoted and discussed in the following pages. I have verified most of this information in the fields of clients.

Many people, especially scientists, find it difficult to acknowledge and grasp the reality of things not readily evident to the physical senses. As a rule, people do not extend their senses into the realm of gifted sense unless they are blind, deaf, or disabled in some way. They are then expected to develop extraordinary sensitivity to the environment.

Only relatively recently have people come to acknowledge the presence of microorganisms, entities so small that a microscope is required to see them. Microorganisms often are discussed in the context of illness, when one's physician prescribes an antibiotic or medication to exterminate the microbes causing the illness. Even so, most people have never seen a microbe and do not know what one is. Of all the microorganisms in existence, only a small fraction are associated with disease, yet these get most of the attention.

In agriculture, microorganisms are one of the three things uninformed farmers most want to eradicate; weeds and insects are the other two. Seldom does anyone acknowledge the contribution microorganisms make to soil fertility and structure, crop production and quality (nutritional value), and the natural checks-and-balances system of every growing crop.

> The biogeny of soil is the most significant indicator of its fertility. As soon as the activity of a microbial population begins in a rock, the first signs of fertility are manifested. The degree of soil fertility is determined by the intensity of the life processes of the microbial population.
>
> It is impossible to solve problems of pedology, not to speak of agriculture and plant growing, without taking into account the microflora of soil. Plants are a very strong ecological factor, selecting certain species of bacteria, fungi, actinomycetes and other inhabitants of soil. *As a result of wrong agricultural practices and crop rotation, the soil becomes infested with harmful microbial forms* [emphasis added]. By use of suitable plants in the crop rotation, one may change the microflora of soil in the desired direction, and eliminate harmful organisms, in other words— restore the health of soil.[2]

We are generally taught in soils classes that microbial populations depend primarily on exposure or contamination. In other words, if one introduces a microorganism into the soil, it will remain there until the soil is sterilized chemically or by steam. Such teaching focuses primarily on disease organisms. What academicians neglect to teach is that soil microorganisms live in a symbiotic relationship with the plants growing in the soil. The microbial population is determined by root excretions, that in turn are determined by crop nutritional management practices.

> Plants create and form microbial societies, which affect the microbial population of the entire root system and harvest remains. During their life, plants excrete through their roots various organic and mineral substances which attract microorganisms. The species composition of the microflora, just as the soil does, has a definite and considerable influence on the growth and development of plants, and consequently on the crop yield.[3]
>
> In general, many different forms of organisms may grow in the rhizosphere of plants, both useful and harmful; those which facilitate the nourishment and development of plants, and, on the contrary, those which inhibit and poison them. The prevalence of these other

organisms depends on soil-climatic condition, on the manner in which the farm is handled, and on the whole agrobiological complex.[4]

Farm management, not chemical warfare, determines whether the soil and crop are healthy or infested with disease.

The vegetative cover, as a whole, is a powerful determining factor in the microbial biocoenoses of soils. In the zone of the root system, only those organisms which assimilate the root excretions of the plant in question more quickly, can develop, supplanting other, less well adapted, species.[5]

The symbiotic relationship between plant and microbe occurs in the rhizosphere, which consists of the third- and fourth-order roots and root hairs, along with the organically complex soil/microbe medium immediately surrounding these roots and root hairs. This zone can be likened to the digestive system of animals.

Increased accumulation of microbes in the root soil was first observed by Hiltner in 1904. He proposed the term "rhizosphere." In investigating the root system of various plants, Hiltner came to the conclusion that the accumulation of microbes in this area was not accidental and that it was caused by the biological activity of the roots.[6]

The microflora of the root zone is of great importance in plant nutrition. Growing near or on the roots, microorganisms, together with the plants, create a special zone—the rhizosphere. Soil in this zone differs in its physical, chemical, and biological properties from that outside the rhizosphere. The interactions between microbial species and between microbes and plants result in the formation of plant nutrient compounds. Substances present in the soil are subjected to a greater or lesser extent of processing before their absorption by the roots. *The plants do not absorb those compounds which are characteristic of soil outside the rhizosphere but rather they absorb metabolic products of the rhizosphere. The rhizosphere microflora prepares various organic and inorganic nutrients for the plants* [emphasis added].[7]

The preceding statement refutes the assumptions and findings of chemical plant feeding studies. These assumptions and findings are taken purely out of context.

> The role of the rhizosphere microflora reminds one of the digestive organs of animals. Microorganisms in the final account serve the same function in the plant nutrition as the digestive system of animal organism.[8]

Like the metabolism of nutrient complexes by microbes in the digestive tracts of animals, the rhizosphere microbes metabolize multitudes of nutrient products for use by microbes and plants. This relationship between plant and microbe is necessary for the health and efficient growth of both.

> In the rhizosphere, iron, manganese, and other metals occur in combination with organic compounds formed by microbes. Amino acids, organic acids, and other metabolites of microbes form stable complex compounds. They are utilized by plants and used as a source of specific organometallic nutrients. These are found in greatest concentration in the rhizosphere and are preserved in the soil for long times.[9]

In contrast, inorganic nutrients are readily leached.

> Research by Weinstein and others (1954) on plants in solutions both with and without microbial metabolites showed that mineral, specifically iron, uptake was faster with the microbial metabolites and showed that plants take up iron in a mineral compound but as an organomineral formed under the influence of microbes.
>
> In the rhizosphere, various minerals, rocks, limestone, marble, etc. are decomposed at a faster rate. This process is not only caused by root excretions (CO_2 and other acids) but also by the microflora of the rhizosphere. The more intense the growth of microbes, the faster the decomposition process of substances. Certain compounds, for instance, tricalcium phosphate, do not dissolve in the sterile rhizosphere of plants, but when soil microbes are added to the vessel the substance becomes available to the plants.[10]

The above cited work is of particular interest because it points out that the availability of plant nutrients is a biological phenomenon as much as or more than it is a chemical phenomenon. Carey Reams repeatedly stressed this, and Dan Skow still does.

As mentioned before, the plant's root excretions determine

the microflora species and populations. These root excretions are significant in quantity.

In 1928, a scientist named Demidenko concluded that the total root excretion comprised 27% of the plant mass.[11] This lends credibility to the contention that soil should get better, not worse, from farming. And why not? The root mass itself is very significant.

> According to a study done in 1937, 1938 on winter rye, the root hairs on the roots of one plant number to 14.3 billions. These root hairs are mainly located on the roots of the third and fourth orders. Totaled, the length of all the hairs of one plant equals more than 10,000 km. The total root mass of one rye plant in the fourth month of growth had a total length of 11,250 km and an area of 6,388 square meters.
>
> Detailed studies of the root system of fescue plants showed that about one third to one quarter of the root mass was living and that the weight of the aerial part of these plants constituted about one quarter of the total weight of the roots.[12]

Upon field inspection of various crop roots around the United States and other industrialized nations in which chemical agriculture is practiced, I have found the root systems of these crops to be far from ideal. The aerobic zone of the soil ranges from nothing to only a couple of inches. The depth of the aerobic zone determines the primary volume of the plants' rhizospheres. It takes oxygen to grow extensive third- and fourth-order roots and root hairs. Primary and secondary roots may be growing outside of the soil's aerobic zone, but their collective mass and volume are minor in comparison to the finer roots and root hairs that proliferate primarily in the aerobic zone.

It is interesting that, in modern institutional rhetoric, most of the biological work of the previous six decades is ignored. Most notable is the contention that disease organisms in the soil prevail where they have "infested" unless they are eradicated with toxic chemistry. It is not understood or suggested that the toxic-chemical practices are themselves the primary cause of the pests' overbearingness. Just as spraying for lygus bugs in cotton guarantees boll weevil infestation crises later in the season, with their accompanying pesticide sprays because the weevil's predators were killed by the lygus bug spray, so do soil fumigation and mismanaged fertility guarantee continued crises, plus they kill the natural predators in the soil, namely the *non*-sporeforming bacteria. These are the beneficial microbes; the sporeforming bacteria are the pathogenic or disease microbes. The latter cannot survive in a healthy rhizosphere.

Sporeforming bacteria are scarce in the rhizosphere. "They are especially scarce during the period of vigorous vegetative growth of plants. Usually these bacteria begin growing abundantly at the end of vegetation, especially on dead, decomposed (or sick) roots."[13]

Soviet scientists made this observation decades ago, before extensive toxic chemistry sterilized the root zone of the soil. They were observing true normality, rather than the commonality that is seen today. The fact that numerous pathogenic microbes are present in modern day soils and cropping environments indicates that an unhealthy situation exists.

One of the more important natural protectors and plant symbionts (companion organisms in which both plant and microbe benefit each other) is the mycorrhiza group of fungi. "Many plants do not grow well without mycorrhiza fungi, and some do not grow at all."[14] Under healthy conditions, these beneficial fungi completely encase plant roots, both feeding plants and protecting them from parasites and pathogens. Every nutrient that passes into or out of the roots is mediated by the mycorrhiza. This is totally ignored by the chemical, inorganic nutrient theorists of conventional agriculture who paint starry eyed theories of soil/root nutrient exchange.

Other organisms that are found in healthy rhizospheres include algae, yeasts, and actinomycetes. All of these organisms metabolize biotic products for the nutritional use of plants and other microbes.

> Research accomplished in 1946, 1953, and 1954 showed that various algae live in the rhizosphere of plants in considerable quantities . . . The number of algae in the root area was two to three times

higher than outside it, mainly diatoms, green and blue-green algae.[15]

Researchers today will not find the same microflora as they did in the 1930s, 1940s, and 1950s because of the pesticides and toxic fertilizers that have been used since then. As a side note healthy soils contain natural pest controls that are more effective than man-made pesticides without the "morning after" price tag.

When diatoms die, they form the material called diatomaceous earth, which is a common natural pest control product. In healthy soil, this product is formed automatically as part of the diatom life cycle. Commercial sources of diatomaceous earth are readily available. The product works well for controlling many insect pests.

Getting back to algae, several companies market blue-green products, accurately claiming the beneficial effect of such materials, particularly when herbicides have reduced or eliminated the natural algae populations.

> Sometimes one . . . observes three elevations in the growth curve of microbes: The first small elevation is seen in the early stage; a second great elevation ensues before and during flowering; the third elevation occurs before ripening.[16]

These elevations would seem to correlate to optimum fertilizing times.

> Upon the ripening of plants, the total number of microorganisms in the rhizosphere decreases. During this period the quantitative ratio between the different representatives and groups changes, the number of sporeforming bacteria, fungi, and actinomycetes increases, and new organisms appear. At the same time, the total amount of nonsporeforming bacteria decreases, some species disappearing altogether.[17]

This seems to mean that sporeforming pathogenic bacteria naturally appear only when it is time for the annual plant to leave or be harvested! The plant growth stage determines microbial composition, which determines the nutrient make-up available and needed by the plant.

> The prevailing group in the rhizosphere, regardless of the conditions of growth and the age of the plants, are the non-sporiferous [nonpathogenic] bacteria. The second place [largest group] among rhizosphere microflora is occupied by mycobacteria.[18]
>
> Microorganisms of soil display a direct, very essential influence on plants, a positive or negative one, depending upon the species and external conditions.
>
> Among the rhizosphere microflora there is a large number of species which affect the growth of plants with products of their metabolism. Some microbial species are active producers of various biotic substances—vitamins, auxins, amino acids and other substances essential to the growth of plants. Other species are antagonists of phytopathogenic bacteria, fungi and protozoa. Microorganisms of these species strengthen the immunization properties of tissues, thus protecting the plants from infections.[19]
>
> A great number of vitamin-producing bacteria are found in the rhizosphere of plants... They comprise 40-80% of all soil bacteria, and their number varies depending on the species of the plant, the stage of its growth, and upon external conditions.[20]

This would seem to validate vitamin fertilization and foliar feeding.

> Yeasts of the genera *Torula, Mycotorula,* and some others are widespread in the soil. These organisms are powerful producers of vitamins and various other biotic substances.[21]
>
> Between 1946 and 1950, it was determined that separate components of vitamins—thiamine, nicotinic acid, para-aminobenzoic acid—present in the air, are used by microorganisms. Biotic compounds enter the air and soil atmosphere from the soil and plants. Vitamins given off into the air by plants are utilized by soil bacteria and by the plants themselves. The air of forests and meadows is the richest in volatile vitamins," according to Krasil'nikov.[22]
>
> Between 1950 and 1955 it was demonstrated that the composition of the microflora of the rhizosphere of various plants differs with respect to vitamin requirements. In the root area of some species of plants, bacteria prevail which require vitamin B_1 and B_2 and, in the rhizosphere of other plants, bacteria requiring biotin, vitamin B_{12}, cysteine, methionine, etc., are prevalent."[23]
>
> In 1942 it was demonstrated "that bacteria synthesize vitamins in amounts exceeding those required for their metabolism. The excess of vitamins is released into the environment."[24]

> It is known that algae secrete various organic substances—metabolic products. Biotic substances are among these products.
>
> [In 1940, a research worker] found ascorbic acid in the secretions of such algae as *Hormidium borlowi, H. flaccidum, H. nitens* and *H. stoechidium.*[25]
>
> Vitamin B_{12} was found in many bacteria and especially in actinomycetes . . . 90-95% of the actinomycetes isolated from soil synthesize vitamin B_{12}.[26]
>
> [In 1933-1934 researchers] added yeast extract to aqueous and sandy cultures of peas. Flowering occurred 5-10 days earlier and the crop was 50% higher than in the control plants. These authors also observed that such effect could not be obtained by growing the plants in soil rich in humus. [The properties of yeast extract are already contained in soils rich in humus.][27]

Two key factors pointed out in the literature merit elaboration. First, in reference to the various biotic substances and the microbes that produce them, an interdependence among microbes and between microbes and plants is needed for optimum health and production. Of particular interest in this regard is the fact that *all* the nutrients and nutrient metabolites are biological in nature, and are not in inorganic chemical forms. This is a significant point in light of the present day mind set toward the use of chemicals in agriculture. Plant feeding and soil fertilization with inorganic chemicals are man-made practices outside of what should and actually does occur in nature under normal healthy conditions. Common health and normal health should not be considered the same thing.

> It is obvious that under natural conditions [life in the soil] the microbial metabolism and the synthesis of biotic substances would differ from that under laboratory conditions [on artificial nutrient media].[28]
>
> It is known that many (if not all) algae may, under certain conditions, grow and develop on synthetic media with mineral sources of nutrition, as well as on organic media containing different nitrogenous and carbonaceous complex compounds. It was shown by special experiments that these organisms, and especially the green unicellular organisms, which usually grow on pure mineral nutritious media, grow much better upon the addition of organic substances to the solution. They can assimilate carbonaceous and

> nitrogenous substances.[29]
>
> The majority of soil microorganisms are partially auxoautotrophs; i.e., they require only some biotic substances of 163 yeast cultures, 87% required biotin, 35% thiamine and pantothenic acid and 12% required only inositol.[30]

These substances must be present in the soil, in order to have yeast balance in the microorganism community.

"There are many bacteria which can synthesize only a fraction of a vitamin's molecule."[31] They must obtain the remainder from their environment. None of these nutritional biotic substances are found in typical N-P-K fertilizers, or in any herbicides or pesticides.

> In 1941 and 1943, researchers found that a certain substance among the metabolic products of the fungus *Aspergillus niger* enhances the copulation of the yeast Zygosaccharomyces. The active principle of this stimulant . . . consists of two substances: an acid closely related to glutamic acid, and riboflavin. [One researcher in 1965] caused an increased copulation and the formation of zygotes in mucor fungi by the addition of beta-carotene to the medium.[32]

This demonstrates the interdependence between soil microbes and the benefit of biological vitamin fertilization.

"Brewer's yeast did not grow at all with corn, wheat, and beans, but wild yeast of the genus *Torula,* on the other hand, grew beautifully and multiplied."[33] Man-made substances do not always satisfy the need.

The other notable point derived from the literature is the difference between what is *normal* in a biological system and what is *common* in a not-so-biological system. Most of the literature on soil and plant nutrition is based on research in invitro (in the laboratory or out of the natural system) or in sterile or relatively sterile conditions. These conditions differ from those in the real world of agriculture, yet the observations made therein are propounded as universal fact. All living organism nutrition is primarily biological, especially that associated with agriculture. Any contention otherwise is not scientific.

> [In 1954, two workers] studied the uptake of inorganic radioactive phosphorus by bacteria. Under sterile conditions the radioactive phosphorus was taken up at a slower rate and in smaller amounts. The plants showed only 4% of the total radioactivity when grown under sterile conditions; in the presence of bacteria the radioactivity of the plants increased to 13%. Experiments with labeled (^{32}P) lecithin gave analogous results. In the absence of bacteria 1.9% of the radioactive phosphorus was taken up, while in the presence of bacteria 16.6-18.6% of the phosphorus was taken up, i.e., 8-9 times more. . . . Bacteria have a considerable effect on the uptake and accumulation of phosphorus compounds in the tissues of plants. Under the influence of the bacteria, a formation of quite different phosphorus substances takes place.[34]
>
> The assimilation of mineral and organic nutrients by plants proceeds at various intensities and depends not only on the composition and properties of the compounds in question but also on the quantitative and qualitative composition of the microflora around the root system. Studies show that under conditions of sterile growth of plants these substances are not absorbed to the same extent as in the presence of microorganisms.[35]

Of interest in the group of biotic substances in soil and plant nutrition are amino acids and enzymes, neither of which is present in N-P-K fertilizers, anhydrous ammonia, herbicides, or pesticides. These substances are, however, found in fish and seaweed products. Amino acids and enzymes are critical for proper protein synthesis and nutrient metabolism. Closely associated with them are the numerous vitamins and organic acid/humic acid compounds.

"The amount of amino acids in soils varies, it depends upon the properties of the latter and upon climatic conditions. The more fertile the soil, the more amino acids it contains."[36] Plants absorb the amino acids that microbes synthesize and excrete. Plants can use these amino acids directly in the synthesis of proteins.[37]

It also has been shown that plants can obtain vitamins from microorganisms. It has been demonstrated that vitamin B_1 was absorbed, using radioactive-labeled vitamin B_1. The greatest activity was achieved from *Pseudomonas autantiaca* cultures.[38] In 1941 it was revealed that vitamins have a favorable effect on the fertilization of plants.

> It was found that the sexual organs are rich in vitamins, especially the pollen. Processing of such pollens with carotene increases its capacity to germinate. The lower the vitamin content, the sharper the reaction to the addition of carotene. Pollens rich in this vitamin stimulate the germination of pollen which contains small amounts of the provitamin if they are left to germinate together. Other vitamins (C_1, B_6, B_1, B_2, PP) also have an effect on the germination of pollen.[39]

Several investigators have found that vitamins and auxins are absorbed through the leaves and roots, giving favorable results.

> It should be noted that even the roots which can grow in vitro do not grow in the same manner as when attached to the plant. The requirement for biotic substances is well pronounced in seedlings. The seed embryos of some plants develop better and quicker in the presence of certain vitamins added to the substrate.[40] Pantothenic and ascorbic acids act favorably on pea embryos.[41]

Vitamins C, B_1, and PP consistently seem to enhance the germination of many seeds. Other vitamins that enhance germination include H, pantothenic acid, and B_2. Further studies revealed that using vitamin PP and adenine had a marked benefit on Bukhara almonds; vitamin PP especially from micorrhyzal fungi or thiamine, vitamin B_6, and nicotinic acid stimulated orchid seeds; vitamin PP, adenine, and thiamine stimulated cotton; and nicotinic acid, vitamin B_1, and vitamin C stimulated carrots.[42] The bottom line is that, in the presence of microorganisms, organic substances that are rich in vitamins have a favorable effect on plant growth.

Interestingly, the fat-soluble vitamins (A, E, and K) often suppress crop growth when they are used as supplements. Vitamin K suppresses the growth of fungi, some bacteria, and the roots of higher plants. This correlates to Dan Skow's observation that animals fed moldy feed need vitamin K supplementation. Vitamin PP antagonizes the action of vitamin K.

> The role of vitamins in the fixation of CO_2 is very great. Vitamins

also play a considerable part in the formation and transformation of proteins. It has been shown that vitamins B_2, B_6, B_{12}, PP, and H participate in the formation of amino acids and their transaminations. The shortage of vitamin B_6 leads to a decrease in the formation of amino acids from organic acids and ammonia.

Vitamins play an immense role in respiration. It was shown that enzymes participating in respiration consist of proteins and a coenzyme. The latter consists of vitamin B_2 and phosphoric acid.[43]

Because of the enzymatic activity of the roots, they can supply plants' demand for phosphorus at the expense of the organic phosphorus compounds if these compounds are present in the medium. These enzymes split glycerophosphate, saccharose, glucose phosphate, and ribonucleic acid. In addition to enzymes, plants excrete into the soil a number of other biologically active compounds. These substances are sources of direct or supplementary nutrition for soil microorganisms and enhance their growth and accumulation in the soil.[44]

The amount of excreted enzymes and their activity varies among the different species of plants. Enzymes include catalase, tyrosinase, phenolase, asparaginase, urease, invertase, amylase, cellulase, protease, and lipase, nitrate reductases. Traces of lipase were detected in only four of twenty-three species of plants evaluated (dandelion, touch-me-not, nettle, and pine).[45]

Two scientists in 1903 studied plant assimilation of many organic compounds. They found that plants use glucose, saccharose, glycerol, dextrin, starch, and potassium humate. Plants assimilate glucose, saccharose, lactose, and levulose.[46]

In 1955 it was revealed that the absorption of methionine is more rapid and pronounced in the presence of vitamins B_1 and B_6, which are products of yeasts. Yeasts are very effective metabolizers of phosphate. It was found that "about 43% of the radioactive phosphorus from yeast was taken up by plants in the first days of their growth."[47]

Biotin and thiamine [have been found] in root secretions of corn and peas. . . The more of these substances are secreted, the more

> intense is the growth of the plants. Corn secretes more biotin, and peas more thiamine... Peas, wheat, and corn secrete more biotic substances in their early growth period than in the period of fruiting. Plants grown in aerated solutions secrete more biotic substances than those grown in poorly aerated solutions.[48]
>
> ... riboflavin can be found in the soil in varying amounts, according to soil fertility, vegetative cover, etc. In soils under forests the authors found 500 micrograms of riboflavin and in soils under plow and in fertile soils about 10 micrograms per 100 grams of soil. After the application to the soil of organic fertilizers such as straw, grass or sugar, the amount of riboflavin increased.
>
> Some amino acids can be listed among the growth factors of lower and higher plants. Plants synthesize these amino acids in the same way they synthesize vitamins, but, nevertheless, the addition of small doses of amino acids has a positive effect on the growth of plants.[49]

This is why the addition of fish, seaweed, and compost is beneficial to soil fertility.

> The enzymatic activity [of the soil] is closely correlated with the activity of microorganisms. Any increase in the amount of the latter leads to enhancement of enzymatic soil processes. . . the enzymatic activity of the soil is an index of its fertility.
>
> There are data in the literature indicating that plant roots excrete various enzymes into the soil, such as catalase, tyrosinase, amylase, protease, lipase and others.[50]
>
> In 1949 a scientist detected the presence of the following enzymes in the soil, catalase, tyrosinase, phenolase, asparaginase, urease, invertase, amylase, and protease, noting that their accumulation depends on soil cultivation.[51]
>
> . . . the following acids can be detected in the soil: 6-7% aspartic acid, 5% glutamic acid, and 18% of other amino acids, totaling 31.9% total nitrogen. According to the author, 66-75% of soil nitrogen is not in the humus but in microbial proteins.
>
> In fertile cultivated soils there are more enzymes than in poor nonfertile soils. The more organic compounds in the soil, the more active is the growth of microbes and the greater the enzymatic activity of the soil (Hoffmann, 1952). The upper layers contain more enzymes than the deeper ones.[52]

Proteins are the most common nitrogenous compounds present in the cells of plants, animals, and microbes. Complex proteins such as globulins are insoluble in water but soluble in dilute salt solutions. This is a major reason salt fertilization gets crop growth. It solubilizes protein compounds for plant nutrition at the expense of the soil integrity.

> In 1940 found that the roots of pumpkins excrete from 9 to 11 different amino acids. [In 1952 it was] observed phosphatides, amino acids, thiamine, biotin, meso-inositol, para-aminobenzoic acid, carbohydrates, tannins, and alkaloids in the root excretions of plants. Sugars, amino acids, vitamins, and other organic compounds were found in root excretions. [It was] noted that the process of the excretion of organic substances is closely linked with external conditions—sunshine, aeration, nutrition, and the pH of the medium. Confirming earlier data . . . [it was] stated that the detected nitrogenous compounds are products of the fixation of free nitrogen, which were not utilized in the formation of protein and plant tissues and not products of protein decomposition.[53]
>
> It was found that the samples of humic acids studied differed from each other in the composition of their nitrogen compounds and amino acids. Alkali extracts contain much of the nitrogen in the form of acid-soluble nitrogen compounds. About 20-60% of the nitrogen does not dissolve after acid hydrolysis. From 3-10% of the nitrogen is in the form of amino sugars. Nineteen amino acids were identified by means of paper chromatography: phenylalanine, leucine, threonine, isoleucine, valine, alanine, serine aspartic acid, glutamic acid, lysine, argine, histidine, proline, hydroxyproline, alpha-amino-butyric acid and others.[54]

It is important that farmers understand the critical interrelationship between microorganisms and successful or failing soil and crop management. Farmers must truly farm the microorganisms, that will subsequently "farm" the soil and crop. To do this, farmers will first realize that chemical "nutrition" falls short of providing the biotic substances necessary for microbial and crop production. They will acquaint themselves with the various biotic substances such as sugar, molasses, compost, vitamins, fish, seaweed, humic acid, herbs, fermentation teas (vitamins, hormones, and so on), and the various groups of microbes and their part in the system. Farmers will know that life processes are biological, not chemical reactions in a test tube. They will know that nutrient feeding and assimilation are almost entirely mediated by microorganisms; thus microbes must be considered in every farming operation, especially fertilization. Farmers will understand that modern inorganic chemical farming produces volumes of "crops" because the inherent nutrients of the soil are readily made soluble by chemical means, bypassing the microbial intermediaries and their biotic contributions. A volume of crop of less than optimal nutritional integrity is consequently produced, at the expense of suppressed bioactivity. This type of agriculture leads to the destruction of soil fertility, infestation by prolific pests, poor commodity quality (nutritional and packing), and full dependence on toxic-rescue chemistry.

Much publicity is given to the availability of inorganic nutrients and the variation of this availability under various pH levels. This information is usually accurate, but it is out of context in reference to real-world agriculture. Such reports are based on the assumption that plant nutrition is fundamentally inorganic; readers are not told that this is true only under man-made situations. Under natural conditions, plant nutrition is biotic and organic. Use of the term organic in this context should not be confused with its use to describe *organic farming*. Almost all of the observations of nutrient flow and availability are different under natural conditions from what they are under the man-made context. The scientists of the former Soviet Union, the Germans, and others have verified this difference.

Using radioactive ^{32}P, a scientist named Kotelev (1955) found that the diffusion of phosphorus in sterile soil was very slow and that uptake by roots was either lacking or negligible. The diffusion of phosphorus in the presence of bacteria was a much more rapid key to the importance of bioactive soil. Chemical agriculturalists do not recognize this, and simply say that phosphorus is immobile in the soil. In addition to inorganic substances, plants assimilate various organic-carbonaceous and nitrogenous substances. Some supply energy, and others serve as biocatalysts. These organic substances are taken up by the roots and aerial parts of the plant and increase the intensity of the biochemical and biological processes in the cells and tissues. They enhance the growth of plants and increase the absorption capacity of the roots, the assimilation of absorbed substances, and other vital functions.[55]

> The biologically active substances of the soil not only enhance the growth and increase the yield of plants but also confer on the plant better nutritional qualities. Plants which obtain vitamins and other organic compounds from the soil in adequate amounts yield crops of higher quality and their seeds are of a higher vitality. The fact that plants can grow in pure mineral nutrient media in the absence of microorganisms cannot serve as proof of the uselessness of the latter in the nutrition of plants.[56]

Plants can indeed be grown in mineral media and yield seeds without the participation of microbes. However, under such sterile conditions, in mineral media without the addition of composts or metabolic products of bacteria, these plants lose their viability and eventually die out. "We cannot understand the assumption of certain authors that microorganisms by their assimilation of inorganic substances preclude the utilization of the latter by plants."[57]

> The presence of bacteria in the substrate causes formation and accumulation of amino acids of a different nature to those formed in the absence of bacteria. Some amino acids (serine, glycine, alaninc, valine, cysteine) cannot be detected; in the control sprouts of the barley grown under sterile conditions, only their traces can be found. These amino acids are present in considerable amounts in plants grown in the presence of bacteria. Different bacterial cultures have a varying effect on the formation and concentration of amino acids in plant tissues. [Investigators] grew corn under sterile conditions, according to the method of Shulov, in the presence and in the absence of bacteria in the solution. They found that when bacteria were present in the solution, organic compounds of phosphorus and nitrogen were found in larger amounts than when the plants were grown without bacteria. The microbial metabolic products increase not only the uptake of these substances [phosphorus and nitrogen compounds] by the roots but also the synthetic capacity of the roots.
>
> Microorganisms and their metabolic products affect the process of nitrogen transformation in the root. In the presence of microorganisms the rate of metabolism of amino acids in the roots increases

as well as the process of transformation of the inorganic to organic nitrogen. In the presence of microorganisms the uptake of inorganic and organic compounds—microbial metabolites—is increased.

Under sterile conditions, without bacteria, the uptake of phosphates proceeds at a lower rate than in the presence of bacteria.[58]

Positive action of humus and organic fertilizers on the growth of plants cannot be explained by the action of the mineral elements of nutrition present in them. [a summary of] the experimental data of 60 years' work of the Rothamstead Station, said that, although plants can grow satisfactorily and reach full development on inorganic nutritious substances only, under natural conditions, however, their nutrition takes place in the presence of organic substances. In the Rothamstead field experiments none of the combinations of the artificial fertilizers is as effective as manure in maintaining crops from year to year. The extensive field experiments of the German investigators Gerlach, Hansen, Schultz, Wagner, Schneidewind, and others, lead us to the conclusion that, whatever the amount of the mineral fertilizer may be, if using mineral fertilizers only, one cannot reach as high yields as by the simultaneous introduction of manure.

[In 1933] Ressel and his co-workers concluded that no mixture of artificial fertilizers can be as effective as manure (and organic substances) in maintaining steady high crops year after year. . . The fact of positive action of humus and organic fertilizers on the growth of plants cannot be explained by the action of the mineral elements of nutrition present in them.[59]

Herein lies the difference between farming the soil and mining the soil. The farmer may obtain a crop using salt N-P-K fertilizers and anhydrous ammonia with the accompanying toxic chemicals, but the crop comes as an exploitation of the soil. It is a mined commodity, rather than the fruit of an ever-maturing biological soil. The latter requires stewardship, compassion, and professional management. It requires that the farmer regularly walk the fields, test the soil and plants, execute timely operations, and feed microbes and plants alike. It requires the possession and use of a refractometer, field meters, a trained eye, and a genuine appreciation of one's fellow inhabitants. With such tools and the information gleaned from accumulated observations, the farmer will understand how the real farming world operates, giving him the foundation to manage it for the maximum benefit of all concerned: his business, his investment, his soil resources, his offspring, his fellow human beings, and nature.

Almost every farmer has had some type of pest affecting his crop. Typically, the farmer manages this pest with toxic chemicals, eliminating the obvious symptom only to compound the real problem—poor bioactivity and balance. It usually takes one to five years to get off this chemical treadmill, depending on the conditions that are present. To wean himself from this addiction to chemicals, the farmer has at his disposal the natural phenomenon of biological checks and balances. As pointed out earlier, the microbial population is determined by its environment. As the farmer alters the soil environment, he arranges it to support whatever microbial system survives best in the surroundings he has created. If he chooses to create a soil environment that is most conducive to pathogenic microbes such as fusarium, verticillium, parasitic nematodes, and mosaic viruses, he need only reduce the soil oxygen level, degrade the humus, destroy the soil structure, and maintain a continuous toxicity level; he will then have the perfect soil environment for such pathogens. This is exactly what thousands of farmers do regularly, in accord with the directions they are given by Extension agents of the land-grant universities.

As pointed out in the following discussion of the literature, pathogenic organisms are present because the conditions have been established for them to thrive there. If one abhors the presence of rats in his surroundings, he need only change his housekeeping practices and the rats leave. Killing the rats will not clean up the garbage that attracted them and continues to attract more. So it is with pest microbes. For discussion in the following sections the pest microbes in the soil are classified as inhibitors and the sanitation engineers as the antagonists. Either one can be introduced into the soil, but soil management determines which group gets the upper hand.

> There are organisms in the soil which, with their metabolic products, suppress the formation of toxins and antibiotics and often inactivate them in the medium, if they are formed there. [This is the

key to bioactive soil.]

Antagonists, therefore, can be considered as one of the powerful factors governing soil fertility and plant-crop abundance. . . In soils in which antagonists grow abundantly (bacteria, fungi or actinomycetes), microbes sensitive to them, saprophytes as well as phytopathogens, grow much more slowly, or not at all. This serves as a basis for the use of microbial antagonists in the struggle against harmful microflora, and against organisms causing plant disease.

The first attempts in this direction were made [in 1924]. He treated wheat seeds with bacterial antagonists and then infected them with Helminthosporium fungi. The seeds either did not become infected and germinated normally, or they were slightly affected. Bamberg [a worker] infected wheat seeds in order to protect them against smut. [Another] used the fungus *Trichoderma lignorum* for the protection of citrus saplings from Rhizoctonia. This fungus, according to other authors, also protects cucumbers and peas from Rhizoctonia and wheat from fusariosis. [Still others] observed a protective effect of the actinomycete-antagonist—*A. praecos*—in relation to the agent of scab in potatoes—*A. scabies*. No scab was observed upon prolific growth of the antagonist.[60]

[In 1953] 31 cultures of actinomycetes [were isolated]. Twenty-two of these inhibited the growth of the fungi *Thielaviopsis basicola,* which causes root rot of tobacco plants, and *Fusarium* sp., which causes the "black foot" disease of citrus saplings. [Two workers, in 1948] had a collection of actinomycetes that inhibited 33 species of phytopathogenic fungi. Kublanovskaya [in 1950] noted actinomycete-antagonists to the agent of wilt of the cotton plant—*Verticillium dahliae* and *Fusarium vasinfectum.* Actinomycete antagonists were found in soils which were active against phytopathogenic fungi—*Helminthosporium sativum, H. victoriae, Coletotricum circinans, Verticillium albo-atrum* and others.

In our studies we found actinomycetes in soil, which suppress phytopathogenic fungi: *Fusarium lini, F. solani, F. vasinfectum, Helinthosporium sativum, Alternaria humicola, Rhizoctonia solani, Botrytis alii, Deuterophoma tracheiphilus, Trichoderma lignorum, Monila fructigena,* and also fungi of the genera *Penicillium, Aspergillus, Cladosporium, Verticillium* and others.

Among fungi there are many antagonists. Antagonists have been described against agents of various diseases: *Fusarium, Peziza, Rhizoctonia, Ophiobolus, Botrytis, Monilia, Sporotrichum, Pythium, Phymatotricum, Phytophthora and Aclerotium.*[61]

More than thirty species of fungi that inhibit more than fifty fungi belonging to different genera and families were reported in 1936. Later in 1953, 170 fungi antagonistic to other fungi from soils were isolated.[62] These findings refute the approach of university experts, who advocate using chemical pesticides as the only feasible method of controlling pests in commercial agriculture.

> [One researcher] found that among soil fungi and bacteria approximately half (out of 86 studied) suppressed the growth of the fungus *Helminthosporium sativum*, and about 12% inhibited *Fusarium lini*. From various soils, 14 cultures of fungi, 29 strains of actinomycetes and 31 strains of bacteria [were isolated], which actively suppressed the growth of the fungus *Pythium debaryanum*.[63]

This indicates that several scientists have discovered similar information and fulfills chemical scientists' appeal for legitimate scientific data.

> The antimicrobial action of the antagonists manifests itself not only under laboratory conditions on artificial media but also under natural conditions of habitation in the soil.[64]

So much for the fumigation for fungus and nematodes. Why haven't the university experts acknowledged this work?

> In the Soviet Union . . . Khudyakov (1935) and Novogrudskii (1936) established the lytic effect of mycolytic bacteria on phytopathogenic fungi. These bacteria were subsequently exclusively tested fighting plant infections under laboratory conditions, and in open fields as well. The results were positive in many cases.
>
> Korenyako (1940) [found that] the mycolytic bacteria suppressed the growth of the fungus and lowered the morbidity of the cotton plant by 60-90% [when troubled by *Verticillium dahliae*]. [These results were corroborated by two workers in 1951 and 1953.] They showed that mycolytic bacteria in mixtures with other bacteria suppress phytopathogenic fungi more vigorously.[65]

In 1955 a researcher named Kuzina used mycolytic bacteria against cotton-plant wilt caused by verticillium. Kublanovskaya, in 1953, used actinomycetes as antagonists to fight cotton-plant wilt caused by the fungi *Verticillium dahliae* and *Fusarium vasinfectum*. His tests showed that 52.6% of the untreated plants were infected, as compared to 18.1% of those treated with variety #8517, and 23.3% of the untreated plants as compared to 3.3% of those treated with variety #108-F.

Chinese scientists Yin, Chen, Yang et al. (1955) corroborated Kublanovskaya's data. They prepared compost from soil with cotton cake and grew actinomycetes in it which were antagonists of the wilt fungus *Verticillium dahliae.* The ripened compost was used for the treatment of the seeds. The latter were sown together with the compost. The incidence of the wilt disease decreased by 50-75%, the cotton crop increased by 13-45%.

Mitchell et al. (1948) observed vigorous growth of antagonists—actinomycetes and bacteria—when those were introduced in the soil together with a composted plant mass. The phytopathogenic fungus causing root rot in the cotton plant perished. The positive role of composted preparations saturated with microbial antagonists was noted by Sanford (1946-1948). He observed in his experiments a drop in the morbidity of potato tubers, pine saplings, etc. Similar data are given by some other investigators.[66]

Four cultures of the fungus *Trichoderma—T. glaucum, T. lignorum, T. koningii,* and *T. album* [were tested]—as antagonists of *Rhizoctonia,* which affects potatoes. Control plants had 54% mortality, whereas the treated plants had 70% to 80% mortality.[67]

Microbial antagonists play an important role in soil improvements. To enhance the growth and activity of these antagonists, certain plant residues should be introduced into the soil as sources of nutrition.[68]

There are many other studies in the literature which confirm the positive effect of microbial antagonists in the struggle against phytopathogenic fungi, bacteria and actinomycetes. All these studies show that microbial antagonists can be used in agricultural practice for the improvement of soil. For this purpose the soil should be enriched with the appropriate antagonists.

By selecting, by means of special experiments, those plants in whose rhizosphere the needed antagonists grow abundantly, and by using these plants in the crop rotation, one may remove or suppress the growth and the harmful activity of the pathogenic microbe.[69]

Several researchers used an Azotobacter culture for the inoculation of oak seedlings in a steppe zone. . . This microbe increases the

> percentage of acorn germination and enhances the growth of oak seedlings.[70]

Another worker studied more than 200 strains of sporiferous bacteria and found very few activators among them.[71]

> . . . certain species of *Ps. sinuosa* increased the percentage of germinating (grape) seeds to 80%, while in the control plants, only 10-12% of the seeds germinated by the 45th day. These bacteria also enhanced the growth of seedlings and roots. In the control plants, the buds swelled on the 16th day and, in those treated with bacteria, on the fourth day. The highest activity was shown by *Azotobacter chroococcum* and the non-sporiferous bacterium *Bact. album*, strains 2 and 3.
>
> Upon introduction of pure cultures of antagonists, one must take into account their adaptability, their growth in the soil, and their activity. If the antagonists lose their activity in the soil or in a substrate which is not suitable for them (which often happens), or if they do not grow or grow but little, their effect will be small or will not express itself at all.[72]

This is why nutrition is so important when microbial materials are introduced into the soil.

"It is also known that plants can absorb various organic substances from the medium, including antibiotics, through their roots."[73] This can be beneficial from the standpoint of potential natural resistance to disease, but it can be potentially detrimental when antibiotics are applied to the soil through manures. These antibiotics can be taken into the plant and then passed on to consumers, creating potential hazards.

Several workers found that the sap contains antimicrobial substances. "There are indications that the cell sap from varieties which are resistant to the infection is more toxic than that of susceptible varieties.[74]

> The antimicrobial properties of vegetative juice are explained by the presence of various substances in the cells. . . An opinion was voiced that plant immunity is caused by the presence of elements of mineral nutrition in the sap. Some of them, potassium, copper, cadmium, etc., create a nonfavorable milieu in the plant tissues for growth of microbes.[75]

Plant nutrition is the key!

"The formation of antibiotic substances is more intense in sterile soils."[76] Antibiotic substances are produced most by the inhibitor or pathogenic microbes.[77]

> The phytopathogenic fungus *Deuterophoma tracheiphila* causes the poisoning of citrus plants by its metabolic products. An antagonist (*A. griseus*) of this fungus was found among the actinomycetes, which produces antitoxic substances and inhibits the growth of the fungus. The antibiotic grisein, obtained from a certain actinomycete, also had suppressive effects on this pathogen, which is the cause of a disease of citrus plants. . . It may be assumed that for any toxin of microbial origin an antitoxic can be found.[78]
>
> On the surface of the aerial parts of plants one finds different microorganisms—bacteria, actinomycetes, fungi, yeasts, algae, and protozoa. . . on healthy seeds there are almost no fungi.[79]

So much for modern-day experts' contention that the fungal disease syndrome is a threat on all seeds.

> In all cases, lactic-acid bacteria were found on all the plants in great quantities. The presence of such a large microflora on the surface of plants cannot be explained by their being carried over mechanically from the air. The accumulation of certain specific species speaks against it. The latter evidently grow and multiply on the plant surface. Consequently, they must find there sufficient quantities of food substances necessary for mass reproduction.[80]

Microbes are present only if there is the correct food source. Pathogens are attracted by a different food source than are beneficials.

Many sporiferous bacteria (anaerobic) have toxic or herbicidal properties on many plants, suppressing growth and lowering the percentage of germinating seeds.[81] "If seeds of plants are kept in a solution of an antibiotic for two to four hours before sowing, the seedlings are completely colorless, without the slightest sign of the formation of chlorophyll."[82] Heavy feeding of antibiotics to animals, passing out through the manure, certainly must have adverse effects on crops. Crops prefer aerobes, and aerobes dislike antibiotics.

A suppression of chlorophyll synthesis is caused by aureomycin, terramycin, and other antibiotics.

Certain strains of gray and pigmented actinomycetes synthesized substances which inhibit the formation of chlorophyll in the leaves of grapevines. . . . The phenomenon of chlorosis as an effect of streptomycin was observed in cereals [in 1947 and 1956]. . .death of the growing tip of flax is not connected with boron starvation, but is the result of poisoning by toxins formed by bacteria.[83]

It is obvious that the importance of microbial inhibitors (pathogens) in soil toxicosis will be mainly determined by the degree of their growth and activity.[84]

Among the inhibiting factors of great importance are the phages: bacteriophages and actinophages. For example, root-nodule bacteria become inactive when phages multiply abundantly in the soil.[85]

Inhibitors are found primarily among the fungi, i.e., Fusarium, Penicillium, Trichoderma, Verticillium, and Pythium.

The total number of inhibitors among actinomycetes is comparatively small, on the average 5-15%, most often among the orange *A. aurantiacus* and the gray *A. griseus*. One finds inhibitors among sporiferous bacteria considerably more often. Inhibitors among non-sporiferous bacteria are encountered much less frequently than among sporiferous bacteria.[86]

As can be seen from the data given, the greatest number of inhibitors was found in soils cultivated to a limited extent. Mineral fertilizers do not diminish but, on the contrary, they noticeably increase the content of inhibitors.[87]

The word *mineral fertilizers* in the preceding excerpt refers to inorganic chemical fertilizer void of bioactivity—in other words, those fertilizers perpetually used in chemical-based agriculture.

Toxins introduced by inhibitors may, under certain conditions, accumulate in considerable quantities and endow the soil with toxic properties (toxicosis). Toxicosis expresses itself in the suppression of the growth and development of higher plants, and in the lowering of crop yields. The phenomenon of toxicosis is frequent under monocultures. In such cases, one speaks of the soil exhaustion as the reason for the suppressed growth of plants.[88]

Other important microbes of the soil are the mycolytic bacteria.

> These bacteria are characterized by their ability to dissolve the mycelia of fungi. Each species in this group of bacteria dissolves certain forms of fungi—saprophytes and phytopathogenic forms. Some plants maximize the populations of these mycolytic bacteria, such as lucerne, while other plants minimize their populations, such as cotton. Some of these bacteria dissolve fungi of the genus Fusarium, and others dissolve fungi of the genus Helminthosporium. . . the limited number of phytopathogenic fungi of the genus Fusarium in soils with tea plantations was due to the abundant growth of the mycolytic bacterial antagonists.
>
> Plants may favor the growth and accumulation in the soil of the microbial antagonists of phytopathogenic microorganisms and even viruses. Two workers observed a lower rate in the root-rot disease in strawberries when they were sown after soybeans, and an increased incidence of the disease when they were sown after clover.[89]

Plants also have an effect on pathogenic microbes affecting humans and animals. Certain plants have toxic effects on such pathogens, some have no effects, and some have enhancement effects on such microorganisms. A rescuer named Arkhipov (1951, 1954) studied the bacterium causing anthrax and found that:

> Certain plants (garlic, winter wheat, rye, onions, rhubarb, and vetch) completely remove the anthrax bacillus from the soil. Lucerne, spring wheat, hemp, and the castor-oil plant exert a weak inhibitory effect. *Ornithopus sativus,* carrots, radishes, rape, watercress, and others have no effect whatsoever. Finally, such plants as potatoes, *Eusgropyrum,* horseradish, radishes, and turnips stimulated the growth and accumulation in the soil of the above microbe.[90]

Special substances . . . stimulate the growth of other plants in root excretions. [The scientist] Timonin (1941) established the presence of substances in a variety of flax resistant to fusariosis (Bizon var.), which activated the growth of the fungus *Trichoderma viridis,* an antagonist of the organism causing fusariosis. In the strain that was sensitive to fusariosis (Novel var.), substances were found in the root excretions that stimulated the development of the fungus Fusarium, the cause of fusariosis of flax.[91]

[The scientist] Mechnikov found microbial antagonism with the lactobacilli in relation to the putrefactive bacteria and certain colon-type bacilli. On this basis he devised a method of changing the intestinal flora and sanitation of the human and animal intestines.[92]

> Microbial antagonists acting upon various pathogenic bacteria against cocci (staphylococci, streptococci, pneumococci, and diplococci), against organisms causing intestinal infections (dysentery, paratyphoid, typhoid, and cholera), against the tubercle bacillus, diphtheria, pest and anthrax brucellosis, tularemia and gas gangrene have been described. A great number of antagonists were described acting against pathogenic fungi, yeasts, protozoa, etc.
>
> Antagonists of viruses and tumors can be found among actinomycetes and bacteria. [Two scientists in 1949] found 11 cultures [of actinomycetes] that suppressed the growth of phytopathogenic strains of Actinomyces scabies.
>
> [Two others, in 1945-1947] found that among the actinomycetes which they studied 21% inhibited growth of the fungus *Pythium graminicola*, which causes root necrosis in a number of plants. The actinomycete-antagonists of *Chalaria quercina*, which causes wilt in the oak, and actinomycete-antagonists of *Cerasto mella ulni*, which affect woody plants, such as the elm, etc., have been described by Stallings 1954.[93]

There are many antagonists among the non-sporeforming bacteria.

> . . . described the species *Pseudomonas phaseoli* which inhibited the fungus *Fusarium oxysporium*. Bacteria which dissolve the mycelium of the fungi *Fusarium graminearum, F. culmorum, F. scirpi, F. lini, F. herbarum, F. equiseti, Sclerotinia libertiana* and others [were isolated and studied in detail in 1935]. These bacteria were called mycolytic bacteria. Subsequently, many other investigators detected these bacteria in the soil.[94]
>
> Fungi which attack nematodes are described in the literature. They were first described by M. S. Voronin in his work *Mycological Studies* published in 1869 and by Sorokin in 1871. In a series of papers Soprunov (1954) showed that these fungi differ in their species composition and are very widespread in soils. The majority of them

> belong to the Hyphomycetes, to the genera Trichothecium, Arthrobotrys, Dactylaria, Dactylella, etc.
>
> These fungi "catch" the nematode with their hyphae and poison it with their metabolites. Attempts were made to use these fungi in the struggle against phytopathogenic nematodes. The introduction of this fungus into the soil lowers the incidence of plant disease. In the struggle with nematodes which affect cucumbers, the fungi-antagonists, or as they are called the predatory fungi, noticeably decrease the incidence of disease . . . Artificial enrichment of soil by predatory fungi lowers the morbidity of cucumbers 1.5-7 times. Tendetnik (1957) used the predatory fungi for exterminating pathogenic larvae, the ancylostomes in mines, and also to destroy strongyles in the manure of infected animals.[95]

As is evident from the literature, soil microbiology and biocontrol of microbial pests are far from novel. Extensive work was done on the subject in the 1930s, 1940s, and 1950s. Thus, rather than enlisting universities to research, repeat, and rediscover what the literature already contains, people might better read the available literature, saving billions of dollars of public and private funds that could be better spent for new search and discovery. However, this approach would seem to indict the university system because it would reveal that academicians possessed such information all along but denied its presence and validity, to promote further the sale of toxic chemicals. In this chapter, I have quoted and paraphrased existing literature on institutional work that has already been done. These sources are available for everyone's use.

Thus far, the interaction among soil, plant, and microbe has been discussed. I reviewed the various biotic substances imperative to healthy soil and plant systems and pointed out the fundamental checks and balances provided by nature in controlling disease problems and microbial populations. Further, I introduced the point that farm management plays an important role in maintaining a cropable biological system, particularly related to fertilization practices. Beyond the literature and institutional studies, the most important factor to farmers and, consequently, to everyone else is the applicability of such information to crop production and sustainability. As pointed out in the literature, one's farm-management practices determine the biological system present in the field. Certain practices encourage the inhibitor, anaerobic, pathogenic microbes to dominate, whereas other practices encourage the antagonist, aerobic, beneficial microbes to dominate.

Perhaps the most profound statement Krasil'nikov made in his book is that "the soil possesses the ability to rid itself of pathogenic bacteria entering it."[96] He continued:

> To a certain degree, microbial antagonists regulate the formation of microbial coenoses [colonies] in the soil in general. They play an important role in the improvement of soils, in the so-called process of self-purification of soils. The removal of harmful pathogenic and phytopathogenic flora and fauna is accomplished by microbial antagonists.[97] [Again, so much for fumigation.]

—8—

MICROBIOLOGY IN THE FIELD

THE KRASIL'NIKOV book, *Soil Microorganisms and Higher Plants,* makes it clear that it is up to us to figure out what practices must be exercised to facilitate the working of nature's systems.

An important aspect of successful farm management is knowing what to expect in the soil and what observations might be made out of context or misread due to a lack of understanding of actual field microbiology. The next step is to get to the field and dig. Here you will find the real classroom. Take your magnifying glasses and your curiosity.

> One found by simple microscopy that the microorganisms grow diffusely in the zone of the roots in the form of isolated colonies or small foci. The colonies are located between the hairs on the surface of the roots, or in their vicinity. In cases where there was a great deal of moisture covering the roots and the root hairs, the bacterial cells spread, often occupying considerable segments along the roots.
>
> Isolated colonies, as well as profuse areas of microflora, mainly consist of nonsporeforming bacteria and mycobacteria. One also finds

colonies of actinomycetes and mycelia of fungi. Very seldom does one see small colonies or single cells of sporeforming bacteria.

Nonsporeforming bacteria and mycobacteria form well-defined, compact colonies, located most often on the surface of the roots, between the root hairs, and also at a certain distance from them. Colonies of actinomycetes often consist of proactinomycete elements, or of entangled mycelia. Fungi also grow around roots in the form of single hyphae and mycelia.

If one prepares imprints of the root area of microflora growing in sand, and not in soil, one does not encounter this picture. Upon the microscopic examination of these imprints, one can observe only mycelial threads of fungi and actinomycetes, occasionally with sporangiophorees on branches. One seldom observes single colonies of actinomycetes. Around the root branches and hairs bacterial cells are also encountered, but in limited numbers and, as a rule, they appear as single cells or in pairs and very seldom in colonies. When such a preparation is being grown, by covering it with a thin layer of an agar medium, a considerably larger number of bacteria and mycobacteria appear than can be observed under direct microscopy. The great majority of cells are not detected by microscopy. This is due to the fact that the cells of bacteria and mycobacteria, and possibly certain actinomycetes, exist in a fragmented state in the form of small granules hardly discernible from soil particles. These granular elements, stained with erythrosin, are encountered in the soil of the root zone in great numbers. They probably form the major part of the mass of the rhizosphere microflora. These elements, when transferred to a nutrient medium, grow out, giving rise to cells of normal size, which are detectable under laboratory conditions.[1]

This accounts for some of the erroneous statements made by university people who do not take this "bion" form of bacteria into account and go strictly by their microscopic observations.

The transformation of mycococci into mycobacteria, mycobacteria into proactinomycetes, and the latter into actinomycetes was experimentally proven.[2]

Adaptation of microorganisms to drugs and antibiotic substances is very strongly manifested. Adaptability of bacteria to penicillin, streptomycin, aureomycin, terramycin and many other antibiotics is widely known. Various species of sporeforming and nonsporeforming bacteria, cocci, mycobacteria, actinomycetes, fungi, yeasts, protozoa, and even insects, become adapted to antibiotics. Numerous cases of adaptation to antibiotics of various pathogenic microbes causing enteric diseases, anthrax, tuberculosis, diphtheria, pests, skin diseases, etc. were described.[3]

With regard to adaptive and directed variability:

> If yeasts which do not ferment maltose are cultivated on a medium with this sugar as the sole source of carbon, then after some time (it may be protracted after many passages) they acquire the ability to ferment this sugar. A new enzyme is formed in the yeasts—maltase; the cells are physiologically altered, they become new variants. Such a physiological rebuilding may be induced in bacteria, actinomycetes, fungi and other microbes with respect to many sources of nutrients, carbon, nitrogen, organic and mineral sources.[4]
>
> Adaptive changes were also observed with respect to nitrogen sources of nutrition. Strains of Clostridium may be trained to decompose casein and gelatin by adopting the same method of cultivation. Some strains of Azotobacter, Azotobacter chroococcum, do not grow on protein media (meat-peptone agar), but by successive "training" they start assimilation. New forms or variants are obtained which develop well on media with organic nitrogen. If Azotobacter is cultivated for a long time on a medium containing nitrogen, it loses the ability to fix molecular nitrogen.
>
> Many saprophytic bacteria from the genera Pseudomonas, Bacterium, Bacillus and others, which do not grow on media with mineral nitrogen, acquire the ability to assimilate it during the process of adaptation.
>
> Under the influence of the nutritional substances of the medium, the cells of the microorganisms elaborate suitable enzymes.[5]

The concept of "training" microbes comes from the old European school of microbiology.

> In the soil the cells of bacteria and actinomycetes occur in other forms and states, and, consequently, the ways of reproduction may sharply differ from those observed by us in artificial laboratory media.
>
> The formation of buds, and regenerative bodies inside the cells, then the fragmentation of the cells into tiny germ elements and the branching off of small particles from the extremity of the cell—all these methods of reproduction take place in particular pathological states of organisms under unfavorable growth conditions.[6]

The last statements resemble those made in several other studies in microbiology. However, although they have been extensively replicated, they are ignored by the general institutions and "peer" publications. Wilhelm Reich's work is perhaps the most notable, yet that of Gaston Neissens and Royal Rife is certainly equally valid. Farmers are advised to read *The Cancer Biopathy* by Wilhelm Reich and *The Galileo of the Microscope* by Christopher Bird for additional corroboration on microbiology as it actually exists.

As under laboratory conditions, sterile soils elicit microbial systems that are the antithesis of optimum cropping systems unless one is farming antibiotics, pathogenic nematodes, and disease organisms. These undesirables are not demonic artifacts sent to torment the farmer, but rather the result of his farm-management practices, which created a sterile, anaerobic environment.

In studies done with Azotobacter, plants were classified according to their effect on Azotobacter.

> This type of plant classification is only relative and only holds true for sterile cultures, where the microbes are subjected to a one-sided action by root excretions, with the exclusion of other external factors.[7]

This simply showed the fallacy of the conclusions of modern in-vitro research, which assumes that sterile laboratory situations apply equally to the field. On the other hand, these sterile laboratory conditions do apply to chemical agriculture, which sterilizes the soil.

> Wheat acts negatively on the growth of Azotobacter under conditions of sterility (poor soil). Even in the same region and in the same soil, the effect of wheat may differ, depending on the extent of the cultivation of the soil. Of great importance for the growth of Azotobacter is the agrotechnical cultivation of soil.[8]

University personnel tell farmers that they cannot generate much nitrogen bacteria activity without legumes. However, research in 1942 revealed that "root-nodule bacteria of lucerne grew equally well under lucerne and under cotton."[9] The sap of corn enhances the virulence of the root-nodule bacteria of peas. Perhaps this is due, in part, to the sugars found in corn sap. "Root-nodule bacteria of lucerne grow well under timothy grass, cotton, and rye grass."[10]

Lucerne sap inhibited the bacteria of its own species to the same extent as it inhibited that of clover, beans, and other plants. Root sap was less bactericidal than the sap of the aerial parts and, in many plants, the root extract had no toxicity for bacteria whatsoever.[11]

The growth of bacteria and mycobacteria, isolated from fallow soil or from soil outside the root zone, in the majority of cases, is inhibited by the roots of many plants, while most of the bacteria of the rhizosphere are not inhibited at all.[12]

Investigations have shown that plants not only determine the microflora of the soil during their growth but also determine them by means of their dead residues, especially those of roots. It was established that these residues, depending on the species of plant or, more accurately, on their chemical composition, are decomposed by various forms of microbes. The qualitative compositions of the microflora of the decaying roots of wheat and clover, cotton and lucerne, differ considerably. . . . The increase in the total number of microorganisms paralleled the multiplication of the cellulose bacteria. The cellulose bacteria are regularly destroyed by anhydrous ammonia!] Actinomycetes, mycobacteria, and coccoid forms appear somewhat later, after the development of the nonsporiferous bacteria. In a half-decayed mass of roots, actinomycetes often cover the root particles with a white coating of aerial myclium. Nonsporiferous bacteria prevail during all stages of root decay, but their species composition varies.[13]

The antimicrobial properties of antibiotics express themselves differently in the soil than in artificial media. So much for the in-vitro studies of the "expert" agriculturalists. Inactivation of antibiotics in soil is probably mainly byproducts of microbial metabolism. It has been shown that, with increase of the latter, destruction of antibiotics is enhanced.

Winter and Willeke introduced penicillin into composted soil rich in humus and into loamy soil poor in organic substances. In humus soil, where there was abundant growth of microbes, the antibiotic disappeared after 2-3 hours. In soil poor in humus the same antibiotic was preserved for 12 hours. If the soil microflora is removed by sterilization, penicillin in such soil is preserved for more than 3 days, while in nonsterile soil it disappears on the second day.[14]

Adsorbed antibiotics cannot be washed out with water or with a number of organic solvents, even upon prolonged treatment. This can be both good and bad.[15]

We now come to the fundamental characteristic of a fertile, biologically active soil. It contains biologically active carbon, particularly active humus. From a humus chromatograph, one can determine the status of the soil's humus. Active humus is "alive" and an integral part of the biosystem. Stable or inactive humus is not generally a factor in the biosystem and can sometimes be a liability because of its hydrophobic (water-repelling) properties, its possible content of toxic residues, and its required energy of activation. *Humus becomes inactive upon repeated sterilization by toxic materials and from continuous dehydration by mineral salts.* Humus is a byproduct of the microbial system. It depends on the microbial system for its creation and maintenance. Once established, active humus provides innumerable byproducts that are beneficial to both the microbes and plants.

> Lockhead and Chase (1943) attempted to elucidate the nature of the activation substance of humus. The experiments showed that the ash of humus and composts had no effect whatsoever on the growth of plants and microorganisms. Of 63 microbial cultures, only one grew in the presence of ash obtained from composts. The extracts obtained from composts and humus of the soil, however, had a positive effect on their growth. The acetone extract was the one most effective. In its presence 26 cultures grew well, 26 cultures adequately, and only 11 cultures did not grow. Alcohol and ether extracts also gave good results.
>
> Allison and Hoover (1936) noted a positive effect of humic acids on the growth and activity of root-nodule bacteria. They discovered the presence in humin of a special substance which they called "Factor R" or "Coenzyme."[16]
>
> Scharer and Preissner (1954), on the basis of their experiments, reached the conclusion that the more complete the mixture of fertilizer employed, the higher the vitamin content of the plants. The amount of vitamins does not always correspond to the weight index of the crop yield. It is not infrequently observed that the vitamin content is high at relatively low crop yields.[17]
>
> Vegetative composts, manure, and humic substances not only activate plant growth and increase their yields but also improve their nutrient value, which is a matter of great importance. Plants grown in fields fertilized with manure are richer in vitamins and other valuable substances than those grown in nonfertilized fields. McCarrison (1926) found that seeds of millet and wheat, harvested from [manure] fertilized fields, contain more vitamins than seeds from fields under mineral fertilizers. Animals fed on fodder from fields fertilized with manure were more resistant to infections, and their

appearance was healthier than those fed on fodder from nonfertilized fields.

Clark (1935) found more vitamin B and C in cultures of duckweed grown in nutrient solution supplemented with humus, than in a similar culture grown without humus.

Nath and co-workers (1927, 1932) have shown that organic substances of manure and composts, as well as cod-liver oil and some other substances, enhance the growth of plants and increase their vitamin content. Cereal seeds grown on fertilized fields have a greater viability and the percentage germination is higher than in seeds from fields fertilized with mineral fertilizers.

Hurni (1941, 1945) grew plants in sand in the presence of a full mineral fertilizer and found that in these conditions the formation of thiamine was less than during growth in the presence of humus substrate.[18]

According to our observations and the available data in literature, the active principles of humus and composts are not the mineral nutrients present in them but the organic substances and the biologically active metabolites of microbes. Mineral substances applied in amounts equivalent to the composts do not have an effect comparable to the latter.

Mineral elements obtained by burning manure, composted peat or soil humus do not produce an effect comparable to that obtained by the application of organic substances... In other experiments ash was added to the nutrient solution. The ash was obtained by burning equivalent amounts of these substances.[19]

Again, the biological test outperformed the ash test. Interested students can perform similar experiments, using various cultures of plants grown over a given period and then weighed and analyzed.

The group composition of the compost microflora varies with the increase of the maturity of the compost. In the first days of incubation on nonsporeforming bacteria of the genera Bacterium and Pseudomonas, and fungi, grow abundantly. The fungi grow only on the surface. Mycobacteria can be detected in large numbers in the composted peat. At the end of the incubation (maturation) of the preparation, the bacteria and fungi decrease in number, their place being taken by actinomycetes. The latter attain such vast numbers that the peat lumps are covered with them. This covering is of a white flour color and is visible with the naked eye. The maturing of the compost can be judged by the intensity of growth of the actinomycetes.[20]

Many authors have noted the positive effect of small doses of

> humus. The detailed studies of Khristeva (1948) had shown that solutions of humic acids exert a direct effect on higher plants. In negligibly small concentrations (0.001% and 0.0001 %) they enhanced growth and increased the yield of wheat, oats, barley, sugar beet, tomatoes and other plants. . . . The action of humic fertilizers was tested by the author in different soils. . . . In all cases the effect was positive.[21]

Therein lies the explanation and justification for biological fertilization, including such materials as vitamins, humic acids, enzymes, and amino acids.

> According to Khristeva, humic substances find their way, in small amounts, into plants, there stimulate the phenol-oxidase system, and participate in the general metabolism of the plant. The physiological function of humic substances is in the promotion of plant respiration. As a result, an increased influx of nutrients, activation of synthetic processes, and better growth of the root and aerial parts takes place.[22]
>
> Humin and humic acids are decomposed by bacteria and fungi, especially by actinomycetes. Many actinomycetes grow well, bear fruit, and form antibiotic compounds on media containing humic acids as a sole source of carbon and nitrogen. Many forms of bacteria also grow on humic acid substrates.[23]

The toxicity to pathogenic microbes also has been studied. A researcher named Peitsa (1952) found that, in forest soils, the strongest toxicity against *Pythium ultimum* was found in extract of humus from under pine, next from under beech, and weakest of all, from under birch. Plant and microbial growth is not the only beneficiary of active humus. The physical structure of the soil also is correlated directly to the humus system.

> [Russian scientist P. A. Kostychev] showed that certain substances which serve as a cement are required for the soil particles to combine into aggregates. Such substances are the products of microbial metabolism and of the disintegration of animal and plant residues. According to his experimental data, the process of structure formation takes place only under conditions optimal for microbial proliferation, when the biological transformation of the organic mass, with the consequent formation of cementing substances, is possible.[24]

A major benefit of optimum structure is soil porosity. Adequate soil porosity allows for the proper movement of air, particularly oxygen and carbon dioxide, and water.

> The aerobic process takes place in the surface layers of the soil, where plant residues are decomposed by microbes, bacteria, and fungi. The bacteria decompose the residues with the formation of humic acid, which, under conditions of full aeration, is fully mineralized.[25]

Under anaerobic conditions, humic acid undergoes denaturation.

The foregoing studies demonstrate the essential, if not exclusive, role that microorganisms play in formatting the soil structure.

> The most important component of the soil atmosphere is carbon dioxide, the final decomposition product of organic matter. The intensity of the biochemical processes taking place in the soil can be judged by the amount of carbon dioxide released. The formation of CO_2 depends to a large degree on microbial metabolism. Everything that favors growth of microorganisms increases the generation of CO_2.[26]
>
> The oxidation-reduction potential of the solution determines the direction and character of chemical and biochemical reactions and the solubility of biologically important components of the medium, as well as the products of microbial metabolism.[27]
>
> There are indications that plant roots not only release, but also actively absorb CO_2. The amount of CO_2 taken up from the soil may be of the same magnitude as that coming from the atmosphere or may even exceed it. The intensity of CO_2 absorption from the soil depends on its concentration. The higher the concentration of CO_2 in the soil, the quicker it finds its way into plants via the roots.[28]

Now that the important factors of soil fertility and its dependence on the biological system have been examined, the reasoning behind actual management practices will be considered. In effect, these practices are the "housekeeping" routines that determine the house's inhabitants.

> Cultivation of podsol soils exerts a large influence on their toxic

> properties; Azotobacter grows better in chalked soils than in non-chalked soils. Katznelson (1940) studied the viability of Azotobacter in acidic soils of America and tested various organic and mineral fertilizers with and without neutralization of the soil. The author reached the conclusion that there was no strict correlation between soil acidity [pH] and the viability of Azotobacter.... Proliferation of Azotobacter is conditioned, not by acidity, but by the ratio of the anions CO_3 to PO_4.
>
> However, the main factors which cause soil toxicosis are, in our opinion, in many (if not in all) cases, excretion products of plants and microbial metabolites.[29]

Although tilling greatly influences the soil, soil properties depend primarily on fertilization. If it were not for the seasonal influences of weather and frost, some farmers would be out of business much sooner. Soil toxicity changes with the seasons of the year, as do other properties of the soil.

> Azotobacter perishes more quickly in summer soils than in winter ones, while in spring soils, it proliferates most. Similarly, seeds of beet germinate with greater energy and in greater numbers in spring soils (April-May) than in summer-autumn ones (August-October).[30]

This lends some validity to astrological shifts and patterns relative to agriculture.

In studies done by Korenyako, some species of actinomycetes (*Azotobacter globisporus*) actually grew more abundantly during the winter months than during the summer and autumn. In addition, certain biochemical processes leading to detoxification of the soil take place in the winter.

"Under the influence of frost, a noticeable change in the chemical and physico-chemical properties of the soil takes place. . . The soil as a whole becomes more fertile."[31] This is what saves the chemical farmers from quicker disaster.

Notice that every discussion of fertilizer, tillage, or weather in this book also includes the subject of microbes. Microbes are the key, the caretakers of the soil. Farmers often forget the essentials required by living organisms. Like animals in the mammal family, micro-animals need suitable quarters, fresh air, and water. It is amazing how many farmers are dumbfounded when biological materials do not work while the farm experiences drought.

In further research on rhizosphere studies, a worker observed . . .

> an abundant growth of bacteria in the form of large colonies around the roots and root hairs; colonies were often located at the hair tips. At a certain distance from the roots, . . . amoebae, Infusoria, and nematodes grew. Only when there is a high moisture content in the substrate do bacteria cells grow in extensive areas.[32]

Obviously, one cannot build soil without water.

Eventually, farmers will understand that the micro-animals in the soil require the same respect and nurturing that all animals need for optimal production. Carey Reams used to tell his students to study animal nutrition to understand soil nutrition, and vice versa. He, of course, was not referring to textbook study. He meant field study.

A very important management practice used by successful farmers is foliar or leaf feeding of the crop. In the late 1950s, the Atomic Energy Commission in cooperation with Michigan State University undertook a study of foliar feeding and concluded that such feeding was often nine times more effective than soil fertilization with the same material. Modern academicians contend that foliar feeding is a waste of time and money and that their "research" proves it. It is understandable why academicians today make such misinformed assertions. They neglect to use a refractometer in selecting a nutritional spray and do not understand nutrient, carbon, and microorganism balance in the soil. The literature shows that plants do, in fact, assimilate materials (some of them undesirable) through the leaves.

> It was experimentally shown that a number of toxic substances penetrate the plants via the roots and spread to the tissues. Toxins and antibiotics may enter the plants through their leaves.[33]
>
> One should assume that the metabolic products of the epiphytic microflora behave in a certain manner in the plant tissue, having a definite effect on them. The ability of leaves to absorb various substances was known for a long time.[34]

On this basis, methods of extrarhizal feeding have been elaborated, as have methods of introducing substances with the aim of changing certain physiological functions—shedding of leaves, arresting of flowering, and so on.

Microbes on the leaf surface also play an important role in plant protection and nutrition.

> It was also established that plants can absorb various microbial metabolites, vitamins, antibiotics and other compounds through the leaf surface. As was indicated above, these substances not only enter the plant through the leaves, but they can be introduced by this route in large quantities for the purpose of feeding as well as for fighting bacterial and fungal infections.
>
> Among the epiphytic microflora there are numerous antagonists which produce antibiotic substances which suppress their competitors, and among them also phytopathogenic microbes. Growing abundantly on plants, such organisms may fulfill a protective role, removing or suppressing infectious agents originating from without. If we were to change the composition of the epiphytic microflora on the surface of the green parts of plants at will, and form certain coenoses of antagonists there, this would prove to be of great value to plant and fruit growing.[35]

Plant nutrition and fertilization will determine what microbes are present on the leaf surface, regardless of exposure.

One only needs to use some common sense to determine what toxic chemicals do to the microorganisms on the leaf. As pointed out in the literature, Lactobacillus, or lactic-acid bacteria, are predominant microbe occupying the aerial parts of the plant. A competent animal or human nutritionist will tell you that toxic chemicals are not components of the conditions Lactobacillus favor. Fungal diseases are a major factor in many vegetable and fruit crops. Lactobacillus is one of the most effective controls of yeast mastitis and Candida, both fungal infections, in "animals." If the Lactobacillus population on the aerial parts of the plant is tended to, these bacteria are the best first line of defense against disease. Crop nutrition and toxic chemistry will determine the Lactobacillus establishment.

Although toxic chemistry is fairly self-explanatory, the crop nutrition referred to throughout this book is far from mere

N-P-K inorganic chemistry. By definition, nutrition must include vitamins, enzymes, amino acids, proteins, carbohydrates, and minerals. Academicians insist that plants do not need to be fertilized with such sophisticated materials because they are able to synthesize these substances for themselves. Plants are able to synthesize these materials if they have the necessary ingredients to do so. Microbes are an integral part of that need, which the use of toxic chemicals curtails.

We use many vitamins, thanks in part to the encouragement of Dan Skow. We use vitamin B_{12} because "it is estimated that approximately four-five million bacteria (per gram) which depend upon growth-promoting substances in soil, may utilize vitamin B_{12} as an essential nutrient."[36] Where will these bacteria obtain vitamin B_{12} if the microbial makers of B_{12} particularly actinomycetes are reduced or killed by toxic chemicals? According to Dan Skow, vitamin B_{12} is also an antidote for the blood-thickening effect of Cygon insecticide. If Cygon has that effect on people, might it do similar things to other living creatures? No vitamin B_{12} is contained in N-P-K fertilizers.

Sugar also is used in our fertilizer mixes. Sugar is the main source of energy that is readily usable by living systems.

> To accumulate 150 million bacterial cells in the rhizosphere solution, we evidently require approximately 50 grams of glucose or some other equivalent substance.[37]

Substances usable by bacteria are excreted in a larger amount than those usable by yeast—approximately 2.5:1.

> West and Wilson (1939) observed biotin and thiamine in the root excretions of flax, and sugar in the excretions of certain cereals. Brown and others (1949) proved the presence of pentoses or closely related compounds (alpha-ketoxylose) in the root excretions of grasses.[38]
>
> Green isolated leaves of plants grown with artificial, sugar-containing nourishment always accumulate more organic substances in their tissues and have a higher turgor than plants nourished with mineral elements [only].[39]
>
> Plants grown in soil well fertilized with manure or compost had

more active sap than the sap of plants taken from unfertilized soil. Eaton and Rigler (1948) observed increased resistance of roots of cotton plant to *Phymatortrichum omnivorum* upon treatment of the seeds with carbohydrates. In these cases, according to the authors, intense growth of bacterial antagonists in the rhizosphere of the plants was observed.[40]

To produce sugar through photosynthesis, the plant must have phosphate. The mobility and availability of phosphate depend on microbiology, both during the living processes of microbial translocation and during microbial death and decomposition.

At the end of the enzymatic lytic processes of bacterial cells, processes of solubilization of their residues by enzymes of other microbes ensue. The process of bacterial cell destruction is rapid. The metabolism of living cells is an endless sequence. The elements absorbed from the substrate are soon excreted. Experiments with radioactive elements have shown that phosphorus (^{32}P), for example, appears in the substrate after a few minutes.[41]

"Experts" often scoff at the suggestion of using very small amounts of nutrients for fertilization. We might use 5 to 40 mg of B_{12}, 1 to 20 pounds of sugar, an ounce of vitamin C, a pint of humic acid, and so on. Yet these same "experts" insist on using similar quantities of pesticides and active ingredients designed to regulate plant growth. They also point out the importance of water dilution and soil moisture in obtaining the desired results from these products.

Other products that are gaining popularity in the professional farming arena are microbial or fermentation "teas." The "experts" often call these materials "snake oils," but whatever they are called, they work just fine.

Filtrates of microbial cultures show a positive effect only when used in small amounts. When the amounts are large, their effect on the growth of isolated roots is negative. The roots do not grow or grow very poorly; deviating from the normal, they thicken, swell, do not branch, become brown too soon, and die.[42]

Similar observations have been made with plant seedlings. This refutes the traditional notion that more is better. Most university research on biological products is done without taking this into consideration, and the same people criticize the use of minute amounts of biological materials.

Many types of biological materials are patented and are used by nonagricultural industries, as well. One such product is generally called "manure tea," which was patented in October 1959 by James Martin. This product or variations on it are used by several biological fertilizer companies throughout America and by Alpha Environmental, Inc., in Texas, a company that specializes in cleaning up oil spills.[43]

Soil and plant nutrition cannot be separated from microorganism nutrition. If the microorganisms are fed, they will take care of the plants' needs, including their need for nitrogen. Farmers must grasp the purpose for which they are growing the crop. Is it to be consumed by animals or people? If so, its nutritional integrity is of primary importance. Thus, the microbes and their nutrition must be considered.

> The vitamin content of plants varies within a wide range, depending on external conditions of growth. It varies according to the soil and climate conditions. Fertilizers have a great effect on the quantity of vitamins present in plants.[44]
>
> Application of fertilizers causes a certain increase in crops but only after the first sowing. The authors obtained the strongest effect after the application of lime and manure. Upon repeated sowing with the same amount of fertilizers, the crop decreased. When nutrient substances were added to the soil solution, only the growth of the aerial parts improved, but the roots remained undeveloped.[45]

Sterile N-P-K farm management simply does not effectuate a sustainable planet or agriculture.

> The soil studies carried out for many years in the Wareham forests [in England] by Rayner and Nelson-Jones (1949) showed that the obstacle standing in the way of afforestation of certain soil zones is not the lack or insufficiency of nutrient elements but the presence of special organic substances. These substances retard the growth not only of young saplings . . . but also that of many microorganisms.

Rayner has shown that by introducing organic fertilizers, manure, or compost into these poisoned soils, the toxicity diminishes or even vanishes. According to her observations, this decrease in toxicity takes place due to the activation of the microflora in the presence of fresh organic substance.[46]

There are different points of view concerning the cause of soil toxicosis, but they can all be reduced to two basic ones.

According to one opinion, soil toxicosis is caused by the accumulation of special toxic substances as a result of growing plants against the rules of agrotechnique. According to the other point of view, the existence of toxins in the soil is denied, and the fatigue of soils is explained by lack of nutrient substances, as a result of their unbalanced withdrawal from the soils in case of monocultures.[47]

Nutrition must be considered full cycle. The organic materials are only as good as the nutrition of the living system that synthesized them.

The amount of growth factors in manure and animal feces also varies according to the quality of the fodder. The microorganisms are the main factor in the enrichment of the soil with these substances.

Biological agriculture, with its foundation resting solidly in nutrition and microbiology, is the way for the professional farmer. It is the only way for a healthy, sustainable planet. As stated in the literature, *biological agriculture is the only methodology validated scientifically*. Sterile N-P-K mineralization misses the mark.

The poorer the soil in organic substances, the less fertile it is, and the weaker is the influence of the root system on the quantitative composition of the microflora of the rhizosphere. At a depth of 90-110 cm in the rhizosphere of clover, growing in podsol soil, the number of bacteria is 2,000 times higher than outside the rhizosphere. . . In the deep layers of soil there are usually very few bacteria, while in the rhizosphere of plants, even at the depth of two to three meters, they grow abundantly.[48]

"The differences in the quantitative relationships [of microbes between rhizosphere and outside the rhizosphere] depend on the species of plant and the soil-climatic conditions."[49] Some or many soils have very poor rhizospheres because of the toxic soil management used on them. Some fields have virtually no rhizosphere as a result of poor soil management. These crops lack the most important defense and nutritional component of the cropping system, the microbes.

> Under unfavorable conditions, when agrotechnical rules are not observed, with an incorrect choice of crop rotation, the fields are contaminated by phytopathogenic bacteria and fungi and other harmful microbes—weeds. Especially, after the prolonged repeated cultivation of plants on the same field in monocultures, one observes this effect. The accumulation of an undesirable microflora under monocultures is most often caused by the insufficient growth of microbial antagonists in the rhizosphere, which are characteristic of the given plant under conditions of normal growth.
>
> Many outbursts of epiphytotic diseases, as for example fusarial infections of cereals, cotton, saplings of woody plants, as well as cotton wilt, are, in our opinion, caused by the above phenomenon. This has been proven by microbiological studies. These fungi are removed as soon as mycolytic bacteria begin to grow in the soil.
>
> Noting the great influence of the vegetative cover on the formation of microbial biocoenoses in soil, one should not forget the importance of the soil itself as a substrate and the effect of the activity of man on external conditions. The physico-chemical state of the soil determines to a great extent the direction of the microbiological processes, and to a similar extent the development of different microbial species. The distribution of Azotobacter in soil not only depends on the plants, on their root excretions and decomposition products, but also on soil acidity and the presence of phosphorus, calcium, molybdenum, and other nutrient elements.
>
> The cultivation of soil, the liming of soils, the use of fertilizers and other factors favor the growth and accumulation of Azotobacter.[50]
>
> The essence of the action of [mycorrhizal] fungi consists in supplying the plants with nitrogenous and carbonaceous elements of nutrition in some cases and, in others, in the supply of auxiliary nutrients or biotic substances, and more correctly with both. There is a great deal of data in the literature on the significance of mycorrhizal fungi in the nutrition of plants. . . It was shown that mycorhizal fungi take up and transmit various nutrient elements. Kramer and Wilbur (1949) and Mellin and Nilson (1950, 1952) have shown that fungi transmit ^{32}P and ^{15}N from the external solution in the tissues of the roots and stems of pine.
>
> We tested 130 different bacteria, mycobacteria, and actinomycetes isolated from various soils. Among them were the following: 32

> strains of Azotobacter, 33 strains of root-nodule bacteria, 40 of Pseudomonas, 10 of Bacterium, four of mycobacteria, six of actinomycetes, three of Bac. mycoides, and two of Bac. subtilis. [51]
>
> A large increase in the length of the roots was observed in the presence of filtrates of Azotobacter, root-nodule bacteria, and bacteria of the genera Pseudomonas and Bacterium. Metabolic products of certain actinomycetes were quite active. Sporiferous bacteria often showed a negative action, inhibiting the growth of roots.[52]

Without question, microbiology is an integral part of any progressive agricultural production. With its proper management, we may participate in an exciting, prosperous process, growing ever bountiful, fully gratifying commodities. Without its proper management, we thrash about in a never-ending, terminal chemical war. The discussion of the past two chapters was not intended to be exhaustive or the final authority on the subject. It was merely intended to emphasize that biology is the basis of true agriculture. Although to some it might have seemed redundant, it was intended to verify that the documentation supporting biological agriculture and the premise that nutrition, not pesticide and herbicide deficiency, determines insect and disease infestation is extensive and substantive rather than sparse and insignificant, as the toxic-chemical proponents would like us all to believe. Crop production is an inexhaustible gift, scrutinized and protected by the most quality-control-oriented inspectors on earth, the microbes. Man simply needs to observe, learn, and abide by the rules of nature that allow the planet to function.

For serious students of microbiology, I suggest participating in a laboratory seminar presented by the American College of Orgonomy in Princeton, New Jersey. The college is listed in the telephone directory. Such a seminar will be one of the most enlightening four days you will ever spend.

—9—

CLAY CHEMISTRY

CLAY MINERAL IS ONE OF THREE PARTICLES that make up a soil's texture, yet the term clay that is used to describe soil texture includes all particles less than 0.002 mm (2 microns or 0.00008 inches). The clay mineral is the topic of interest in this chapter, particularly because of its effect on soil tilth and fertility through its high cation exchange capacity (CEC) and nutrient carrying potential.

There are two basic types of clays, expandable and nonexpandable. The following is a list of some clays and their corresponding CEC values, which will be referred to later in this chapter.

TYPE OF CLAY	CEC VALUE
1. Expandable (swelling) clays	
• Smectites: montmorillonite, saponite, hectorite, beidellite)	80-150
•Vermiculite (clay-like silicate)	100-150

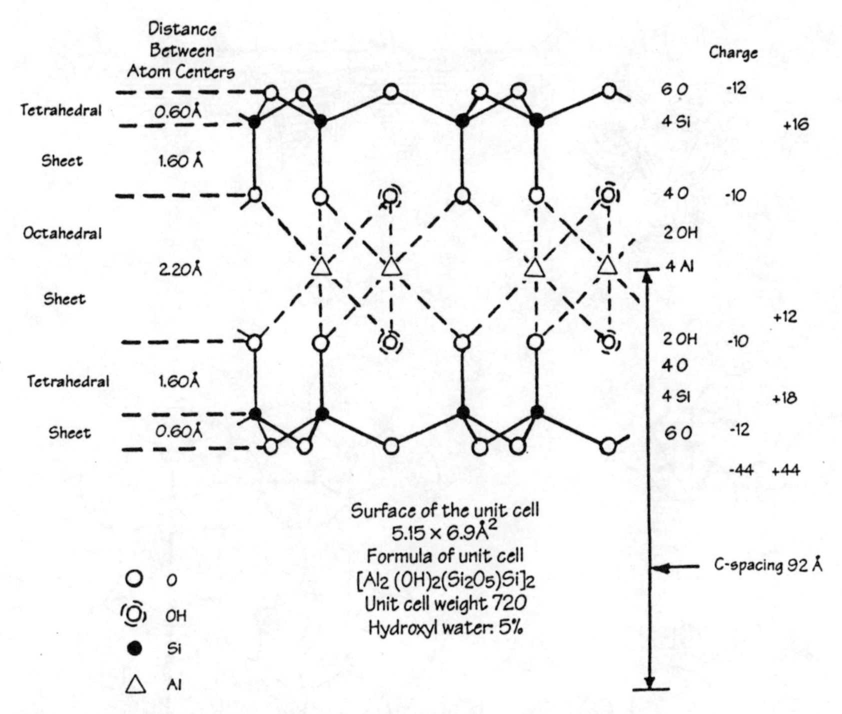

Figure 1. Atom arrangement in the unit cell of a three-layer mineral (schematic). (From *Manual of Drilling Fluids Technology: Sources of Mud Problems*. [Houston, Texas: Baroid Industries].)

2. Nonexpandable (nonswelling) clays

Illite	10-40
Chlorite	10-40
Kaolinite	3-15

Figures 1 through 4 illustrate the structure of montmorillonite clay. Thirty-five to 40 million single sheets of this clay stacked on top of each other would be only about an inch thick. As shown in figure 3, the mineral cations attach to the flat outer surfaces of the clay sheets, whereas the anions (not shown) attach to the edges of the clay sheets. The surface area and the available negative charges on that surface determine the CEC of the clay. As illustrated herein, expandable clays have a considerably higher CEC than do nonexpandable clays. Ex-

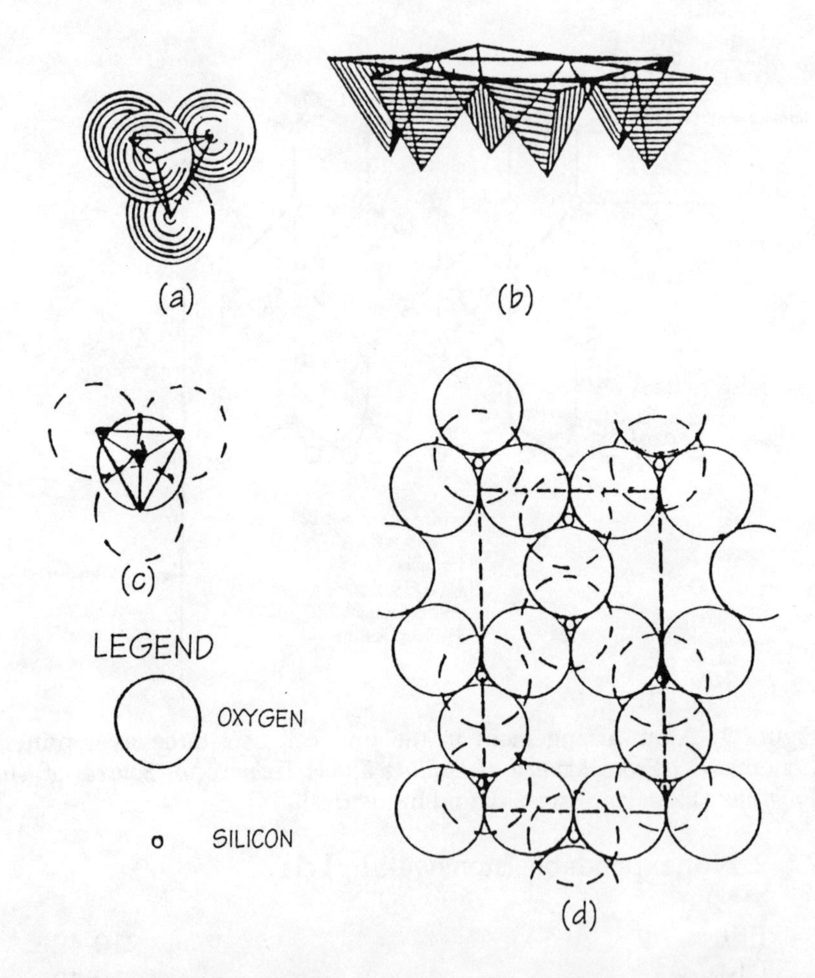

Figure 2. Structure of the tetrahedral sheet. Tetrahedral arrangement of Si and O. (b) Perspective sketch of tetrahedron linking. (c) Projection of tetrahedron on plane of sheet. (d) Top view of tetrahedralsheet (dotted line: unit-cell area). Large circles represent oxygen; small circles, silicon. (From *Manual of Drilling Fluids Technology: Sources of Mud Problems* [Houston, Texas: Baroid Industries].)

pandable clays are also more pliable or elastic and sticky than nonexpandable clays. Farmers prefer expandable over nonexpandable clays because of their greater fertility potential.

Clays expand by adsorbing water between and around the clay sheets. As seen in figure 3, the sodium ion (Na^+) collects a larger water envelope around itself than does the calcium ion

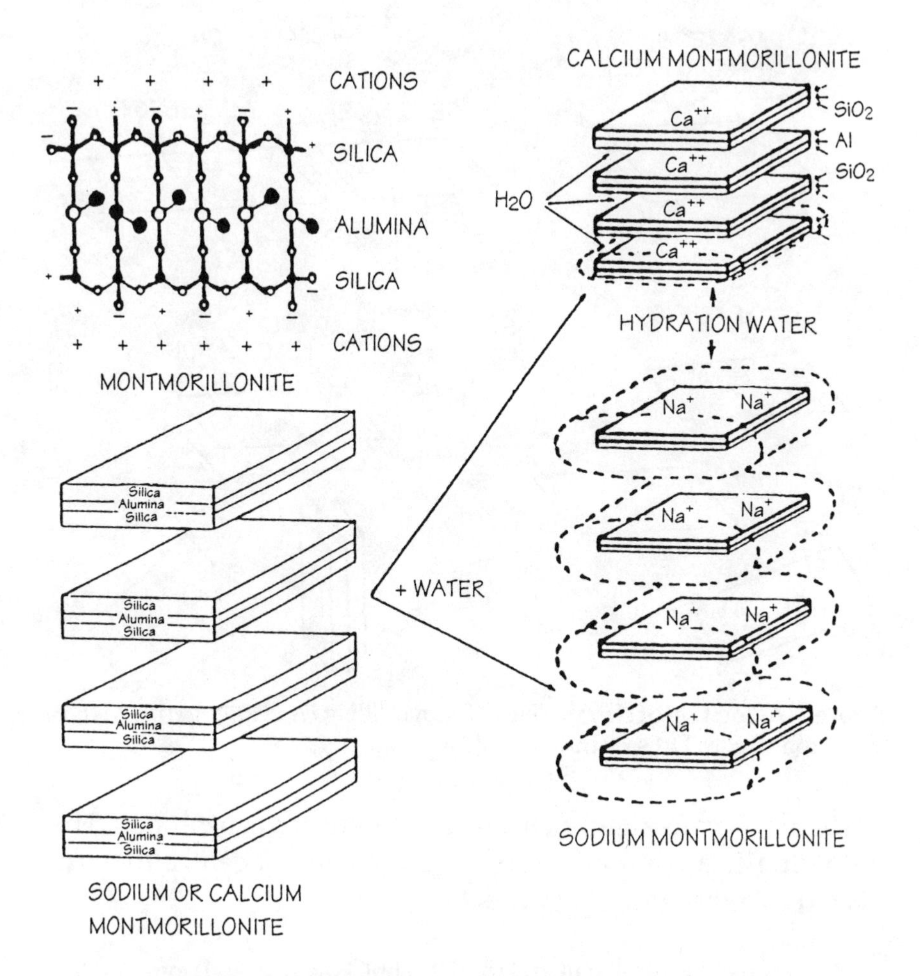

Figure 3. Water of hydration of calcium and sodium montmorillonite. (From *Drilling Fluid Engineering Manual* [Houston, Texas: Dresser Industries].)
anual [Houston, Texas: Dresser Industries].)

(Ca^{++}). As a result, the clay retains more water, and less free water is available to plant roots and microorganisms. This situation is all too familiar to farmers combating high sodium conditions. The saltier the water around the clay particle, the greater and tighter the aggregation (face-to-face association) of the clay sheets (see figure 7 on page 146). This leads to progressively worse soil compaction. Of importance to this discussion is the fact that different salts—chloride salts, in

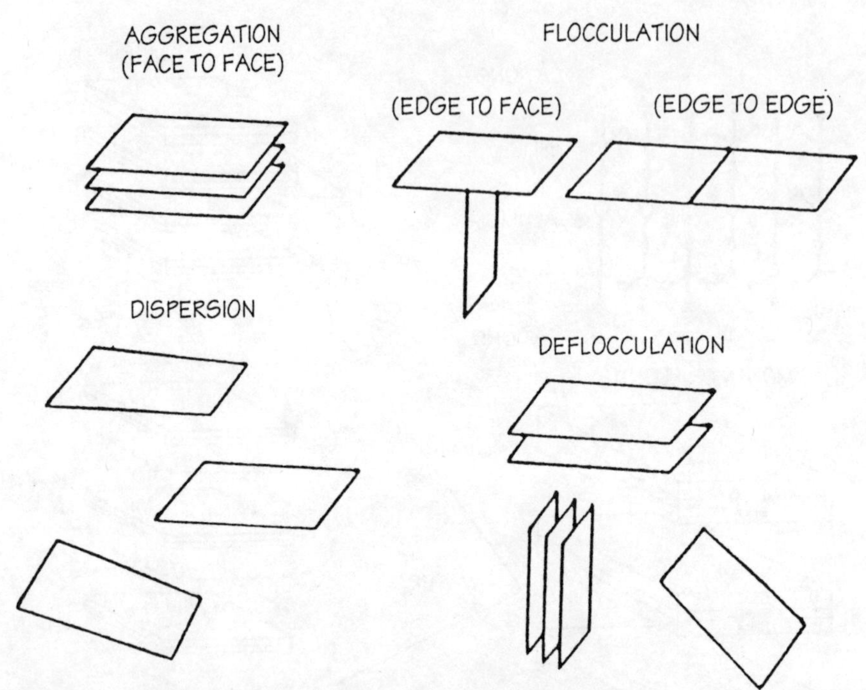

Figure 4. Assocition of clay paricles. (From *Drilling Fluid Engineering Manual* [Houston, Texas: Dresser Industries].)

particular—cause aggregation at different concentrations. The following is a comparison of chloride salt concentrations required to aggregate clay particles:

> Sodium chloride: 400 meq/l (8.20 lb./bbl. NaCl), 22800 ppm
> Calcium chloride: 20 meq/l (0.89 lb./bbl. $CaCl_2$), 1100 ppm
> Aluminum chloride: 10 meq/l (0.156 lb./bbl. $AlCl_3$), 426.6 ppm

Any time the soil dries out to the point that the salt concentrations reach these levels, complete or nearly complete aggregation occurs in that local environment. It is also important to realize that this is the level at which complete aggregation occurs. This does not mean that no aggregation occurs up to these concentrations, for more and more aggregation does transpire as the concentration of salt increases. Many farmers experience the hardening of their soils as the soils dry out. This is one possible contributor to that problem. As the soil dries, the relative salt concentration increases, resulting in greater soil-particle aggregation and increased soil hardness.

CLAY AGING

Aging is the process by which clays lose their expandability over time through their reaction with various elements in the environment. Ancient sedimentary deposits are largely void of montmorillonite or expandable clays, whereas modern soils are rich in such clays, according to *Physical Geology*.[1] This is because, with age, expandable clays like montmorillonite tend to become nonexpandable clays such as illite and chlorite. This propensity has been defined in *Sources Of Mud Problems*.[2] Chlorite clays are silicates of iron, magnesium, and aluminum; they are similar to montmorillonite, yet nonexpandable. Illite clays are similar to montmorillonite but with aluminum (Al^{+++}) substituted for 10% to 15% of the silicon (Si^{++++}) making up the clay, which results in a -1 charge on the clay particle (the difference between +4 of silicon and +3 of aluminum, 3 - 4 = -1) and a hole measuring 2.8 Å in the clay sheet. Due to this size of hole and the -1 charge, potassium (K^{+}) is "fixed" or locked tightly in the clay sheets. This produces a very strong bonding together of the clay sheets, resulting in a nonexpandable clay. The condition can be reversed, but with difficulty.[3]

The processes of aluminum substituting for silicon and potassium fixing or locking into the resulting hole (the conversion of montmorillonite to illite) are both natural and man-inducible. These processes continuously occur in the oceans. Clay particles enter the ocean through the rivers, react with the aluminum and potassium in the sea water, and settle to the ocean floor. This process seems to account for the deficiency of the expected potassium level in sea water and the fact that nearly all sedimentary clay is nonexpandable.[4] In oil-drilling situations, this process is desirable for preventing shale instability in drilling holes, which occurs due to hydration of the bore hole's sides. Converting the hole surface clay to a nonexpandable material diminishes the possibility of the bore hole washing out.

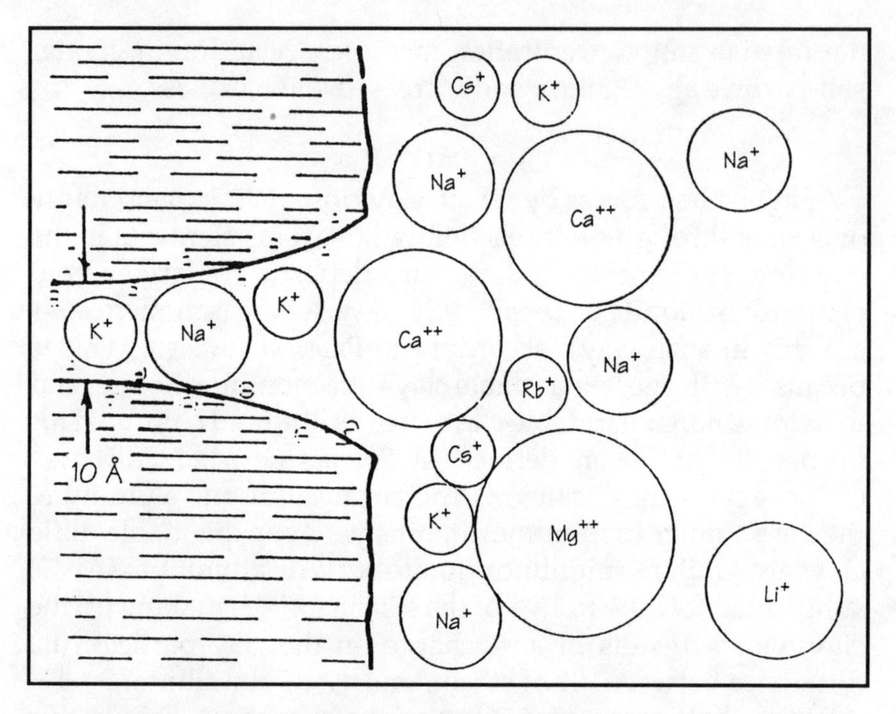

Figure 5. "Frayed edge" of illite with hydrated cations (to scale). (From *Manual of Drilling Fluids Technology: Borehole Instability* [Houston, Texas: Baroid Industries].)

Figure 5 shows how the potassium ion, being the ideal size, fits into the clay crevice and, through ionic forces, effectively seals off the clay. Other ions, particularly calcium and magnesium, are so large that they would only widen the crevice, furthering the water absorption by the clay and resulting in wash out. The following chart shows the sizes of the various cations that can be compared to the 2.8 Å hole size in the clay. An angstrom (Å) equals 10^{-10} or 0.0000000001 meters, or 0.00000000357 inches.

	ION DIAMETER, (Å)		HYDRATION ENERGY*
	NOT HYDRATED	HYDRATED	KCAL/MOL
Li^+	1.20	14.6	124.4
Na^+	1.90	11.2	97.0
K^+	*2.66*	*7.6*	*77.0*

Rb^+	2.96	7.2	71.0
Cs^+	3.34	7.2	66.1
Ca^{++}	1.98	19.2	377.0
Mg^{++}	1.30	21.6	459.1
NH_4^+	2.16	—	72.5

**Hydration energy corresponds to the amount of energy required to deform or change the water envelope around the mineral ion.*

Because the potassium ion is near the ideal size of the holes in the clay sheet and its water envelope is easily deformable, the clay sheets are drawn snugly together, with a relatively low concentration of K^+ required.

In the drilling industry, potassium chloride (muriate of potash or KCl, the most widely used potassium fertilizer in the industrialized world) is used to bring about this potassium fixation, resulting in the conversion of the expandable clay to a nonexpandable clay. "Expandable clay becomes nonexpandable through the reaction with potassium chloride."[5] In fact, "potassium chloride gives maximum shale stability at lowest mud weight," according to *Borehole Instability.*[6] In this process, the CEC and the pore space of the soil both decrease appreciably, as shown in figure 6.

Because the ammonium ion is 2.16 Å, it can also become "fixed" although not as tightly—like the potassium ion, creating a stable, nonexpandable clay sheet. This simply depends on which cation is most available at the time. This fixation, coupled with the fact that anhydrous ammonia burns out the humus, explains why anhydrous ammonia was used in World War II to harden airplane runways, and can be used effectively to harden the soil.

It is important to note that, when a clay devolves from an expandable clay to a nonexpandable one, the resident CEC declines from 80-150 to 10-40. As a result, a significant number of nutrient and nonnutrient cations will be dumped into the soil solution with no exchange sites on which to attach. These abandoned cations are available to plant roots, microbes, or leaching water. This is good for present crop growth, but it is

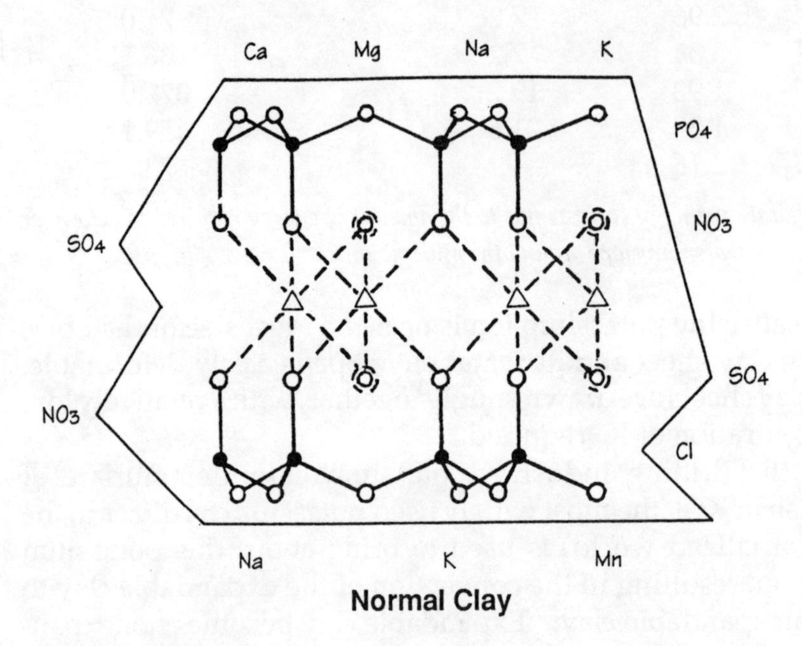

Normal Clay

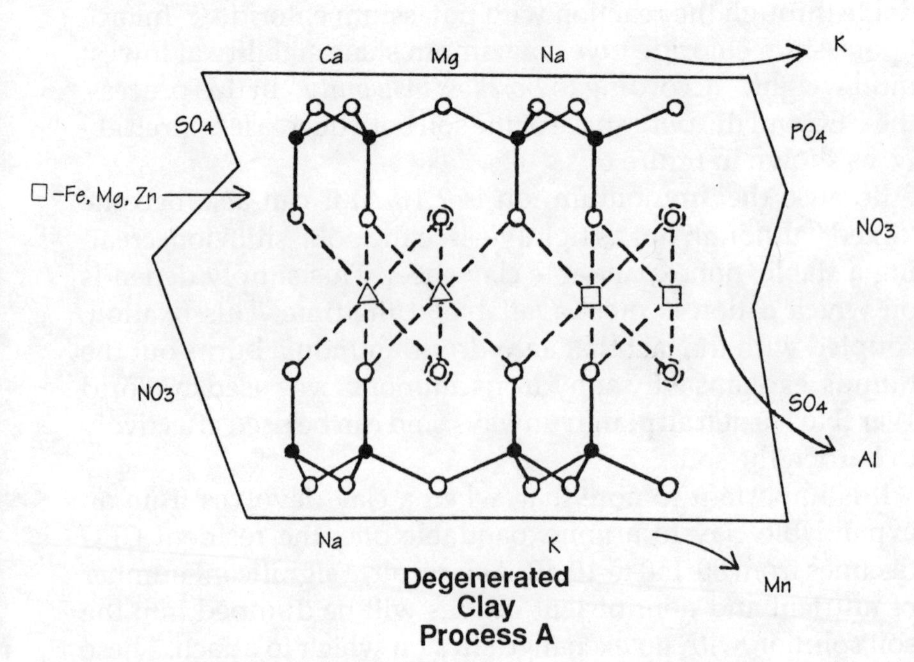

Degenerated Clay Process A

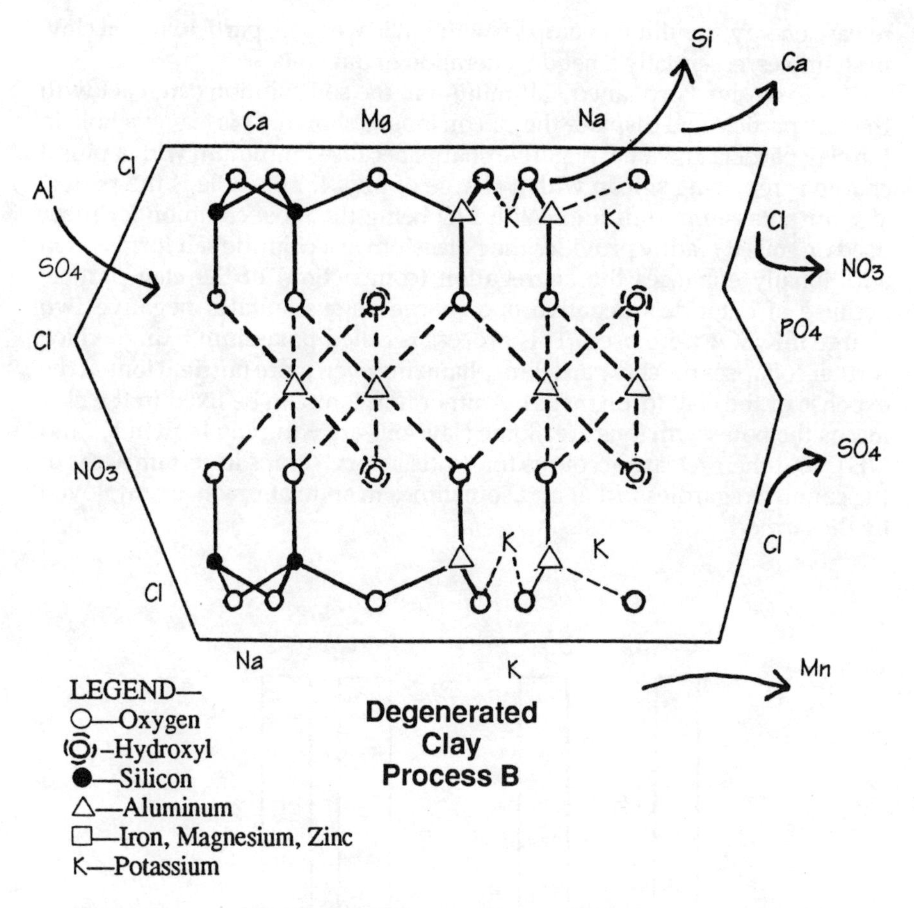

(□) Fe (high iron applications, i.e., $FeSO_4$, molasses, or ore—must be alert not to apply too much iron ore in attempt to increase soil magnetic susceptibility.)

(□) Mg (high magnesium applications, particularly dolomite.)

(□) Zn (high zinc application is common as status quo practice by many farmers even though they don't really *need* it.)

Figure 6. Schematic of clay evolution.

These three metals can replace the aluminum(△) in the clay sheet. (A)This collapses the clay sheet at these points, reducing CEC, which "orphanizes" the nutrients attached to the clay surface, leaving them free in the soil solution. This provides a "punch" for crop growth but we are left with a degenerated clay particle. Further, aluminum ions are "orphaned" into the soil solution creating problems inherent of aluminum, one being degeneration of other clay particles as described next. (B) This entire process does release energy, resulting in crop growth but leaves soil, particularly in clay, in shambles, essentially a net degeneration of our soil.

This excessive "orphaned" aluminum in the soil solution can react with the clay particle and displace the silicon ions as shown. This leaves a hole in the clay particle and a net negative charge because aluminum with a plus 3 charge is replacing silicon with a charge of plus 4. This hole is the perfect size for potassium to fit into. With KCl being the most common fertilizer used on soils it readily provides the potassium in a chloride salt form which additionally enhances the aggregation (compaction) of the clay particle because of chloride's negative one charge versus sulfates negative two charge if K_2SO_4 were used. This process is called potassium fixing, which further collapses the clay particle orphanizing even more nutrient ions at the expense of the clay (orphanage). Ammonium ions can be fixed in the clay just as the potassium ions are. Some clay soils are naturally high in K^+ and NH_4^+ and their release accounts for "satisfactory" crops in certain areas of the country regardless what and sometimes in spite of practices employed by the farmer.

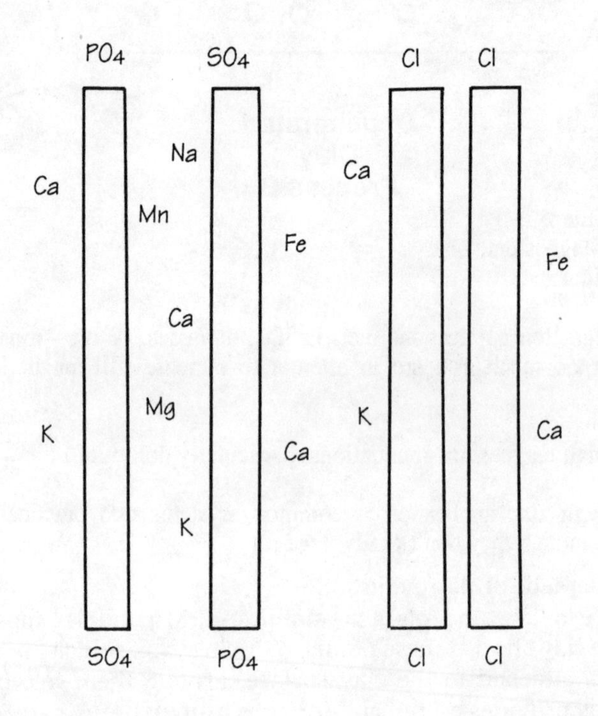

at the expense of net long term capacity—like living on principal rather than interest. In effect, assets have been destroyed or liquidated to feed the crop. Sooner or later, one realizes that the interest (available nutrients) dividend is minuscule because the principal (CEC) has been depleted. This accounts for both the apparent result from the application of muriate of potash and anhydrous ammonia and the eventual soil hardening, poor fertility, and declining response to equal amounts of fertilizer. Accompanying these characteristics will be soil compaction (nonexpandable clay), erosion (nonexpandable clays are less sticky), reduced water-holding capacity (nonexpandable clays take on less water), and limited nutrient reserves (fewer exchange sites and higher potassium mean fewer varieties of cations held by the clay).

These conditions can be reversed, although not necessarily with ease. The petroleum industry, desiring to leave the expandability of the mud clay unchanged, yet wanting a less viscous mud, uses the principle of anion exchange to thin the mud. Anion exchange uses negatively charged ions like phosphates, sulfates, chlorides, plant tannins, humic-acid lignins, and lignosulfonates, which act as dispersants (see figure 4 for a schematic of clay dispersion) without substantially altering the expandability characteristics of the clay as redrawn from *Drilling Fluids Engineering Manual.*[7] Polyphosphates rather than orthophosphates are used. The latter, alone, tend to flocculate the clay. Orthophosphates are used to reduce calcium and magnesium availability in the mud, which in effect reduces the flocculation due to calcium and magnesium. In such cases, the orthophosphate is combined with a dispersant like humic-acid lignin, lignosulfonate, chloride, or sulfate.

The above-mentioned dispersants are used instead of the polyphosphates in higher-temperature and salty conditions because phosphate dispersants function poorly under such conditions.[8] This sheds some light on why there appears to be an opening of the soil after gypsum or sulfuric acid is applied. The SO_4^{--} anions cause dispersion of the clay colloids in a thinning action. Unfortunately, this does not address the non-

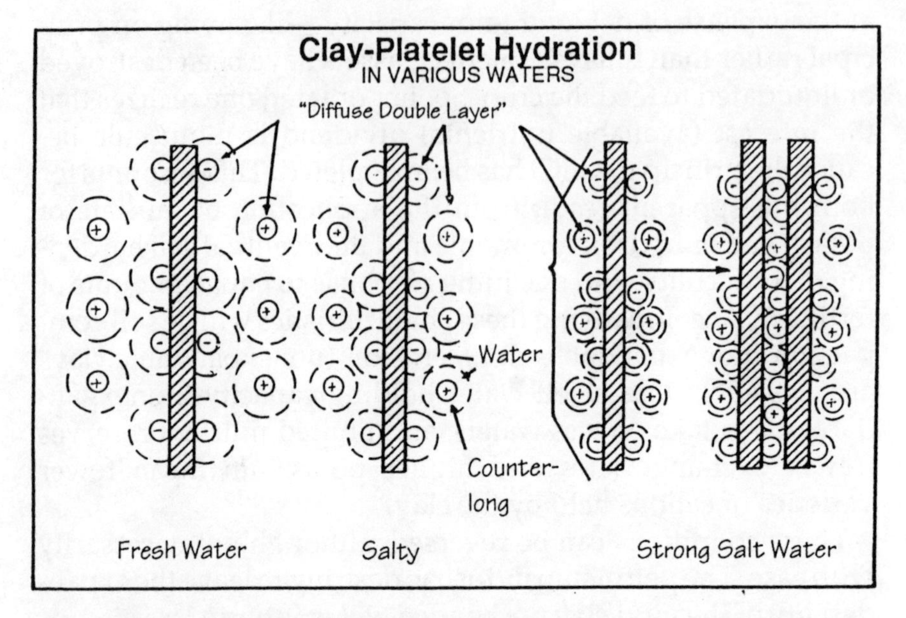

Figure 7. Clay behavior in different waters. (From *Sources of Mud Problems.*)

expandability of the clay, except for the calcium in gypsum, but when acid phosphates are added the calcium is largely precipitated or chelated out of solution, rendering its flocculating characteristic ineffective. With the continued application of the SO_4^{--} anions for this thinning purpose combined with the chlorides from $CaCl_2$ and KCl, the soil solution becomes highly concentrated with anions, that effectively and progressively reduces their thinning characteristic. In effect, the clays continue to become nonexpandable by potassium ion fixation and become aggregated due largely to chloride salts. The chlorides and sulfates oxidize and reduce, respectively, the organics; and the soil becomes less and less manageable, just as would a drilling mud.

Unfortunately, it seems to be human nature to look for magic solutions rather than to determine and eliminate the underlying cause of a problem. The material that is often seen as filling the role of panacea is humic acid. As with the application of sulfuric acid and gypsum, some people reason that if a little is good, more must be better; thus, larger and larger quantities

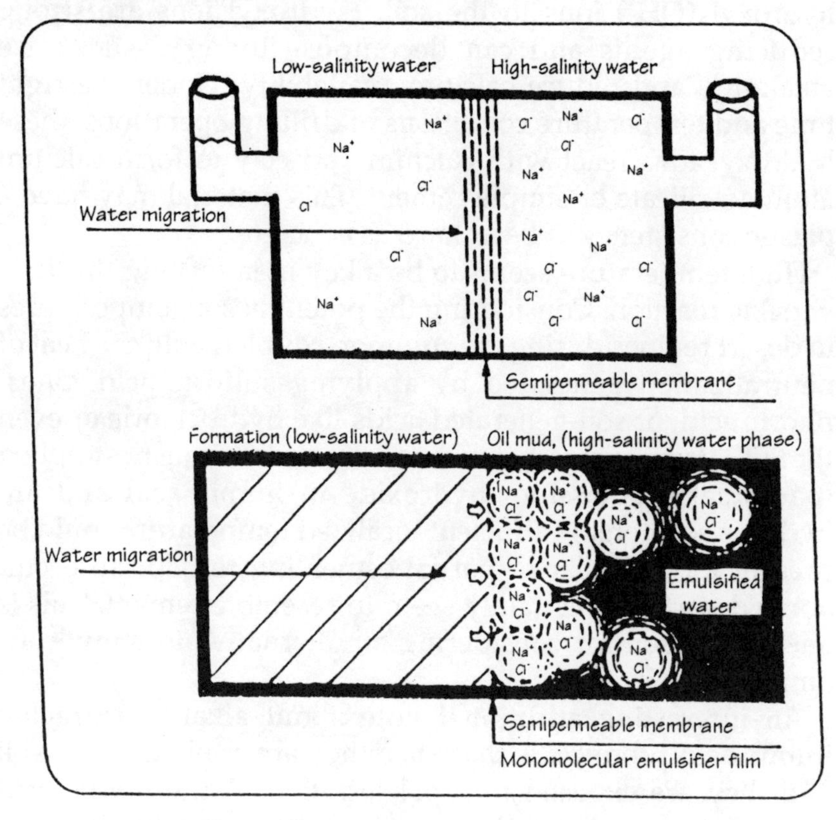

Figure 8. Osmotic mechanism. (From *Sources of Mud Problems*.)

of humic acid are applied with the intention of loosening the soil. It does loosen the soil at first, and perhaps for some time. However, there are two major problems with this approach.

First, the humic acid, like the other dispersants, is addressing a symptom but failing to address the underlying issue—clay expandability. Humic acid is an effective nutrient chelator and dispersant that works well in salty conditions. It definitely is useful in mud engineering and soil fertilization. There is a catch, however, and this is the second potential problem. Most commercial humic acids are made by reacting caustic soda (NaOH) or caustic potash (KOH) with leonardite to extract the soluble humic acids. When this liquid is applied, it adds hydroxyl (OH^-) ions to the soil. Hydroxyl ions are strong reducing agents and can decompose living tissue, raise alkalinity, and reduce calcium availability. Under the right time and temperature conditions of drilling operations, these hydroxyl ions react with calcium and clay to form calcium alumino-silicate or simple cement. This material may have a plastic consistency or be as hard as concrete.[9]

High temperature seems to be a key in launching this irreversible reaction. Considering the potential soil temperatures in desert regions during the summer, coupled with the heat of neutralization generated by applying sulfuric acid, phosphoric acid, or soil-generated acids like hydrochloric or even the SO_4^{--} from excess gypsum and the various caustics applied (particularly potassium hydroxide in humic acid and anhydrous ammonia), sufficient localized temperature could be present for a similar or partial cementing reaction to occur. Some desert soils certainly seem to resemble cement. This is another reason for considering an alternative to anhydrous ammonia.

An interesting additional note about alkaline extracted humic acid products is that once they are applied to the soil and they are exposed to a pH less than 7, the humic acid precipitates and has little or no activity or benefit. The acid soluble fulvic acid component of the humate is the only component that remains active to give soil/crop benefit.

The items that cause high alkalinity or high pH, such as KOH and NH_3, lower the temperature stability of the soil or mud, alter soil tilth, and decrease clay stability. The important issue in this discussion is really the judicious use of products like humic acid and careful selection of materials to be used, noting the product quality and base.

THE EFFECTS OF SALT ON CLAY

Salts, especially the chloride salts, alter the behavior of clay. Of particular importance is that salts, especially NaCl, cause the clay to cluster in lumps, making it more difficult to till. As mentioned before, the calcium and aluminum chlorides—or any multivalent chloride salt—are even more effective at solidifying clays through aggregation. Consequently, pore space diminishes, resulting in decreased water-holding capacity and oxygen levels. Figure 8 shows how salty mediums cause water to migrate to them rather than away meaning water will not readily be given up to plants or microbes from the salty soils.

The higher the clay and calcium content of the soil, the more aluminum and calcium that will be available to form chloride salts when muriate of potash (KCl) is applied. The excess K^+ can be fixed to form nonexpandable clay, and the Cl^- can form $CaCl_2$ and $AlCl_3$ salts. As a result, the farmer will need larger equipment to pull his anhydrous applicator, further solidifying the soil. More pesticides will then be needed to combat the only organisms that can survive this man-created anaerobic condition, and the farmer will need to apply more of these same fertilizers to achieve a sufficient "blasting" effect to release enough energy to raise a crop. Although farmers are told this is high-tech, scientific farming, the effect of such methods is to destroy fertile soils.

Anhydrous ammonia and muriate of potash, both important fertilizers, are the two most widely used fertilizers in American agriculture. According to scientists, they are two of the most effective clay-aging materials available.

REVERSING THE AGING PROCESS

To reverse the conditions described above, several things must be done. First, farmers must stop using KCl, NH_3, H_2SO_4, excessive amounts of acids and caustics, and salts. Then we can begin to discuss how to reverse many undesirable conditions: compaction, salinity, clumping, poor water retention, poor available nutrient balance, low oxygen content, and sparse aerobic microflora activity. Calcium, nonacid phosphate, humus, oxygen, water, microbes, and energy are the key components of this regeneration of the soil.

—10—

CAREY REAMS' TESTING AND EVALUATION METHODS

The Reams soil test was developed to reflect, in the test values, the characteristics actually observed in the field. These characteristics include soil compaction and tilth, weed and pest problems, crop quality and yield, and overall stability of soil and plant nutrients. No other testing system can make such a claim.

CAREY REAMS[1] realized that traditional soil testing did not give an accurate picture of a soil's fertility level. He understood that pH did not indicate calcium availability, rather, it measured electrical resistance in the soil. He observed that soil pH would fall into place naturally once the various nutrients were balanced. Thus, he contended that pH was an effect, not a cause. Reams perceived that calcium needed to be addressed as a fertilizer rather than a soil amendment.

Because of the drawbacks inherent in traditional soil testing, Reams adopted a system that closely resembled the biologically soluble level of major nutrients. Reams understood that just because a nutrient was present did not guarantee that it was of any value, analogous to being in the middle of the ocean and suffering from a lack of water. He tested calcium, phosphate, potash, nitrate and ammoniacal nitrogens, ERGS (conductivity in micromhos or microsiemen), and various trace elements. Using this test method, now known as the Reams test, which makes use of the LaMotte testing kit and the Morgan procedure, Reams established the following nutrient levels for a minimally balanced soil:

Calcium	2,000-4,000 lbs.
Magnesium	285-570 lbs.
Phosphate	400 lbs.
Potash	200 lbs.
Nitrate nitrogen	40 lbs.
Ammonium nitrogen	40 lbs.
Sulfate	200 lbs.
ERGS	200-600 micromhos
pH	6-7
Sodium	20-70 ppm

The test values alone contradicted what traditional thinking claimed was indicative of fertile soil with optimum nutrient levels and ratios. Reams developed his ratios by observing nature and evaluating the soil in conjunction with such observation. Consequently, using the Reams soil test, many soil characteristics can be identified before one sets foot in the field.

For example, if the calcium level is less than 2,000 pounds per acre, there will be possible energy-reserve deficiencies, weakened skin and cell strength, bruising susceptibility of fruit, soil compaction—especially if there is a narrow calcium-to-magnesium ratio (7:1)—weakened stems or stalks, and grass-weed problems. Further related to the calcium-to-magnesium ratio is the fact that a narrow ratio reduces nitrogen efficiency, requiring additional applications of that nutrient.

When the phosphate-to-potash ratio is less than 2:1 for row crops and 4:1 for forage crops, it will be difficult to sustain crop refractometer readings above 12 brix at the crop's weakest point. There also will be less than maximum production and crop vigor, as well as broadleaf weed problems and the possibility of insect and disease infestation.

The nitrate nitrogen levels indicate the potential growth status of the nutrient reserves in the soil. If this level gets too high, there will be problems with blossom drop and in getting fruit to set. High nitrate nitrogen levels also increase the potential for frost damage and winter kill, especially if the phosphate levels are less than desirable.

A low ammoniacal nitrogen level indicates poor biological activity and stability. Reams noticed that this level was practically nonexistent unless very active microbial populations were present. One can briefly get some ammonia nitrogen after an application of ammonia fertilizer, but Reams was looking at a consistent level.

The nitrate nitrogen levels on the Reams test are relatively easy to achieve with applications of chemical nitrogen. The ammoniacal nitrogen, however, will not remain until a very active microorganism system is established. The ammoniacal nitrogen seems to be one of the last factors to come into line when regenerating a soil.

Sulfate, the next item on the test, is not to be confused with elemental sulfur. Elemental sulfur can cause rot at maturity of fruit and can tie up or interfere with calcium. Sulfate, on the other hand, can help enhance calcium availability, is needed in certain protein and enzyme complexes, and sometimes can aid in mellowing the soil. However, it is possible to apply too much sulfate, which seems to be happening in some areas where reductionists are attempting to "hammer down" soil pH with large amounts of gypsum and sulfuric acid. This practice causes additional salt problems, calcium demand, and microbial stress.

ERGS (energy released per gram of soil), measured in micromhos or microsiemen, represents the amount of energy available to the growing crops and microorganisms. The reading must be interpreted in relationship to the inherent conductivity of the base soil due to salts and nonnutrient minerals. If the overall reading gets above 1,000, there is generally a salt problem, energy loss and waste, and increased potential for root burn and nematode proliferation. If the ERGS level drops below 200, little or no crop growth is occurring. Late-season crop finishing is directly correlated to the ERGS level.

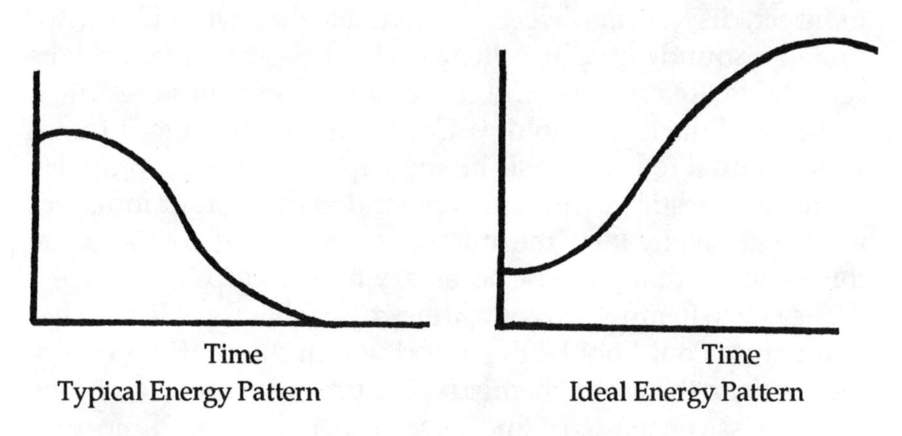

Typical Energy Pattern — Ideal Energy Pattern

Soil pH is an indicator of energy resistance in the soil. It varies throughout the growing season and is a reflection of what types of microorganisms are flourishing. Extremes in pH can indicate problems—problems with vegetative growth if pH is too low, or with fruiting if pH is too high. pH will vary throughout the growing season and should be monitored to track this change. Maximum nutrient exchange occurs between 6 and 7 pH.

pH also is a handy indicator in checking foliar sprays. Ideally, the final spray will be between 6 and 7 pH. Some people contend that foliar sprays should be between 4 and 5.5 pH because research has shown that plant sap is close to this level. It is, under inferior nutritional standards and low refractometer readings. It is also easier for the chemical people to get higher-analysis spray solutions when the pH is this low, but that does not mean this is ideal for the plant or the efficiency of the spray.

Sodium is a fairly ubiquitous element, yet it can often become problematic when in excess concentrations. As sodium concentration surpasses 70 ppm, the soil will become increasingly clumpy and compact, exemplify poor water-exchange characteristics, require greater calcium levels for balance, and show excessive ERGS levels.

Reams observed that if he took care to balance the soil sufficiently to achieve these test values, his crops would be free of insect, disease, and weed infestations; they would be nutritionally sound, give excellent yield, be profitable, and be repeatable. Reams knew he could not achieve these results if he ignored the microbiology. Consequently, he taught that it was essential to learn basic biology applied to agronomy. He found that destitute microbes responded to sugar or molasses and calcium. In fact, the microbes responded to the same things he postulated to be necessary for the crops.

The key to Reams' program, though, was energy. He realized that nature could not be described within the confines of any mechanistic theory of chemistry. Nature is energetic and thus encompasses chemistry and every other science. Philip Callahan has contended and repeatedly proved this for years. Most people, trapped in the reductionist mind-set, find it difficult to make the distinction between what to apply literally and what to apply conceptually of Reams's teachings.

The major conceptual aspects of Reams's teachings involve the use of fertilizers. Reams advocated applying several tons of high-calcium lime and a ton of soft rock phosphate per acre, as well as several tons of chicken manure. These recommendations are conceptual relative to today's applications. They were developed several decades ago in different conditions from today's. Experience has shown that if smaller amounts of these materials are applied, we often get better results. The challenge in most areas is determining what to use to get the calcium and phosphate in line. Mechanistic thought advises farmers just to apply the calcium or phosphate or whatever nutrient is deemed deficient, according to the symptoms. Common sense advises that farmers first determine whether quantity of nutrient is the issue. Perhaps sufficient quantity of a nutrient is present but the quality is insufficient. Perhaps we have a case in which the microbes need some energy, e.g., sugar or molasses, a vitamin, an enzyme, or a trace element. Perhaps we simply need a little "charcoal lighter fluid" in the form of liquid calcium to make more reserve calcium available.

Reams used soft rock phosphate rather than acidized or hard rock phosphate. Although he was not opposed to hard rock phosphate, he preferred to use soft rock phosphate because it was colloidal. Colloidal particles are the key to biological systems. They do not tie up as readily as do noncolloidal materials. Reams found that, over the long term, the only way to achieve the phosphate availability of 400 pounds per acre in a 2:1 ratio with potash on the Reams soil test was by using soft rock phosphate.

Reams used calcium carbonate, never dolomite. He observed that sufficient magnesium would be available if he balanced the calcium, phosphate, and microorganisms and then applied fertilizer quantities of sul-po-mag. Magnesium, he found, interfered with nitrogen. Large amounts of magnesium require large amounts of nitrogen and vice versa. An excess of magnesium relative to calcium also causes the soil to compact, thus further degrading the microsystem of the soil.

In traditional agriculture, plant-tissue testing is done in addition to soil testing to evaluate the need for nutrients. Reams placed little credence in plant-tissue analyses for two reasons. First, they test symptoms, not causes; plants are reflections of the soil. Second, they are evaluated using suboptimum health standards. Farmers may find that their crop possesses adequate levels of nutrients according to the tissue analyses, yet the crop still has a low refractometer reading, insect and disease infestation, poor shelf life, and so on.

Tissue analyses do not reveal or address causes, only symptoms. A prime example is the tissue-analysis indication that there is a deficiency of magnesium and potassium. It is possible that there is a deficiency of these elements, but correcting these deficiencies usually does not mean applying magnesium or potassium. More often than not, magnesium and potassium deficiencies in tissue analyses are a result of insufficient calcium availability in the soil, especially in areas where there are large quantities of magnesium and potassium in the soil, such as in the western United States. That is not to say that applying magnesium or potassium does not relieve the symptoms. It often does, but the symptoms later reappear.

For tissue analyses to be of value, the standards that the farmer is seeking to achieve for his crop must be increased to represent the actual crop quality that is found when plants are nutritionally sound and not dependent on chemicals to protect them from insect pests.

At present, there are no standard correlations between tissue analyses and refractometer readings. In establishing these correlations, distinctions must be made between leaf, vein, and petiole evaluations. The lower the nutrient balance, the greater the variation will be between the parts of the plants, both in the refractometer readings and the nutrient analyses.

Multiple nutrient interactions also must be considered. A student of Dan Skow's discovered an example of such an interaction. He noticed that if the phosphorus levels of grapes remained around .35% or greater in the leaf, the refractometer reading in the leaf remained at 12 or above. Interacting with this, however, was the magnesium level in the leaf. If it remained above .3%, the above-mentioned phosphorus rule applied. This seems logical when one remembers that magnesium regulates nitrogen in the system. If the magnesium level decreases too much, there will be an excess of free nitrogen in the system; this free nitrogen carries water with it, resulting in a diluted nutrient concentration, a lower refractometer reading, and lower plant health.

This type of observation and record keeping needs to be done on all crops so that the results of such testing become more valuable and useful. In all likelihood, it will be the farmers who eventually do just that. The important question remains. What will it take to correct the nutrient imbalances, not just address their symptoms?

Several competent consultants use tissue testing extensively, fertilize according to the test results, and achieve reasonable successes. Investigation of these successes usually reveals that a biological carrier material, e.g., fish, seaweed, fermentation byproducts, cultures, and plant growth regulators, was applied with the fertilizer. It seems, though, that the same nutrient deficiencies appear season after season because the consultants fail to address the real causes of these deficits. The biological materials are successful in concealing a multitude of symptoms. Addressing nutrients in a symptomatic manner is like taking an aspirin for a headache. The symptom is temporarily eliminated, but it will reappear later if the cause is not addressed. Headaches are not caused by a lack of aspirin.

Using the Reams soil test, we can predict accurately whether soil compaction is present in the field. This can be determined by evaluating the calcium-to-magnesium ratio. If this ratio is less than 7 pounds of calcium to 1 pound of magnesium, compaction will occur. Even at a 7:1 ratio, if there are more than 70 parts per million (mg/liter) of sodium, there will be compaction. As these ratios come into line, compaction decreases until it ceases to be a problem. People often blame compaction on heavy equipment and frequent traffic across the soil. These things do cause compaction of soils with calcium-to-magnesium ratios of less than 7:1. They do not cause compaction of soils with calcium-to-magnesium ratios of 7:1 or more and less than 70 parts per million of sodium. Compaction is a phenomenon of physics (particle attraction/repulsion) and aeration.

Take two magnets and hold them together, north pole to north pole. Then release your grip on the magnets and observe what happens. The magnets separate by themselves. Proper mineral ratios in the soil reflect the same magnetic phenomenon. You can press the soil particles together, but as soon as the compression is released, the particles again repel each other.

Now take a sponge, place it on the floor, and step on it. It compresses. Lift your foot, and the sponge returns to its original form. Pick up the sponge and inspect it closely. Notice that it contains as much air space as sponge material. The air space allows the sponge to be compressed and then to return to its original form after the compression passes. This is what happens in the soil once biological activity and humus are restored. The soil will function like a sponge, even under the heaviest farm equipment. The biological activity and humus are restored in direct proportion to the restoration of the calcium-to-magnesium ratio.

The calcium-to-magnesium and phosphate-to-potash ratios constitute the bulk of information from the soil test. One must remember, though, that the soil test indicates only what was happening when the soil was tested. Traditional opinion suggests that soil be tested only once a year, at the most. Ideally, however, a farmer should test the soil with the Reams test each week of the growing season, charting the variations in nutrient levels.

Initially and every few years, it also is beneficial to compare the Reams test results to those of a conventional soil test from A & L Labs or a similar reputable firm to establish a guideline as to the reserve nutrient levels in the soil. The combination of these two tests provides a directive concerning the approach to take in fertilization. For example, if the A & L test indicated several thousand pounds of calcium but the Reams test indicated only several hundred, we would know that there is poor microbial activity. Initially, our fertilization approach would probably favor those materials that would catalyze the releasing of calcium rather than the building of a calcium reserve. Such materials might be sugar, molasses, vitamin B_{12}, humic acid, fermentation products, enzyme materials, liquid calcium products, hydrogen peroxide, compost, or simply aeration of the soil.

If, on the other hand, the A & L test showed only several hundred pounds of calcium and the Reams test only several hundred pounds, we could assume that there was very little calcium with which to work. In this case, we would apply a few to several hundred pounds of calcium carbonate (high-calcium lime) in either ground or pelleted form, in addition to the catalyst materials previously mentioned, to gradually build the calcium base.

Even in the first example, if economics permitted, we would probably apply a few hundred pounds of calcium carbonate per acre. In traditional practice, calcium is treated as a soil amendment and is applied by the ton rather than by the pound. We are treating calcium as a nutrient and applying it as a fertilizer, in fertilizer quantities. This is not to say that one cannot benefit from applying a ton or two of calcium carbonate to the soil, but this would be our second choice. Keep the quantities low in the spring or just before a crop is planted. This timing will lessen the chance of reducing the yield. Several applications of a few hundred pounds of lime will give better results more quickly than single large applications.

Farmers often ask how they can decrease their magnesium, potash, or other excess nutrients.[1] In some cases, certain nutrients will actually decline when the overall nutrient balance comes into line as the microorganism population is regenerated. One such nutrient is sodium. Often the high sodium levels will actually drop due to soil regeneration. This is due to complexing and perhaps transmutation of the sodium.

To correct the imbalance, raise the other nutrients. If you have a 2:1 calcium-to-magnesium ratio, correct it by raising the calcium. If you have a 4:1 potash-to-phosphate ratio (very common in American agriculture), correct it by raising the phosphate. Sugar is an important component to add to acid phosphates. It helps buffer the phosphate and make it compatible with microorganisms. Especially relative to phosphate is the microorganism activity. It is imperative to stimulate this activity in order to get the 2:1 phosphate-to-potash ratio on the Reams test.

It is advisable to couple any soil test with field history and characteristics to further correlate the soil-test nutrient levels to their meanings. The more complete the picture formed from these data, the more effective will be one's fertility recommendations. Accurate record keeping is essential, as is soil testing at least once during the growing season to establish nutrient status under load. Nutrient draw from the soil is greatest during the latter part of the growing season. This is when we want to know how the soil is performing "under load." An analogy would be to evaluate the capacity of a water-well aquifer while the pump is pumping full capacity, versus while the pump is idle. No single item will show you the entire situation. All items must be combined with astute field observation and common sense. No number is perfect unless all the numbers are perfect.

A good example of this and the value of routine testing is the case in point on a well-managed farm in Wisconsin the summer of 1991. Two corn fields were soil tested weekly and managed the same under irrigation. One field yielded over 200 bushel per acre while the other yielded only 155 bushel per acre.

Early in the season both fields looked identical. Late in the season the low field's calcium and phosphate levels gradually increased on the soil test yet the ERGS readings consistently declined. The high field's calcium and phosphate levels continued to decline while the ergs levels remained in the desired range. What happened? We do not know exactly, but have an idea. The high field's calcium and phosphate continued to be used up thus creating minimally sufficient energy to keep the crop expanding. The phosphate and calcium in the low field lost their assimiliability and consequently were not usable by the soil system for energy production or the crop for grain fill. The high field calcium and phosphate apparently remained assimilable, consequently, yielding more grain fill. Here we have the situation where we must consider not only going from unavailable to available nutrient activity but also from available to assimilable nutrient activity. This is, in all probability, a microbial issue. This is very similar to the problem encountered when feeding cattle. Nutrients may be in the feed ration but getting satisfactory nutrient assimilation and feed conversion are determined by nutrient quality and digestive system health. To reiterate, we are truly farming the microbes in the soil.

REAMS SYSTEM OF SOIL TESTING, AS CURRENTLY UPDATED BY BOB PIKE AND DAN SKOW

• • •

SOIL TESTING INSTRUCTIONS: MAJOR NUTRIENT AVAILABILITY USING MORGAN EXTRACTING SOLUTION

Assuming that the farmer wants to know the availability to rootlets of the major plant nutrients of the soil, i.e., ions of calcium, magnesium, phosphorus, potassium, nitrate, and ammoniacal nitrogen, so that ratios of specific elements may be amended, the following testing method may be used. After these ions have been measured and amendments made, the farmer may want to monitor and control the minor elements, such as iron, sulfate, boron, manganese, copper, and so on.

The Morgan extract (UES) is a weak organic acid solution that acts on soil particles to dissolve nutrients that are likely to be made available by the exudant from plant rootlets. This test is often referred to as testing for water soluble nutrients.

PREPARATION OF SOIL EXTRACT/FILTRATE

Use multiples of 2.5 ml of soil to 7 ml of Morgan universal extracting solution to obtain sufficient filtrate to perform desired tests. The following table lists quantities of filtrate for each ion, as well as the extraction efficiency for each ion:

Ion	Amount(ml)	Efficiency(%)
Ca	.50	50
Mg	.50	50
P	.50	4
K	3.00	50
NO_3	(one drop) .05	80
NH_4	.05	50
Total	4.60	

It is therefore apparent that, for each atom of phosphorus in the filtrate, there are 24 atoms still remaining in the soil. These must be factored when reference standard solutions are produced.

At least two measures of soil or 2 × 2.5 ml. = 5.0 ml and 2 × 7 ml of UES will be required to obtain the 4.6 ml of filtrate required if no repeat or dilution tests are necessary. For convenience of measurement, it might be better to use 10 ml of soil to 28 ml of extracting solution.

To simplify the measuring out of the soil sample, it is convenient to collect the soil in a plastic bag in which the sample may be mixed and loaded into the measuring vessel without touching the soil with fingers.

1. Use a modified syringe (cut off tip at "0" mark of 0-60 cc plastic syringe) and adjust plunger to dispense 10 cc of soil from plastic sample bag. Usually the face of the plunger should be set at 1 division greater than desired amount, i.e., at 11 divisions to obtain 10 cc or ml sample. Push soil out of syringe body into coverable plastic cup.

2. Use a constant-volume dispenser to measure and dispense 28 cc or ml of the Morgan universal extracting solution to the plastic coverable container.

3. Cap, and shake 60 to 80 times, i.e., until soil is completely wetted by the solution. Allow to settle for 2 to 3 minutes.

4. Swirl contents, remove cover, and pour contents slowly into funnel lined with a folded filter paper cone. Funnel may be supported by a test tube, a flask, or a disposable plastic cup rack. Allow approximately 5 minutes for filtrate to collect in container.

5. When all solution has run through the filter, a suitable pipe may be used to transfer the filtered soil extract per the following instructions.

NITROGEN TESTING: NITRATE OR NO_3

6. Add 1 drop of the filtrate to each of two wells of your porcelain well plate.

7. Into one of these wells, drop 4 drops of NO_3 nitrogen test solution (5148).

8. Wait 4 to 5 minutes, and compare blue colors to those in the phosphorus color chart #1312. Use the following conversion table to convert P numbers to NO_3 -N pounds per acre as N.

Phos. No.	NO_3 -N (lbs./acre)
10	2
25	8
50	15
75	25
100	40
150	60
200	80

AMMONIUM NITROGEN (NH_4) TEST

9. In the other well position, drop 4 drops of ammonium test solution (5103).

10. In approximately 30 seconds, compare the colors with those of the NH_4 color chart (#1302).

11. The following table will convert the card values to NH_4 -N pounds per acre as N.

NH_4 Card Value	Ammonium -N (lbs/acre)
Very Low	10
Low	20
Medium	30
High	40
Very high	50

USE CAUTION TO PREVENT CONTACTING ANY PART OF THE HUMAN BODY WITH ANY OF THE CHEMICALS SUPPLIED WITH THIS KIT. BE SURE TO IMMEDIATELY WASH SKIN OR EYES WITH LARGE QUANTITIES OF WATER IF ANY CHEMICALS CONTACT THE BODY.

CALCIUM AND MAGNESIUM TESTING USING DIGITAL TITRATOR

12. Using a precision pipe set to .5, transfer .5 ml of filtrate to 125 ml flask marked Ca test.
13. Also transfer .5 ml to second 125 ml flask marked Mg.
14. Using graduated cylinder, transfer 49.5 ml of distilled water to both flasks.
15. Using pipettor and clean tip, transfer 2.0 ml of 45% KOH to Ca flask.
16. Add 1 CalVer II pillow to Ca flask.
17. Set digital titrator counter to 0. Carefully place a clean teflon-coated magnet in flask. Then titrate while magnetically stirring using the Digital Titrator with .08 molarity sodium EDTA cartridge and tube attached. Slowly introduce titrant until pink color turns to blue. A white background is helpful to see and point. A reference sample of distilled water is also helpful to see what the end point looks like.
18. Read number of digits dispensed . . . × 13 = pounds/acre calcium.

TOTAL HARDNESS (TH) or Ca PLUS Mg (Mg = TH - Ca)

19. Take the Mg flask from #14 above and add 1 ml of Buffer Solution Hardness 1 to the contents of the flask.

20. Add 2 drops of the MgVer Hardness Indicator to flask.

21. Use same procedure as above and titrate to obtain blue color end point using #14364 cartridge and magnetic stirrer and magnet in flask.

22. Read digits from counter . . . × 13 = TH. Magnesium = (TH - Ca) x .6 = . . .

PHOSPHORUS TEST

23. Using the pipettor, transfer .5 ml of the soil filtrate to the Hach 1730 tube. Add 1 drop of the Phosphate Test Reagent 2 (#5156). Rotate or gently swirl to mix filtrate and molybdenum in #5156 reagent.

24. Add 1 drop of the V-M Phosphate Reducing Agent (#4411) to above mix and then rotate and swirl to further mix ingredients.

25. Add 5 ml of distilled water to above and swirl to mix. Blue color is now ready to be read in the inner position of the Hach Phos color wheel. Tainted distilled water may turn the solution a yellowish tint. This solution must be discarded and the test redone with different distilled water.

26. Place a reference tube (#1730) filled with 5 ml of distilled water in outer position of color wheel tester.

27. Turn color wheel to match colors in two windows while looking toward a diffuse daylight sky.

28. Read color wheel and multiply times 100 to obtain pounds/ acre phosphorus. To convert to P_2O_5, multiply above by 2.3 to obtain fertilizer equivalent. It is believed that P_2O_5 should be 400 pounds per acre. This is equal to 174 pounds per acre of phosphorus.

POTASSIUM TEST

29. Fill the Potash A Tube (0245) to the bottom line with soil extract. Or, alternatively, using the large pipettor, transfer 3.0

ml of filtrate to the clean (0245) tube.

30. Add 1 Pot Reagent B Tablet (5162) and swirl until dissolved. An alternative procedure would be to use a .05 ml spoon (0696) and transfer the Pot Reagent C powder to the 3.0 ml of soil filtrate and swirl until dissolved. The latter process is faster.

31. Slowly add Pot Reagent C (5162) to the second line on the Pot A tube. Or use pipettor and transfer 4 ml of Pot Reagent C to A tube. Allow the "C" Reagent to run slowly down the side of the tube. Twirl the tube to mix.

32. Place the Potassium B tube (0246) onto the potash reading plate. Use a pipe to transfer the mixture from the "A" tube to the inside wall of the "B" reading tube.

33. Observe the black line down through the "B" tube and continue to add mixture until black line just disappears.

34. Read the potassium in pounds per acre off the scale on side of reading tube. $K_2O = 1.2 \times K$

CLEANING OF LABWARE

If labware is to be cleaned (and not disposed of), rinse item immediately, then submerge in pan of clean water for storing until all apparatus is to be washed. Wash in hot, soapy tap water. Rinse with hot tap water. Spray rinse inside surfaces with distilled water. Store on rack to air dry.

For further information, contact: Pike Lab Supplies, Inc., RFD 2, Box 92, Strong, ME 04983. Telephone, (207) 684-5131 Fax, (207) 684-5133.

COMMENTS ON SOIL TESTING

Attentive soil testers will formulate many questions concerning the dependability of soil testing, whether it be using the Reams method or any of the more common university and state registered laboratory methods. Typical soil testing methodologies calling for testing only once per year, at most, are not too concerned about differences between soil at a two inch depth versus soil at a four inch depth. Soil samples are mixed, and these differences are not considered. An argument could be waged either way. With the recommendation of soil testing more than once a year and even weekly with high value crops using the Reams methodology, it is argued that differences in test results between depths and field areas become statistically significant. I would agree that this is very possible. In other words, the variation between two soil tests taken one week apart could be due primarily to sampling differences. This is why we stress that the sampling should be done in the same area of the field, at the same depth, using the same procedures each time. We can still have sampling error, but there is a back up mechanism inherent in the Reams methodology. That back up mechanism is the observation of field characteristics.

Soil sampling and testing, regardless of the consistency and care exercised in their performance, are subject to human and mechanical error. When relied upon exclusively, they inevitably will lead to failure somewhere, sometime. That does not mean we disregard or dispense with such procedures. It simply means we must correlate test results with crop performance and integrity, soil tilth, compaction and erosion, weed pressure and variety, disease pressure, and insect conditions. These field conditions hold true regardless and thus can be depended upon as the final management information.

Soil tests can provide valuable information, particularly with regard to the growing crop. They can help head off impending disaster, but like all man-nature interactions, soil tests entail an element of art. Keep in mind that consistency is most important in sampling and testing. Ratios are important in evaluating nutrient levels on a soil test as long as the ergs are in the satisfactory range and water is adequate. However, if the soil test has nearly perfect numbers but the field is hard, the crop is mediocre, weeds are rampant, and insects are having a feast, your soil test is probably inaccurate. The soil test must be correlated to field observation. The active observer spots a problem or imminent difficulty and tests the soil to confirm his suspicions, not vice versa. He knows intuitively and from experience where the sample should be taken and what it represents. Cookbook testing with corresponding recommendations is only a guideline, not a guaranteed road to success. All in all, the better your testing, the better foundation you will have for management decisions.

—11—

THE TERMINOLOGY OF CAREY REAMS EXPLAINED

EVERY FARMER HAS BEEN inundated with fertility theories, good and bad products, mandated practices, government welfare programs, and university academician solutions. Many farmers are overwhelmed by these choices and resort to their local Cooperative Extension agent for advice. When the Cooperative Extension Service was established in the early 1900s, it was intended to provide farmers with sound, reliable information about such things as fertilizer recommendations, tillage practices, soil testing, and pest control. In association with the U.S. Department of Agriculture, the Cooperative Extension Service provided farmers with valuable resources and solutions such as the sterile screw worm project in Florida, and that practically eliminated that horn fly problem for Florida cattlemen.

Over the years, the Cooperative Extension Service and the U.S. Department of Agriculture have become less sources of solution and more representatives of the petrochemical industry. Ask almost any Extension agent or USDA authority how to grow crops without pesticides, and you will be told that it cannot be done without a tremendous loss of crop output and quality. The only truth to this statement is that it cannot be done with *their* understanding and technology.

Today, revolutionary breakthroughs in agricultural technology and viable solutions to cropping problems come from private researchers, consultants, small entrepreneurs, farmers, and old-fashioned common sense. Common sense is of greatest importance because much of the information and many of the new breakthroughs today are primarily rediscoveries of "old" information. Biological insect control is ancient knowledge. Regenerative fertilization is as old as man.

Intensive monoculture farming is relatively unique to modern man. However, the principles of natural science remain constant. It is man's use of the principles that must change.

Conscious awareness must precede the successful application of the natural principles that are necessary to reverse the degeneration of our soils and produce the necessary quantities of nutritionally sound food to meet the demands of an ever-expanding world population. Modern agricultural practices are performing successful operations, but the patients are dying.

In his book *Nutrition and Your Mind,* George Watson explained that nutrition governs the way people think. Change a person's diet, and you will change the way that individual thinks. We observe the degeneration of the attitudes of society and blame everything but the real cause. Agriculture has degenerated the food chain to the point that the food does not contain enough minerals to feed an active, creative mind. It is difficult to be conscious of nature, your fellow man, or yourself, for that matter, if your mind does not have enough energy to function beyond the basic survival instincts. The less energy the brain has, the more primitive the individual's behavior.

Many people have realized this self-destructive cycle of empty food production and consumption, altered their diets accordingly, and realized the consciousness necessary to turn the system around. Some may choose to call this change spirituality, being born again, or being touched by the Holy Spirit. Nevertheless, the result is that they have come to understand the symbiosis of man and nature, are applying that understanding, and are doing their best to teach others. There is a problem, however, in teaching a concept that involves expanding the student's consciousness or awareness.

Knowledge can be taught, but the intuitive understanding or consciousness of nature is multidimensional. Knowledge is an inner spiritual intuition and appreciation. It is not gained solely from textbooks. The best current example of this is the work of Philip Callahan. Many scientists have textbook learning similar to Callahan's, but few possess his consciousness of nature and practical experience to explain how insect antennas are omni-range radar devices, how viruses are self-contained communication satellites, how blood vessels are wave guides, or how ancient "religious" structures such as stone circles are tuned radio or UV/IR transceivers.

Despite his solid academic explanations for these phenomena, Callahan's colleagues often do not believe him due to their lack of conscious awareness. Others contend that his explanations are sacrilegious and cannot be of any scientific value. It is ironic how the religionists condemn the scientific community for teaching that humans evolved from apes and yet turn around and demand that individuals like Philip Callahan produce verification from this same scientific community that their work is not sacrilegious.

These problems are similar to those Carey Reams encountered with regard to his biological theory of ionization. Reams would say, "Get even with these people by praying for them." Reams was a bit different from Callahan in that he was not as meticulous in using proper scientific vocabulary. This alone causes many people great consternation. The concepts

Reams taught do not differ markedly from those taught by Callahan. Further investigation bears this out.

Reams picked up on the work of Charles Northern, M.D., who in the 1930s warned that nutritionally deficient soils would ultimately cause the diseases of mankind. Rudolf Steiner gave similar warnings, but the "scientific" community found no foundation for these assertions. Carey Reams did, however, and he embarked on a life of research, consultation, and teaching. Many farmers have attended seminars on Reams's biological theory of ionization (RBTI) over the years. Some scoffed at it as fraud or nonsense, some never used what they learned, and others set out to prove or disprove Reams' theory. For the last group, it touched their hearts in some unexplainable way—unexplainable because it did not make sense from a textbook approach to soil science, chemistry, or agronomy, yet it worked in the field. Many people have criticized Reams' teachings for that very reason.

People often ask me why Reams did not use accepted terminology. Why doesn't RBTI make sense academically? For some time, I did not know how to answer these questions. I knew intuitively that what Reams intended to convey was correct, regardless of his vocabulary. As I have studied further, I have concluded that Reams might not have been so far off, even with his terminology and explanations of hard chemistry and physics. I say this, having learned that the scientific community has suppressed, altered, or actually destroyed many mathematics, physics, and chemistry teachings. The exceptional work of E. T. Whittaker in mathematics around the turn of the century, which laid the groundwork for "free energy," anti-gravity, time reversal, and directly engineerable physical reality, is neither taught nor discussed in classrooms today. Neither is the work of Nikola Tesla, Albert Abrams, Georges Lakhovsky, Wilhelm Reich, Henry Bastion, Antoine Bechamp, and many others. It appears that Reams' terminology is closer to what should be described than is the conventional glossary. That is what distinguishes the textbook academician and the common-sense natural scientist.

Reams observed nature as an energetic symbiont rather than as a mechanical system fitting nicely within the confines of man's mechanistic sensors. He recognized the fallacy of the conventional scientist's reductionist mind-set. The reductionist approach to science says that the whole is just the sum of the parts, and that is all there is to it. The most obvious problem with this belief is determining who is the authority who inventories the parts in the beginning. Who is to say this inventory is complete?

Reams realized that there is no mechanical, man-made device as sensitive as man's own senses of touch, smell, taste, sight, sound, and, most important, intuition. He repeatedly prompted his students to see what they were looking at. This seemed trivial to me at first, but after further study I realized how profound this statement was. To notice that one field of corn has dry ears and green leaves, whereas the adjacent field has no green leaves, is to recognize the difference between dry down and die down. To notice that one field of beans has a sheen and the adjacent field does not indicates a difference in nutritional balance. To notice that one field has insect damage and the other does not also denotes a variation in nutritional balance. Such an awareness allows one to notice that a particular flock of chickens cackles more than or at a different pitch from another flock, and encourages the sensitivity to observe that one hen house is two degrees cooler than another without looking at a thermometer. An observant individual notices that cows in a particular group hold their ears higher and look more contented than those in another group. These are all examples of "seeing what you are looking at."

The concept of energy is foreign to most nonphysicists. Consequently, and partly because of his own vocabulary limitations, Reams took existing scientific terms and defined them in his own way. This enabled him to teach soil chemistry with an energetic component. It was the only way he knew to convey to his students what he observed and knew intuitively. This act *per se* alienated him from most scientists. However, his critics' assurance wavers when they are asked to define energy, electricity, and life or to explain what happens in the nucleus of an atom. People like Tom Bearden have a pretty good idea, but as Bearden has said, only a handful of physicists around the world really understand these phenomena.

The final straw, however, for Reams' critics was his contention that insect, disease, and weed infestations were indicators of nutritional imbalance. In the eye of the expert, not only was Reams ignorant of true scientific terminology, he blasphemed the petrochemical industry, much to his discredit, as did the late William A. Albrecht. Industry leaders had been successful in suppressing the works of most other humanitarians; they certainly were not going to let Carey Reams gain credibility.

Reams taught very basic concepts: "God is the basis of life. Life is the basis of energy. Energy is the basis of matter." Thus, plants and animals live off of the energy derived from their diet, not off the diet *per se*. What does this really mean? Two analogies may help to clarify this assertion.

If a builder has in inventory all the materials necessary to build a house, does he have a house? No. Work energy is needed to build the structure. When gasoline is used to power an engine, is it the gasoline that actually moves the pistons? No. When the gasoline is combusted, energy is released; this creates tremendous heat, resulting in expansion pressure on the pistons and hence their movement. Without energy, the house would not be built, nor would the engine run.

With respect to plants and animals, when food is presented to them, it must be digested in such a way that net energy is released. This energy allows work to be done so that amino acids, proteins, enzymes, tissues, and cells can be built from basic building materials. It also allows the living system to maintain its temperature and to carry out daily activity.

This concept of energy is clear to a biophysicist, but present it to an agronomist and he is confounded. To convey this principle of energy to his students, Reams used the terms *anions* and *cations*, which chemists define as negatively charged ions and positively charged ions, respectively.

Terms are regularly borrowed from other disciplines to

describe things for which no better word is available. A good example is the word *organic*. Chemists know that *organic* means *with carbon*, but lay people use it to describe farming without chemicals. Thus, the word "organic" now carries that connotation. Reams did somewhat the same thing by using the words *anion* and *cation* to convey a particular concept.

Reams's anion-cation concept is correlated to the growth-fruit phenomenon observed in plant production. The problem is that Reams borrowed these terms from chemistry, so their use in this context has created some controversy. Reams observed that there is a difference in energy quality between the growth and fruiting experiences. This difference correlates somewhat to the male-female energy qualities. I believe it more accurately correlates to the Yang-Yin quality of energy discussed in many energy-medicine contexts, such as acupuncture. Yang energy is associated with growth or expanding energy, the male manifestation. Yin energy is associated with fruiting or condensing energy, the female manifestation. I believe there is no actual difference in the Yin/Yang energies, but rather a difference in the dynamics of the energies. Throughout nature there is a continuous compression-rarefaction (expansion) pulse, as shown in the following figure.

Spiral representing the dynamics of energy.

The spiral in the figure represents the dynamics of energy. Pick any point on the spiral. Move in a clockwise direction from that point, and you are moving in an expanding dynamic.

This dynamic is Yang, or anionic, as Reams termed it. This dynamic is expressed in the rapid vegetative growth of plants. Move in a counterclockwise direction from the reference point, and you are moving in a condensing dynamic. This dynamic is Yin, or cationic, as Reams termed it. This dynamic is expressed in the setting of fruit on plants.

In the natural world, movement in both directions occurs simultaneously. Understand that this spiral is only a two-dimensional figure. Whether you draw it as a counterclockwise or a clockwise spiral to the center, the Yin/Yang concept holds. Once you put it into three dimensions, it is difficult to distinguish spin direction. It then becomes a question of perspective rather than direction. What can be distinguished is the expansion/compression dynamic. Is the sphere expanding or compressing? This holds at every level of the cosmos. Relative to the atom, it is either gaining energy or expelling energy. It does both continuously. Man, with his "crude" instrumentation, isolates the test item at a particular instant and draws a conclusion, often missing the full dynamics over time.

Further evidence of this shift in energy dynamics can be found in the Soviet literature on soil microbiology.

> During the ripening of plants . . . the quantitative ratio between the different [microbial] representatives and groups changes, the number of sporeforming bacteria, fungi, and actinomycetes increases, and new organisms appear [in the rhizosphere]. At the same time, the total amount of nonsporeforming bacteria decreases, some species disappearing altogether.[1]

Wilhelm Reich observed the compression-rarefaction pulse in his study of bions, and human function explained *The Cancer Biopathy*. Bearden mentioned the same pattern in his explanation of the vacuum. If you study the physiology of society over time, you will notice this same compression-rarefaction pattern. Society moves out into new areas, brings up its reserves, stretches out, collects its reserves, and so on. Many students of Reams' concepts interpret the components of his anion (growth), cation (fruiting) classification of nutrients and fertilizers as being absolute and distinctly separable. They are not. This reductionist opinion of nature and nutrition is a carryover from the allopathic chemical agriculturalists who want to compartmentalize everything. This viewpoint fails once one leaves the test tube and goes into the field, yet, unfortunately, many eco-agriculturalists adhere to these basic textbook beliefs.

According to Reams' concept of energy, calcium is classified as the kingpin of growth (anionic) energy and manganese is classified as the kingpin of fruit (cationic) energy. People consequently assume that these materials must be separated and applied at distinctly different times. The point to remember is that every cell of every living organism requires both growth and fruiting energy. Every living cell needs both calcium and manganese. Consequently, manganese is needed to get healthy growth, and calcium is needed to get healthy fruit. The fact is, however, that the expansion (rarefaction)-condensation (compression) pattern constantly occurs in all growth and fruiting, whether it be a new cell or a kernel of corn.

When functioning according to nature's intended design, growth and fruiting occur in such a way that there is perfect harmony between the two, both at the cellular level and at the foliage/seed level. This perfect harmony requires a balance of all the mineral elements, microbes, and other factors; they are all components of the whole.

For example, in an engine you must have the compression stroke in order to get the expansion or power stroke. You cannot have one without the other, yet each can be evaluated separately. So it is in nature. We must have the growth to get the fruiting, but if one or the other is imbalanced, the "living engine" functions at less than peak performance. When a mineral imbalance occurs, the expansion (growth)-condensation (fruiting) pattern is disrupted. There may be a deficiency in the growth-energy nutrients, which would cause a plant to set more fruit than there is vegetative growth to support, like an engine that has too rich a fuel/air mixture for the power (growth) cycle to get maximum benefit from the compression stroke. The fuel is not getting burned or used completely in this case. On the other hand, there may be a deficiency in the fruiting-energy nutrients, causing a plant to put forth tremendous foliage with very little fruit. This is analogous to an engine with high rpm but no lugging power (torque).

When Reams discussed applying a fertilizer or material such as vinegar, superphosphate, or thio-sul to set fruit, he stated that a cationic material should be added. In reality, more fruiting or condensing-energy (Yin) material was needed. If he discussed applying a fertilizer or material such as calcium or nitrate nitrogen (like in forage or leaf crops) to get mostly growth without fruit, he stated that an anionic material should be added. In reality, more growth or expanding-energy (Yang) material was needed.

To follow Reams' teachings, we simply need to observe the expansion-compression cycle and fertilize to suit it. Reams's observations were correct, and they were effective. However, he did not know what words to use to describe those observations, so he appropriated scientific terms to describe these phenomena. This led to some of his conflict with traditional scientists. The underlying difficulty was that the traditional scientists were not sufficiently attuned to recognize or even care whether such phenomena occurred in nature. In their minds, Reams was observing fiction and using their scientific terms to describe that fiction. How dare he contend that nature was energetic, divinely ordered, and noncompartmentalizable!

The next issue that arises with the concept of anions and cations is that Reams assigned anions an energy value of 1 to 499 millhouse units of energy and assigned cations an energy value of 500 to 999 millhouse units of energy. This allowed him to describe his observations of various elements and compounds from an energetic standpoint as they interacted in nature. Until that time, no one had described nutrient interaction in terms of energy.

Reams assigned the different values of energy to anions and

cations based on his observations of what they seemed to cause or create in nature. This does not necessarily mean there is an actual difference in energy values, but is more correctly a description of what the energy seems to be doing, or the energetic state of the nutrient. When a substance is condensed, it often takes on different characteristics than it had before condensation. Water, in the liquid state, has much different characteristics from those of water vapor or steam, yet both states are water. Whereas water vapor and steam can be compressed, water in the liquid state cannot. Water vapor and steam have lower specific heats than water in liquid. Specific heat is the amount of energy (calories or BTUs) required to alter the temperature of a given quantity of the material.

The difference in the energy state is the reason Reams said that anions spin clockwise and cation spin counterclockwise. Anions appear to be reversed from cations because compression and rarefaction appear to be opposites if either is taken out of context, but each is actually the other half of the same cycle. In reality, spin is occurring in both directions simultaneously, as Reams said, but most people missed hearing this.

Reams noticed that two lots of the same kind of material, such as two different sources of high-calcium lime, can have different energy levels. In classical physics and chemistry, this difference is attributed to variations in the orbits of the electrons revolving around the nucleus of the element in question. At this point it might be useful to review the chapter on chemistry and the model of an atom. Classical agronomists, however, do not readily view this phenomenon as being significant to fertilizer effect because they assume that energetics has nothing to do with agronomy or soil science. Further explanation of Reams's observations and contentions regarding energy levels of growth and fruiting materials can be found in the writings of such gifted mathematicians and physicists as E. T. Whittaker, Vlail Kaznacheyev, Richard Feynman, Norman Levanson, John Eric Von Neuman, and Thomas Bearden. An excellent reference list and general explanation of this area of physics is found in *Gravitobiology: A New Biophysics,* by

Thomas Bearden.[2]

Reams's observations relative to fertilizers can be explained in two parts. First, the differences in energy quality between growth and fruiting exist partly because electrons, which are the orbit of the atom or the "planets," are composed of collected photons (particles of light). As a result, the number of photons per electron can vary, which accounts for the differences observed between diverse materials and similar substances from different sources. Photons themselves vary according to the energy and information they contain. Popp documented the role of photons in the communication systems of living things in *Photon Storage in Biological Systems*[3]

The second part of the explanation of the difference in energy quality Reams observed between growth and fruiting materials lies with the nucleus or "sun" of the atom.

> As the nucleus charges up with a new vacuum potential pattern, its EM [electromagnetic] coupling with the electron shells gradually is changed also, producing electron changes (of a wide variety) as well. Everything is continually charging and discharging internally. Any change in the infolded Whittaker[4] structuring of the local vacuum (potential) will gradually activate everything . . . Activated masses—atoms, nuclei, ions, neutrons, and electrons, e.g.—can then exhibit quite abnormal external chemical, electrical, and gravitational interactions as compared to unactivated (normal) "inert" masses.[5]

It is important to remember that, to be healthy and to thrive, every cell requires both fruiting and growth energies simultaneously. To initiate fruit set, the fruiting or condensing energy of the soil and/or plant is increased. When you increase the fruiting energy in the soil, you increase the fruiting energy in the plant. This causes an increased condensing, compressing, cationic (Yin) nutrient energy flow in the plant that triggers bud initiation. If there is sufficient complementary growth energy, the bud will grow into a developed embryo that is ready to be fertilized; upon fertilization, the embryo will grow into a seed.

This process can be likened to what happens with a blinking light. Current flows into the bimetallic thermal switch. As the switch heats up, it deforms until the contact opens and current flow through the switch stops, being diverted to the light and thus lighting it. If there is insufficient current flow, the bypass switch does not get heated enough to open, consequently, the light does not get lit. Likewise, if there is not sufficient fruiting energy, growth energy never gets diverted to the fruiting parts for fruit initiation, growth, and development. Concurrently with the above-mentioned process, the energy dynamic of the soil changes. During the course of growth, the plant transfers energy to the soil and the soil accumulates energy and nutrients—causing the apparent paraconductivity of the soil to decrease slightly—until a "trigger point" is reached. This initiates fruiting. If the proper ratio of fruiting energy to growth energy remains, the growth energy is now directed to the fruit. If there is more growth energy available than is needed to fill the fruit, the extra energy will continue to produce new foliage at the same time, for yet more fruit initiation.

The major regulators or control centers of these energy flows are the genetic code of the plant and the microbiology of the soil. The plant is an antenna, which functions both as a collector and as a sender of energy. As such, plants deposit nutrients into and extract them from the soil. This interaction between plant (antenna) and the living soil system (tuner) determines what kind of crop (music) is produced.

Fruit and growth are incremental. In space-time (our present reality of space and time), they appear to have definite incremental cycles, yet even out of space-time (if time is taken out of the equation) there is "frequency" or "cycling" due to the criteria we set up in space-time. A given interval is assumed in the fruit-growth cycle as we usually experience it, yet time is a cyclical function and, therefore, its interval is not undeviating. In other words, time patterns can be altered, e.g., the time it takes to grow a crop.

As Bearden and other physicists have explained, we can have frequency or cycling without time involvement. Time, in fact, is not the determining factor in the fruiting and growth sequences. Great alterations can be made electromagnetically or chemically (which is fundamentally electromagnetically) that change the time cycles of growth and fruiting.

For example, animal growth hormones cause premature puberty. Birth control pills alter the menstrual cycle. Malnutrition also can alter that cycle, e.g., malnourished cows that will not come into heat. Dan Carlson showed that he could accelerate the time cycle of growth and fruiting in nut trees, getting the equivalent of several years of growth and development in one year, with fruiting in two years.

The branches of apple trees will grow straight up, with no fruit production, if there is too much vegetative growth energy. On the other hand, if there is too much fruiting energy the branches will grow straight out from the trunk, thus setting more fruit than the vegetative growth can support. Apple growers will tie or brace branches at a 45 degree angle to the main trunk in an attempt to achieve a balance between fruiting and growth. In doing so, however, they are handling only the symptoms, not the cause of the problem. In such cases, the trees have an imbalance between the growth and fruiting energies, which can be corrected more effectively by judicious fertilization than by manipulation of the branches.

In non-tree crops, fruiting often is initiated by physical change in the plant, injury of the plant, or cold temperatures. If someone tied your arms together, injured you, or put you in a cold room, you, too, would pull in your defenses and withdraw somewhat. This response is analogous to the energy pattern or state of fruiting discussed above. The compression, condensation, or Yin energy is the fruiting-energy pattern. When such conditions as physical changes, injury, and cold temperature are inflicted on the plant, its energy condenses, thus inducing the fruiting response. This happens by design because nature endeavors to maintain the species. Under threat of elimination, the species will condense or collect its energy to reproduce and carry on. When threatened, many single-celled organisms will condense their energy into endospores, which are almost indestructible. They do this to survive the threat of death, only to germinate and grow at a later time.

Seeds are much denser than foliage. This fact seems to validate further the concept of compression-rarefaction. Reams taught that energy, not time, is the factor that determines growth and fruiting. If the student can grasp these concepts, he will be well on his way to understanding the workings of nature.

Relatively speaking, the growth-fruiting cycle observed in agricultural crops seems to be "long," whereas the growth-fruiting cycle of individual cells in the vegetative part of the plant is a harmonic or fraction of the larger foliage-seed cycle. For example, the pitch of C at 512 cycles per second is a harmonic of the pitch of C at 256 cycles per second.

Everything in nature—life, reproduction, weather, economics, electricity—is based on cycles. As electricity, light, or any electromagnetic energy flows through a medium, it continuously compresses and rarefies. Because our instruments detect these cycles in two dimensions, we see their forms on an oscilloscope as a sine wave. Picture the "Sun" machine in an auto repair shop hooked up to your car's engine. Every firing of each piston creates a squiggly line on the machine's CRT screen. That squiggly line is a type of sine wave.

In our context, the positive side of the wave is the rarefaction phase. At the point where rarefaction shifts to compression, the sine wave crosses zero (the "*x*" axis) in transition between the positive crest (Reams's anion spinning clockwise) and the negative trough (Reams's cation spinning counterclockwise). Each portion of the sine wave could be divided into an expansion/contraction increment. From zero to the peak is expansion. From the peak back to zero is contraction. When a material or substance is compressed, increased stress or potential is created, thus the potential of the seed.

For example, with a steam engine, as more steam is packed into the system, greater and greater pressure (stress) is created. This stress is actually raw potential, which can be transferred into doing work, such as driving the wheels, or altering certain physical conditions, such as exploding the boiler if the pressure becomes too great.

The plot on a graph of this stress due to compression is also seen two-dimensionally as a sine wave. The amplitude of the stress sine wave is twice that of the compression-rarefaction sine wave because the stress is the sum of the crest (compression stress) and the trough (tensile stress). There is tensile stress in something being expanded, and compression stress in something being compressed.

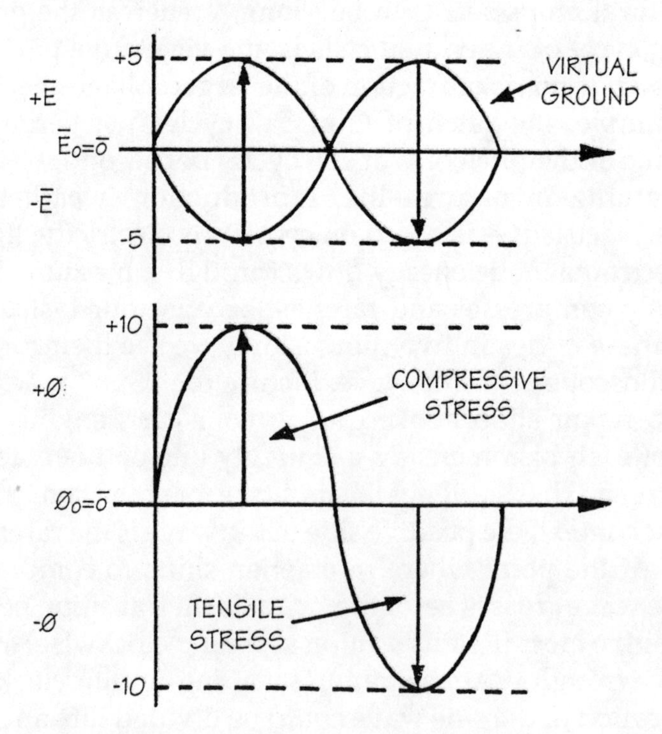

I have included this technical information because Bearden discusses these topics in reference to "free" energy generation, Tesla super weapons, and the alteration of space-time and physical reality through an understanding of these principles.

Another problematic issue with Reams' teaching and terminology is his statement that likes attract. This statement seems absurd to people who have studied basic chemistry and physics because they have been taught that likes repel and opposites attract. Yet instructors can teach only what they, themselves, have been taught; this does not guarantee that what is being taught is "the whole truth and nothing but the truth." Reductionists have concluded that like charges repel and opposite charges attract. However, they are seeing only half of the picture; they are missing the most important half, the gravitational-field half. Reams was observing and referring to the gravitational field, the "normal gravitational potential," which is the internalized (infolded) electromagnetic energy. When people challenged Reams, saying that opposite charges attract at the atomic or electrical level, he would tell them that he was not referring to those levels but rather to what comprises the atomic particles and electricity. This was beyond even his most erudite students because they perceived the make-up of atomic particles and electricity to be protons, neutrons, and electrons. Reams would say, no, he was talking about the cause or composition of these particles. He impressed on students that he was discussing the fundamental basis of energy. Years ago, such concepts were not readily discussed, let alone defined, in classical science (nor are they today, for that matter), so it is understandable that there was great consternation when Reams proposed those ideas.

I believe that anyone who thoroughly understands Reams' concepts and then reads the many papers and books by Thomas Bearden and the references he suggests will recognize that Reams was discussing the very things Bearden has described using the appropriate terminology. It becomes clear that, in discussing basic anionic and cationic energy, Reams was speaking of the electromagnetic "embryo of physical manifestation," the system that gives rise to our physical reality as we generally perceive it. In *Gravitobiology: A New Biophysics*, Bearden defined this as vacuum flux, internalized electromagnetic energy, or electrogravitational potential energy.

Energy is simply the "amount of ordering" or the "amount of coherence" that exists in something.[6] In this medium, the gravitational field, like charges—that is, all masses—attract each other.[7] This system, which Bearden also refers to as the Whittaker potential, being internalized electromagnetic energy, is the information content of the externalized electromagnetic energy.[8] In other words, this gravitational field, the virtual flux, is the blueprint of what is manifested in our physical reality. It is the true "living electromagnetic energy" of all biological systems,[9] the information content of the field. "It is this infolded and hidden EM structure—the internalized or covert EM energy of the potential field—that is deterministically utilized by living biological systems in all their cellular control and regulatory functions."[10] We are actually speaking about the hologram or "cookie cutter" from which our physical reality is manifested. Yes, protons and electrons are opposite charges and they attract, but there is a holographic world from which they come, wherein likes attract and opposites repel.

This assertion of a holographic world was experimentally proven by Kaznacheyev, Devyatkov, and Lisitsyn of the Soviet Union, Valerie Hunt of the United States, and Fritz-Albert Popp of Germany," and related in Bearden, *Gravitobiology*. It was demonstrated by Hieronymus in his experiments on plant growth in darkness, Wilhelm Reich in his bion experiments, and Gaston Naessens in his stomatid research."[11]

In every teaching with which I am familiar, including the Bible, the reader is impressed with the need to look within. Why should the fundamental aspects of physics and physical reality be any different? Reams often said to study nature for your answers. The aforementioned scientists have proved that the answer lies within, whether it be in reference to biological systems or inanimate objects.

Modern reductionist science is based on external observations and measurements. Scientists, technicians, and lay people seek the outfolded, translated electromagnetic energy system (essentially the message bearer), thus overlooking the infolded Whittaker structure, the blueprint (the message)—the body rather than the spirit. In his book *Love, Medicine, and Miracles*, Bernie Siegel explains that physical disease is preempted by the individual's personality and emotions. My book, *Applied Body Electronics,* further elucidates this subject.

The Bible says to seek God within and to see the light in every person, yet most people today insist that reality is that which is discernible to the physical senses. Most people are unaware of or have ignored Reams' most profound counsel: "See what you look at." With this understanding, as Bearden wrote,

> One now can immediately verify the truth of the old adage, "as above so below." Since the external world is just the outfolded content of the inner order, then internally reflecting our macroscopic scientific concepts and models is justified.

Reams was looking within to gain an understanding of the wisdom of nature. He realized that there are no secrets in nature if one looks in the right places. Observing an entire world that was foreign to conventional science, he had to formulate explanations for his students without overwhelming them with complex explanations. He wanted to convey the concept of what he saw, not necessarily its complexity or technical description. Consequently, he appropriated the terms "anion" and "cation." In this context, I believe that Reams' anion was possibly Bearden's antiphoton. Reams said that his anions would penetrate all matter, were not influenced by gravity, and could not be detected with conventional instruments. Reams's cations would not penetrate matter, were influenced by gravity, and were detectable with conventional instruments.

The other possibility for Reams's anion is Bearden's graviton, which has negative mass. The other possibility for Reams's cation is Bearden's traditional atomic particles, which have positive mass.

Further study is required to correlate Reams's terms to Bearden's, but there is little doubt that the two men's conceptual teachings are nearly parallel. Countless times, Reams read the RBTI health numbers of people and animals, which he derived by evaluating their urine and saliva. He reported that these numbers allowed him to "see" the patients' physical and mental characteristics. Most astounding was the fact that he saw such characteristics long before they were detected with traditional medical evaluations. On a reserve-energy scale of 0 to 100, 100 indicating perfect health, Reams said he could detect an alteration in body health using the urine and saliva numbers if the reserve energy dropped only one increment, from 100 to 99. Conversely, traditional medical evaluations could not detect illness until the reserve energy dropped to 30 or less.

I believe Reams was seeing the infolded Whittaker structure, described by Bearden, containing the peremptory information of the living system. All that is needed to perform such a task is what Bearden called a scalar inferometer. The brain possesses such a characteristic. Reams repeatedly said that nature, God, is a God of math. Bearden has pointed out that Whittaker, in his 1903 and 1904 papers, described the mathematics of this infolded information. Such is the need to study the basics. In *Gravitobiology, A New Biophysics,* Bearden wrote,

> With a developed knowledge of the external and internal energy balances of our biosphere, we can clean up the biosphere, permanently rid ourselves and our children of poisonous chemical and nuclear wastes, and restore the vitality of the planet for our living health and enjoyment.

In any event, Reams's observations are borne out in practice. The fruiting and growing cycle of plants can be altered by the fertilizer materials that are used. His concepts work in the field; this we know by experience. To understand this phenomenon requires continued spiritual growth and the naturalist's desire to know.

—12—

WEEDS: CARETAKERS OF THE SOIL–WHAT THEY CAN TELL US

THIS CHAPTER IS INTENDED to give the reader a more accurate and realistic perception of economic weeds. It is not intended as a final treatise on the subject. In fact, I highly recommend reading *Weeds, Control Without Poisons* by Charles Walters Jr.[1] for more in-depth coverage of weeds and their meaning.

Plants, like people and animals, have particular diets under which they function best. Nature always exists in harmony and allows the survival of every species according to its food supply.

If man destroys the habitat for small game, like the rabbit, quail, and pheasant, the population of foxes, owls, and other predators will soon decline. If the population of small rodents, like field mice, were artificially increased, one or more situations would occur. The food supplies for these rodents would decrease, the population of predators would increase, and/or the population of small game would increase due to the overabundance of food for the predator and the consequent reduced pressure on the small game. The latter situation might lead to a lack of food for the expanded population of small game and consequently result in starvation, increased susceptibility to disease, and so on. If the predators increased, in certain areas there might be increased pressure on small game as a matter of convenience for the predators, whereas in other areas there might be a lessened pressure due to the availability of field mice. In any event, food would regulate the types of species and the numbers of each species.

Farmers know that the feed rations of turkeys are different from those of feeder pigs, which are different from those of lactating sows, gestating sows, baby pigs, lactating dairy cows, dry cows, baby calves, feeder cattle, work horses, race horses, colts, veal calves, mink, alligators, trout, and monkeys. One needs to provide a different feed ration for every type of organism in nature and for every stage of growth of each organism.

Farmers also know that corn is fertilized differently from potatoes, which are fertilized differently from grapes. In turn, grapes are fertilized differently from bananas and tobacco. The nutrient ration that is suitable for ocean plants would be deadly for freshwater plants or alfalfa.

Farmers and agriculturalists confront these differences every day and acknowledge them as common understandings. Why, then, is it so difficult for people to grasp this same principle when it is applied to insect pests and weeds? It is commonly believed that healthy soil supports weeds and crops equally well. By this logic, quinoa, blueberries, and alfalfa should all grow well in the same soil. But they don't. It is also commonly believed that insect pests are ravenous, undiscriminating eaters of our food and fiber crops. They are not. Both of these beliefs are fundamentally and logically incorrect. It has repeatedly been shown that insect pests are correlated to the nutritional imbalance of the crop and are readily curtailed or eliminated by raising the nutrient density and refractometer reading of the crop. Insect pests are essentially nature's "sanitation engineers." More information on this topic can be gained from my book, *The Anatomy of Life and Energy in Agriculture, Tuning In To Nature* by Philip S. Callahan, *Mainline Farming for Century 21* by Daniel L. Skow and Charles Walters Jr.[2]

Weeds also play a sanitation-engineering role in some cases. Velvetleaf, for example, works at cleaning up excess methane in the soil. More often, however, weeds seem to fill the role of "construction engineer" and "lighthouse." They can play an important role in altering soil structure, tilth, and composition. They play an essential part in signaling where the soil health is. The belief that healthy soil grows weeds equally as well as the desired crop is based on the misconception that the soil in question is healthy. Evaluating the refractometer reading of the plants, both weeds and crops, growing in the soil tells the observer whether the soil is truly healthy. In this case, one will find that the refractometer readings of both the crop and the weeds are about the same, probably in the 4 to 8 brix range. Neither the crops nor the weeds are well balanced nutritionally at these brix levels, but the conventional soil test and nutrient standard may indicate that this is a "healthy" soil. In any event, *It is not!*

Some people contend that the refractometer is not a valid indicator of plant or soil health. In every case I have encountered in which a consultant, farmer, or fertilizer dealer has made such a statement, I have found that he had been unable to get the refractometer reading of the crop to increase and therefore contended that the concept of using a refractometer to assess plant and soil health was invalid.

Basically, as Carey Reams instructed, sour grass weeds like quackgrass are indicative of calcium deficiencies, at least qualitatively if not quantitatively. Broadleaf weeds are indicative of an improper phosphate-to-potash ratio. Using the

Reams soil-testing method, this ratio should be 2 pounds of phosphate to 1 pound of potash for row crops and 4 pounds of phosphate to 1 pound of potash for alfalfa and grass crops. Succulent types of plants like pursilane are indicative of soils that are deficient in biologically active carbons.

Specific varieties of weeds within the three categories—grasses, broadleafs, and succulents—are indicative of specific nutrient conditions beyond the general group meaning. Ragweed, for example, is generally indicative of a phosphate/potash imbalance, but, more specifically, it indicates a copper problem. Copper is important in the use of manganese and iron, as well as in many metabolic reactions. Copper also seems to be important in controlling fungal disorders. Many people have allergic reactions to ragweed pollen. This reaction seems to be related to a copper deficiency in the mucous membranes.

As mentioned, weeds play a role of "construction engineer" in the soil. In performing this role, the weed directly deposits nutrients and metabolites into the soil or rearranges those nutrients already in the soil. Both of these activities seem questionable to many people who hold the long-standing belief that plants are essentially soil parasites—they simply extract nutrients from the soil. Such a belief is a result of several decades of chemical agriculture, in which weeds and insects are viewed as parasites warranting expeditious chemical eradication, and soil nutrition is seen solely as a function of chemical-fertilizer intervention. In this belief system, nature is strictly a primitive force needing heavy-handed conquest by man. Fundamentally, insects, diseases, and weeds are interpreted as being the results of pesticide deficiencies. These beliefs are very strong paradigms, and as Joel Barker states in the educational film *Discovering the Future: The Business of Paradigms* and Thomas Kuhn says in his book *The Structure of Scientific Revolution*, lay people and scientists alike see only what their belief systems allow them to see. People must look beyond their beliefs in order to make new discoveries and assimilate new knowledge.

The rearrangement of nutrients by weeds, by all plants for that matter, actually occurs in a symbiotic relationship with soil microorganisms. In *Soil Microorganisms and Higher Plants,* it is repeatedly emphasized that every species of plant supports different populations of microorganisms. These relationships are very important because each combination is both indicative of and crucial to the correction of imbalances in soil health. Certain weeds obtain specific nutrients from the soil at various depths and in various circumstances, e.g., compacted soil, water-logged soil, chemically contaminated soil, and high-salt soil. In turn, they manufacture certain metabolites, which they excrete into the soil rhizosphere to be processed by specific microbes. Likewise, these microbes uniquely obtain and metabolize certain nutrients from the soil, which they, in turn, alter and re-excrete into the rhizosphere for use by plants and other microbes. Left to operate naturally, this process regenerates a "sterile" or "toxic" soil to the point where the plant and microbe populations change again and again, eventually to support whatever crop and microbe group we desire. Every organism, whether bacterial or human, is specific to a particular habitat medium. An organism will survive and thrive only if the proper conditions have been established for it to do so.

In his monograph, *Weeds! Why?*, Jay McCaman lists the field conditions under which various weeds thrive. In every case, the field condition conducive to the weed is much less beneficial for the desired crop. The weed, however, is able to obtain nutrition from this soil and, as a result, alter the soil slightly in a positive direction. Quackgrass, an indicator of calcium deficiency, has a prolific root system and is better able to project its roots into a compacted soil than, say, corn or beans. Eventually, this weed would reduce the calcium deficiency and loosen the soil as a result. This might take decades or even centuries to accomplish, but the result is nonetheless valid.

The other role that weeds perform is to deposit nutrients into the soil. This is another concept that many people have difficulty accepting. How can a plant possibly deposit nutrients into the soil? Where is the plant obtaining the nutrients?

Plants actually get about 80% of their nutrition from the air. Most of this nutrition is from carbon dioxide and water, but it also includes cosmic and solar energy and airborne nutrients. The effectiveness of this direction of nutrient flow depends totally on the inherent integrity of the plant and/or seed and the health of the soil.

The soil-plant relationship can be thought of as a radio apparatus. The soil contains the electrical ground and the receiver/ tuner, whereas the plant is the antenna. If the antenna (plant) or the receiver tuner or electrical ground (soil) is faulty or malfunctioning, the entire system is mute (there is no plant growth). If the system is only slightly faulty or malfunctioning, it emits poor-quality music (plant growth) or just static.

If there is a nutrient deficiency in the plant that started with the seed quality, there will be increased resistance in the antenna and it will not function as it was designed to do. If the nutrient deficiency is severe enough, an open circuit occurs and the antenna (plant) will not function at all. In the case of weeds, they are designed according to their genes to collect specific nutrients and energy spectra from the air, to convert them into specific nutrients or metabolites, and subsequently to deposit them into the soil.

If there is an imbalance in the soil, there will be a poor electrical ground. If this imbalance is an excess of an electrolyte or a deficiency of an insulator nutrient, too much current will flow, creating increased resistance and resulting in increased heat and depletion of the soil. If the imbalance is a deficiency of an electrolyte or an excess of an insulator nutrient, we have a brown-out phenomenon or simply insufficient energy to drive the system. Any imbalance reduces the effectiveness of the receiver/tuner. If this mechanism cannot function properly, whatever energy and nutrition make their way into the soil are not processed and used as efficiently as they should be. This results in further malfunctioning of the system or least a lower quality of functioning.

The entire system is designed to be an alternatingcurrent system in which the energy and nutrients flow in both directions. Unfortunately, we usually observe only a direct-current system in which energy and nutrients flow in just one direction. In this case it is, of course, an extractive process resulting in soil depletion. Weeds are differently designed antennas intended to reestablish the alternating current system. If they are evaluated properly, weeds can be extremely helpful to the farmer in deciding what steps he needs to take to reestablish this alternating-current system for his commercial crop.

DESIGNING AND BUILDING "THE SYSTEM"

The agricultural system in America and the world is deteriorating. It has brought about extensive soil erosion, universal environmental contamination, and a degeneration of the health of almost every living species on the planet. Public outcry is growing as people become aware of the problem. Politicians are voicing their concern in order to garner votes. Legislation to curb or banish the current widespread use of toxic chemicals in agriculture is becoming more plentiful.

Caught in the middle of this political dispute is the farmer. One by one, his toxic supports are being eliminated from the market. He is told that he must farm without these panaceas that he has become accustomed to using—on the advice of the United States Department of Agriculture—land-grant universities, and Cooperative Extension personnel. Politicians are telling him he cannot use these materials anymore, but the agriculturalists are not informing him what to use in their place. The farmer will continue to feel lost and frustrated until he realizes that he can solve his dilemma by using his own intuition and common sense. Farming is not a desk job or the work of a laboratory technician. It is a natural experience. Farming entails understanding, appreciating, and coexisting with all life on earth.

The first step in building an agricultural system without toxic chemicals is to change your attitude. Become a farmer rather than a miner of the soil. Decide to leave the farm in better condition than it was last year or when you started farming it. Decide to accept responsibility for the health of your country, your family, and yourself.

Now that we have the appropriate frame of mind, we can proceed to address how to alter the soil so that weed pressure on crops is minimal or nonexistent.

Recalling the general discussion about the three major weed groups (grasses, broadleafs, and succulents), we know that we must alter our fertilization practices in such a way as to change the nutrient status in the soil. Conventionally, we would add a given quantity of a particular nutrient if it was said to be lacking, giving no consideration to its side effects, biological compatibility, or the actual need for it. The most important concept to understand is that, ideally, the soil is a living system. The life of this system depends on the status of the innumerable microorganisms inhabiting the soil.

The key to this is the specific groups of microorganisms that are in the majority. Ideally, there would be a majority of aerobic microbes in relation to anaerobic microbes. The desirable microbes are ultimately responsible for the availability of all nutrients in the soil. As a result, every fertilizer material that is used must be compatible with these microbes if the desired result is to be realized. Also, because the microbes are ultimately responsible for nutrient availability, the real crop is the microbe; it is what really needs feeding. If the microbe is satisfied, it will take care of nutrient availability to the plant. This means that the farmer might often add materials to the soil that are different from what he perceives as deficient or unavailable to the plant. In this case, weeds are the "lighthouses," helping the farmer make these decisions. Often these signals can be difficult to interpret, so it is valuable to use other tools, especially a radionic scanner, in determining the most appropriate material to add to the soil.

At the end of this chapter is a list of common weeds encountered in the United States, their general meaning in the soil, what materials reinforce each weed, and what materials begin to correct the cause of each weed. Keep in mind that the nutritional conditions that resulted in weed infestation occurred over a number of years of corresponding farming practices. Likewise, it will take a few years to reverse these nutritional conditions and to eliminate the weed. The quick-kill approach using chemical herbicides addresses only the symptom. This approach will not remove the underlying cause. This is not to say that a dramatic change in plant growth will not occur; it might, depending on the conditions involved. Also, keep in mind that the list is general; local conditions might warrant applying different materials to achieve the same result.

If one wants to grow certain weeds as commercial herb crops, the same principles apply. Many commercial herb growers are experiencing increased pressure from weeds, diseases, and insects. These problems are due to nutritional imbalances in the herbs. If growers can get the herb's refractometer reading above, roughly, 12, these problems will disappear.

Low biological activity is inherent in every weed problem because that is what causes or at least results in the mineral imbalances. Each weed is keyed to a specific environment slotted for its proliferation. Notice that there are some interesting nutrient interactions in that the nutrients needed to correct the weed problem are not necessarily the same ones the weed is indicating are lacking. This is where the conventional agronomists get tripped up because they are looking strictly for mechanical and methodical cause and effect (symptoms).

The farmer needs to be able to read his fields and crops as if they were complete, detailed logs of exactly what has happened and is happening on his farm. This allows him to be independent of the allopathic system, but it also forces him to get out into the fields and learn to farm. One innovative farmer in Illinois, Bill Fordham, has discovered that he gets the best results from rotary hoeing by timing his hoeing operation according to the number of degree days that have passed.

WEED MEANING

WEED MEANING: What does the weed tell us?
ENHANCING MATERIALS: What contributes to the weed's proliferation?
CORRECTING MATERIALS: What reduces or eliminates the weed?
COMMON CONDITIONS IN THE FIELD: See *Weeds! Why?* by Jay McCaman, and *Weeds, Control Without Poisons,* by Charles Walters Jr.

Black Bindweed (Polygonum convolvulus)

WEED MEANING: Potash excess, narrow Ca:Mg ratio, salt excess, vitamin B_2 deficiency.
ENHANCING MATERIALS: Mg, K, salts.
CORRECTING MATERIALS: Ca, Cu, Mo, vitamin B_{12}.

Black Nightshade (Solanum nigrum)

WEED MEANING: Ca, P, Cu, vitamin B_3 deficiencies; improper P:K ratio.
ENHANCING MATERIALS: KCl, excess K, Zn (majorly)
CORRECTING MATERIALS: Ca, soft rock phosphate, Cu, vitamin B_3.

Common Mullen (Verbascum thapsus)

WEED MEANING: Ca, Co, B, Xenon, Vitamin C deficiencies.
ENHANCING MATERIALS: Magnesium, sodium chloride, diammonium phosphate, magnesium sulfate, vitamin K, zinc, KCl, reduced biological activity.
CORRECTING MATERIALS: Molasses, biologicals, molybdenum, RL-37, vitamin C, selenium, phosphoric acid, seaweed, Biomin calcium, calcium carbonate, vitamin E, hydrogen peroxide, soft rock phosphate, sulfates.

Common Purslane (Portulaca oleracea)

WEED MEANING: Lack of biocarbons or active humus, sodium salts; Ca, sugar, Mo deficiencies.

ENHANCING MATERIALS: Zn, KCl, Na, salts, Mg (minorly) but especially $MgSO_4$.

CORRECTING MATERIALS: Dextrose, white sugar, molasses, Ca, Mo, H_2O_2, Mn, humic acids, enzymes, soft rock phosphate (majorly).

Dandelion (Taraxacum officinale)

WEED MEANING: Ca, P, vitamin A, Fe deficiencies.

ENHANCING MATERIALS: Zn, Mg, KCl, K.

CORRECTING MATERIALS: Vitamin A, soft rock phosphate, Ca, molasses, black rock phosphate (majorly), humic acid (majorly).

Field Bindweed (Convolvulus arvensis)

WEED MEANING: Potash excess, narrow Ca:Mg ratio, salt excess, vitamin B_2 and B_{17} deficiency, specific amino acid deficiency.

ENHANCING MATERIALS: Mg, K, salts

CORRECTING MATERIALS: Ca, Cu, Mo, vitamin B_{12} and C.

Giant Foxtail (Poaceae)

WEED MEANING: Ca, P, Cu, B_{12} deficiencies.

ENHANCING MATERIALS: KCl, chemical nitrogen, salts (manure or chemical).

CORRECTING MATERIALS: Ca, soft rock phosphate, B_{12}, carbohydrates, sulfates in some areas.

Green Bristle-Grass (Setaria viridis)

WEED MEANING: Compaction; excess Mg; narrow Ca:MG ratio; Ca, P, Se, vitamin C, carbohydrate (minorly) deficiencies.

ENHANCING MATERIALS: KCl, Zn, Mg.

CORRECTING MATERIALS: Ca, Fe, Mo, Se, vitamin B_{12}, vitamin C (minorly), sugar, soft rock phosphate, black rock phosphate, aeration.

Johnson Grass (Sorghum halepense)

WEED MEANING: Ca, P, Cu, Se deficiencies.
ENHANCING MATERIALS: KCl, compaction, Zn, certain pesticide interactions.
CORRECTING MATERIALS: Ca, P, Cu, Mo, vitamin C and B_{12}, sugars, humic acid.

Lamb's quarters (Chenopodium berlandieri)

WEED MEANING: P, Ca, Se, Vitamin C deficiencies.
ENHANCING MATERIALS: Potassium chloride, diammonium phosphate, chemical nitrogens, magnesium oxide, magnesium sulfate, paper mill lime.
CORRECTING MATERIALS: Seaweed, Sea Leaf, molasses, selenium, molybdenum, hydrogen peroxide, cobalt, manganese, copper, Micro-cal, Biomin calcium, Dynamin, soft rock phosphate, monoammonium phosphate.

Milfoil (Achillea millefolium)

WEED MEANING: Excess nitrogen
COMMON CONDITIONS: Sewage and nitrogen seepage into lakes.
ENHANCING MATERIALS: Nitrogen sources of all types.
CORRECTING MATERIALS: Calcium carbonate, vitamin E, vitamin B_{12}, molybdenum, cobalt, Biomin copper, sul-po-mag, potash.

Milkweed (Asclepias syriaca)

WEED MEANING: P, carbohydrate, Co, boron (slightly) deficiencies.
ENHANCING MATERIAL: Potassium chloride, poor electrolyte conditions.
CORRECTING MATERIALS: Seaweed, humates, humic acid, molasses, sugar, copper, biologicals, RL-37, cobalt, manganese.

Nutgrass (Cyperus rotundus)

WEED MEANING: Ca, P, Si, carbohydrate deficiencies; compaction;

low water-holding capacity.
ENHANCING MATERIALS: Salts, KCl, sodium, ammonia.
CORRECTING MATERIALS: Sugar, Ca, humic acid, Co, vitamin B_{12}.

Poison Ivy (Toxicodendron rydbergii)

WEED MEANING: Ca, P, Mo, Si, Se deficiencies; negative polarity.
ENHANCING MATERIALS: Diammonium phosphate, magnesium and magnesium compounds, stray voltage.
CORRECTING MATERIALS: RL-37, seaweed, sea leaf, phosphoric acid, monoammonium phosphate, calcium carbonate, violet color, cobalt, selenium, vitamin C, vitamin A, vitamin E, biomin copper.

Quack Grass (Agropyron repens)

WEED MEANING: Ca, P, Fe, Co deficiencies.
ENHANCING MATERIALS: KCl, sterilization conditions, and compacted conditions.
CORRECTING MATERIALS: Cu, Mn, soft rock phosphate, B_{12}, Ca, sugar, molasses, aeration of the soil, sulfates in some areas.

Ragweed (Ambrosia trifida)

WEED MEANING: Cu, P, Ca, vitamin A and E deficiencies.
ENHANCING MATERIALS: KCl, $MgSO_4$, Zn, low bioactivity, chemical toxicity, lack of soil aeration.
CORRECTING MATERIALS: Mn, Cu, vitamin C and B_{12}, Ca, P, sugar.

Red Root Pigweed (Amaranthus palmeri)

WEED MEANING: P, vitamin A, Fe, Co, bioactive carbon deficiencies.
ENHANCING MATERIALS: Salts (manure or chemical), Mg, KCl.
CORRECTING MATERIALS: Ca, P, Cu, Mo, carbohydrates, humates, some areas vitamin C.

Shepherd's Purse (Capsella bursa-pastoris)

WEED MEANING: Salt, narrow Ca:Mg ratio, Ca (majorly), phospho-carbonate, Cu, Mo, vitamin B_3, beryllium (minorly) deficiencies.

ENHANCING MATERIALS: Zn, Mg, salt, chemical nitrogen, KCl, elemental sulfur, basic H in some cases.

CORRECTING MATERIALS: Organic copper, Ca, phosphate/molasses or sugar mix, molasses, Mo, vitamin B_3, B_{12} to some degree.

Velvet Leaf (Abutilon theophrasti)

WEED MEANING: Methane absorption, P deficiency, highly anaerobic conditions.

ENHANCING MATERIALS: KCl, chemical nitrogen, Mg, herbicides, putrid conditions.

CORRECTING MATERIALS: Ca, P, Mo, Si, B_{12}, molasses, humates, sulfates in some areas.

Western Nutgrass (Cyperaceae)

WEED MEANING: Ca, P, Cu, carbohydrate deficiencies; soil compaction.

ENHANCING MATERIALS: Magnesium, potassium chloride, zinc, vitamin K, humates where excess has been applied, diammonium phosphate, sulfates where excess has been applied.

CORRECTING MATERIALS: Sugar, molasses, vitamin C, vitamin B_{12}, biomin calcium, Microcal, biomin copper.

Wild Carrot (Daucus carota)

WEED MEANING: Excess Mg; Fe, Mo, phospho-carbonate, Ca (minorly) deficiencies.

ENHANCING MATERIALS: Mg, KCl, Fe tie-up (cold temperatures).

CORRECTING MATERIALS: Organic Fe, molasses, Mo, Ca, vitamin B_{12}, phospho-carbonates.

Wild Dewberry (Rubus flagellaris)

WEED MEANING: Ca, P, SO_4, Mo, Si, B_{12} deficiencies; acidic soil.
ENHANCING MATERIALS: Zinc, vitamin K, chemical nitrogens, potassium chloride, sodium chloride.
CORRECTING MATERIALS: Sulfates, vitamin C, selenium, boron, copper, molybdenum, iron, phosphate, humic acid, cobalt, sugar, vitamin B_{12}, calcium.

Wild Oats (*Avena fatua*)

WEED MEANING: Ca, P, Mn, Co, Si deficiencies.
ENHANCING MATERIALS: KCl, Zn (slightly), excess nitrates, compaction, certain metabolic byproducts.
CORRECTING MATERIALS: Ca, P, NH_4^+, Cu, sulfates in some areas, Mo, vitamin C and B_{12}.

Wild Vetch (*Vicia sativa*)

WEED MEANING: Ca, P, Mn, Mo, Si deficiencies.
ENHANCING MATERIALS: Potassium chloride, magnesium, zinc, vitamin K.
CORRECTING MATERIALS: Manganese, molybdenum vitamin C, phosphate.

In most of the cases where calcium is listed as a correcting material for the weed, the first choice of material to provide this is high-calcium lime, calcium carbonate, which should preferably be applied in fertilizer quantities, e.g., 100 to 300 pounds per acre, on a regular basis.

—13—

FERTILITY PROGRAMMING

CROP FERTILIZER PROGRAMMING traditionally is viewed as a very mechanical, precise process. A typical soil test is taken in the off season, the crop to be grown is listed, and an arbitrary quantity of N-P-K fertilizer, with perhaps some sulfur and zinc or possibly boron, is recommended according to the predicted yield. The reductionist viewpoint is that so many pounds of each of these nutrients are required to grow so many bushels of corn, boxes of tomatoes, or bales of cotton. No consideration is given to the source of each nutrient, whether there is actually need for such nutrients, the resulting refractometer reading of the crop, the effect on soil health, or the relationship of such fertilization to insect, disease, and weed pressures. If this system worked, we Americans would not have lost more than half of our topsoil, polluted a vast number of our natural resources, or become addicted to using pesticides.

If you entrusted your assets to an investment broker and five years later noticed that your real net worth (principal) had decreased, you probably would no longer leave your assets with that broker, even if he claimed he was getting record yields. In all likelihood, the risks and subsequent losses of principal he created more than offset the "record yields." Relative to agriculture, consumers and enlightened farmers are finally beginning to realize that university and government "experts" have been incompetent in handling their natural resource assets over the past several decades and are demanding a change in management.

Proper programming of fertilizer must take into consideration not only the items listed above, but also the type of seed used, the cultural practices employed, the climate, available water, the farmer's management ability, and, of course, the economics. The farmer must know what his field conditions are, what yields and quality of crop he has been harvesting, and what weed, insect, and disease pressures he has. These items should partially shape his fertilizer program because they disclose what is really happening, regardless of what a soil or tissue test indicates.

The best time to start your next programming task is right now. If you have just planted a new crop, get out into the field throughout the growing season and open your eyes. Take refractometer readings, get weekly Reams soil tests, and dig up roots and note their characteristics. Check weeds and insects. Do all the things discussed in this book. Most important, keep good records!

During this time, formulate a list of characteristics that need to be altered, such as reducing compaction, eliminating hardpan, eliminating grass pressure, eliminating aphid pressure, increasing tuber uniformity and specific gravity if you are growing potatoes, and so on. This list will help you recognize areas in which improvement is needed, evaluate current and past practices for their contribution to the problem, and formulate a plan of action to correct the problem.

Take your current fertilizer program and write it out on

paper. Let's say the program is for corn and looks something like this:

Preplant	100 lbs. of N as anhydrous ammonia
Planting	250 lbs. of 18-18-18 in the row, "X" lbs. of insecticide in a "T" band
Pre-emerge	Broadcast herbicide
Sidedress	100 lbs. of N as anhydrous ammonia

This is a simple program and is not unusual for a midwestern corn farmer.

My first alteration of this program would be to replace the 200 pounds of nitrogen as anhydrous ammonia with 120 to 160 pounds of nitrogen as liquid UN28 or 32% mixed with about 10 pounds of sugar or 2 to 4 gallons of molasses. I would split-apply this nitrogen mixture, just as was the anhydrous. I would then reformulate the 18-18-18, which is generally made with urea, triple super phosphate or diammonium phosphate, and muriate of potash. I would use ammonium nitrate and/or ammonium sulfate, possibly a little urea, monoammonium phosphate, and potassium sulfate. In addition, I would add about 2 gallons of liquid calcium with 1 to 2 gallons of molasses or 4 to 7 pounds of sugar to the herbicide mix. The herbicide rate might possibly be cut using this mixture, especially if some *SprayTech Oil* were used with the herbicide.

Another option would be to rig the anhydrous applicator so that you could inject into the ammonia at the point of injection: 2 to 6 gallons of clean 10-34-0, 4 to 8 pounds of sugar or 1 to 2 gallons of molasses, 5 to 10 pounds of potassium nitrate or sulfate (it is not recommended to use potassium hydroxide in this case because it is highly caustic, thus adding causticity to the already caustic anhydrous ammonia), 1 to 4 quarts of liquid fish or 1 to 4 pounds of powdered fish, 1 to 2 pints of humic acid, 2 to 4 ounces of soil conditioner, and 10 to 20 gallons of water (very important!). This would take the initial caustic harshness out of the anhydrous ammonia by forming aqua-ammonia, while at the same time adding some other nutrients. This makes the most use of a trip across the field. This alteration to the anhydrous ammonia application would be done for both applications of anhydrous ammonia with some alteration in materials, possibly for the sidedress. I would probably add 1 to 4 gallons, depending on soil calcium levels of ammonium thio sulfate, to a sidedress mix. In addition, the farmer would be able to purchase his nitrogen at the cheap price of anhydrous ammonia without damaging the soil as much as he would using straight anhydrous ammonia.

I do not claim that this is the ideal fertilizer mix. The solution being injected will still be very alkaline and cause some solubilization of the soil humus, but it is certainly a step in the right direction of reducing detrimental practices. Also, for the latter program, I would treat the seed with a product like that marketed by TransNational Agronomy to aid in germination, eliminate the dry starter fertilizer (we provide it in the anhydrous ammonia injection mix), and add liquid calcium with a carbohydrate and *SprayTech Oil* to the reduced herbicide spray.

A further alteration would be to apply the herbicide in a band over the row on the planter and then cultivate the middles of the rows. Eventually, all herbicides and insecticides will be eliminated from the program. They do as much as or more than anything else to inhibit the regeneration of the biological system in the soil.

The determining factor in this or any program is what the farmer is willing to do and what he can manage. Once these factors are established, a program can be formulated.

Keep in mind that we want to eliminate the harmful materials and replace them with tolerable ones (tolerable does not necessarily mean ideal). As we do this, the soil will begin to mellow and be regenerated. Eventually, the soil will not need to be rescued from the "garbage crew." Usually, we can eliminate insecticides before we can dispose of herbicides.

I realize that proponents of organic farming might find this discussion a bit compromising, and it is. However, the vast majority of farmers are not ready to convert completely to organic farming, and I doubt they ever will. At issue here is not philosophy, but rather regeneration of the soil and the quality of food and fiber production. Whether one is an organic, a biological, or anything but a straight-line chemical farmer, the key to success is nutrition that begins with mineralization. That is the common ground. Integrated with the materials mentioned above must be well-managed cultural and management practices. The job must get done in a timely and efficient manner.

Many additional material options are available for our corn program. A dry soluble fertilizer blend could be used. These are technical or foodgrade fertilizers that are readily soluble in water and are intended to be dissolved in water before application. They originally were used widely in the greenhouse industry but have since found their way into commercial production agriculture. They are noted for their purity and efficiency. Using such materials, the corn fertilizer program might look something like this:

Preplant	10 gal. N25 (liquid N blend) 2 to 4 oz. of soil conditioner or a biological "soil treat" product 2 gal. liquid calcium with carbon 8 or more gal. of water
Planting	5 to 10 lbs. of 12-50-0 5 to 10 lbs. of 18-18-9-8s 5 lbs. of sugar Several oz. of a trace element 1 gal. of water per pound of fertilizer
Sidedress	10 to 20 gal. of N25 5 to 10 lbs. of 12-50-0 5 to 10 lbs. of 18-18-9-8s or 14-30-15 3 to 5 lbs. of sugar 1 gal. of water per pound of fertilizer

It also would be good to add some fish, liquid or powder, to the planting and sidedress mixes and perhaps another trace element or biological product such as seaweed.

If the farmer desires an organic program, it might look like the following: apply 1 to 3 tons of true compost, a dry blend mix of North Carolina black rock phosphate, soft rock phosphate (Idaho or Florida sources), potassium sulfate, aragonite, possibly some gypsum, digested fish, and kelp (seaweed). Then apply an organic liquid calcium—it cannot be calcium-nitrate based, but will probably be calcium-lignosulfonate based—with sugar or molasses and perhaps a trace element if needed. The sidedress program would be similar to the previously mentioned dry blend. If nitrogen is needed in addition to that provided by the compost (3 tons at 2% nitrogen is 120 pounds of nitrogen), blood meal, leather tankage, or bean meal can be added.

In any program, it is a good idea to split the fertilizer applications and consider foliar feeding if it is feasible to do so. Also, regardless of the program, I would seriously consider applying 100 to 300 pounds of calcium carbonate, either in granular or pelleted form. In some areas, 50 to 200 pounds of gypsum could be applied.

As with any project or situation, there are several ways to achieve the desired goal. Some people would have you believe that the only way to fertilize is with their product or program. There are many successful biological-product companies and consultants across America. The common characteristic of all the successes is mineralization. When mineralization is accomplished, the crop refractometer readings increase. Insect, disease, and weed pressures decrease. Crop quality and production improve. Soil health and tilth improve. And economics improve. The major reason so much organic product is substandard is that mineralization has not been accomplished. Organic fertilizers have only the mineral integrity of the items that constitute the fertilizer. If you are using compost or manure from an animal whose diet is not mineralized as well as it should be, the manure and subsequent compost will lack the same minerals that the animal does.

A typical organic program seen in some parts of this country consists of 10 tons of compost, 10 tons of raw manure, and 1 to

5 tons of gypsum. This program occasionally might be followed by applying a few fish and seaweed foliar sprays directly on the crop. This program produces lush, green growth with substandard fruit quality and low refractometer readings, but it is organic. It can be assumed that the compost and the manure have 2% nitrogen (that is low for manure). Combined, 20 tons or 40,000 pounds of material are being applied to the soil. To calculate the quantity of nitrogen that would be applied, simply multiply 40,000 by .02, which equals 800 pounds of nitrogen. In contrast, chemical farmers generally apply only 200 to 400 pounds of nitrogen. This excess amount of material also adds several pounds of salt to the soil. Finally, this program adds an excess of calcium sulfate in an attempt to lower the soil pH which contributes, along with the excess nitrogen and salt, to the depression of the biosystem. Poor-quality commodities produced by imbalanced fertilization are poor quality by any label, chemical or organic. The underlying problem in this organic program or any imbalanced approach is lack of mineralization.

Programming fertilization needs is partially an art because each field is different. Early in a remineralization program, the materials and quantities used are very similar. As the soil's health improves, the individualization of programs becomes more important. Using the same materials, the same program, and the same cultural practices year after year may increase the soil's health to a reasonable level, but eventually it will cease to improve. If you reach a plateau and do not seem to be able to get beyond it, evaluate your mineralization program for missing or unnecessary components. Consider such things as vitamins (B's, C, K), trace elements other than the norms (Si, Se, Mo, I, Co), colors, yeasts, different seaweeds, spoon-fed calcium carbonate, rock dusts, alternate sources of currently used materials, and the timing and methodology of alternate applications. I have seen several cases in which the programs reached a plateau because less-than-optimum sources of fertilizer materials were used. Programs are only as good as the quality of their individual components. A component may perform well in one part of the country but poorly in another. As a rule of thumb, avoid industrial-waste acids like 10-34-0 made with waste phosphoric acid, which was used to clean metal.

Keep in mind that, for all practical purposes, working with the soil is like working with a young animal's or infant's digestive system until the soil is regenerated to "adult" capacity. A baby is given small amounts of food every few hours. Relatively speaking, the soil is no different. Treat it as if it were an infant's digestive system, and you will be successful. This means no large quantities of anything, no materials that are difficult to digest, and no toxic substances. As with individual babies, you might find that you need a few specific nutrients that are out of the ordinary or that your field is "allergic" to certain substances. Some of you have "sick babies" to deal with, and it is no wonder, with all the junk food you have been feeding them.

Programming a fertility program should begin a year before it is to be applied. Conduct soil tests during the last third of the growing season to determine how your soil is holding up under load. Observe weed, insect, and disease pressures. Obtain refractometer, penetrometer, and crop-characteristic data. Use all this information to begin next year's program this fall.

Keep in mind that desirable fertility programs do not come out of textbooks. At best, such books simply provide guidelines for farmers, gardeners, and turf managers to use. Each individual must learn the principles of nutrition; the meanings of weeds, diseases, and insect pests; the ramifications of compaction and toxic chemistry; and the interactions of microbes, soil, organic materials, and oxygen. Along with this understanding, one must apply common sense, initiative, and sound management to formulate and execute a successful fertility program. People have been lulled into believing that soil fertility is a fixed concept, based on taking a simple CEC test and adding the cheapest N-P-K material available, followed by spraying the life out of everything with toxic chemicals.

As mentioned before, the ideal time to start a fertility program is in the fall. This is the best time to address basic nutrition and to initiate desirable microbial action in the decomposition of crop residue. A late-season Reams soil test gives an indication of the basic nutrient status. Generally speaking, as one moves east of the Rocky Mountains, the reserve calcium level in the soil progressively decreases, and the need for fall application of calcium carbonate increases accordingly. This does not necessarily mean that the quantity of calcium carbonate that is applied at any one time is greater, however. As a rule of thumb, if the Reams test shows calcium at less than 2,000 pounds, calcium would be considered a priority in the fall material application. As the sodium level approaches a level of 70 parts per million or greater, the 2,000-pound desired threshold is raised, especially in areas with medium to high natural levels of magnesium or where dolomite lime has been applied regularly. At less than 2,000 pounds of calcium, I would consider applying 200 to 500 pounds of fine, high-calcium lime, ideally aragonite. In many areas, using such a small quantity of calcium lime is not feasible, so my compromise would be to use 1 ton of fine and 2 tons of coarse high- calcium lime per acre. If the calcium level is greater than 3,000 pounds and budget and ease of spreading small quantities are poor, I should not be concerned with applying calcium the first fall.

The next consideration is the phosphate level. Ideally, we like a minimum of 400 pounds on the Reams test in a 2:1 ratio with potash. In areas west of the Rocky Mountains, I would use the Idaho soft rock phosphate at a rate of 200 to 400 pounds per acre, applied in the fall. East of the Rocky Mountains, I would use either the Idaho or the Florida soft rock phosphate, whichever costs less, at 200 to 400 pounds per acre. Depending on the cost, availability, and feasibility of application of soft rock phosphate for areas east of the Rocky Mountains, I would consider the North Carolina black rock phosphate. This is a hard rock phosphate with 4.8% carbon, about 40% calcium, and a 4% soluble phosphate analysis. It is a very good material applied at 200 to 500 pounds per acre. Obviously, if your budget permits only a 50 pound per acre application of soft rock or North Carolina black rock phosphate, apply 50 pounds per acre.

Next, I would consider the appropriateness of some gypsum ($CaSO_4$) in the fall program, 50 to 500 pounds per acre. Gypsum is often helpful in mellowing the soil, especially in areas of medium to high magnesium, where dolomite lime has been used repeatedly and where salt levels are rising. This is a specialty product and is not effective in raising the calcium reserves in the soil. If potash is significantly low (below 100 pounds on the Reams test), I would consider applying 50 to 200 pounds of potassium sulfate in the fall program. Calcium and phosphate are generally the most needed materials on modern American farms, so these two nutrients should be given priority in the fall program budget.

Next, I would consider adding some compost or humate, if only a few pounds per acre, to coat the calcium so it will stay in the root zone better. If you can apply some type of biological material to help degrade the chemicals—natural and man-made—in the soil, with a little sugar and/or molasses, you will probably see an improvement, not only in soil tilth but also in spring soil test values. Ammonium sulfate could also be added to help release calcium and set up an energy field with conditions for desirable soil temperature stability.

Finally, I would consider applying 50 to 2,000 pounds of granite dust or colloidal clay mineral powder such as *Dynamin*, *Azomite*, or Noble mineral, to reestablish the soil mineral foundation and improve soil magnetic susceptibility. The following are a few examples, to be considered as guidelines only:

Soil Test August 1		*Fall Program*
Calcium	1200	200 lbs. aragonite (high-calcium lime)
Magnesium	500	
Phosphate	50	150 lbs. North Carolina black rock
Potash		300
Nitrate nitrogen	30	50 lbs. soft rock

Ammonia nitrogen		100 lbs. gypsum
ERGS	180	50 lbs. compost or 10-
Sodium	60	15 lbs. dry humate 50 lbs.
pH	6.8	ammonium sulfate

Soil Test August 1		*Fall Program*
Calcium	3000	100 lbs. aragonite (high-calcium lime)
Magnesium	700	
Potash	250	50 lbs. soft rock
Phosphate	20	150 lbs. North Carolina blackrock
Nitrate nitrogen	80	
Ammonia nitrogen		150 lbs. gypsum
ERGS	300	50 lbs. compost or 10-
Sodium	90	15 lbs. dry humates
pH	7.1	

The latter soil test indicates that there probably is not actually 3,000 pounds of calcium effectively available because of the higher-than-tolerable sodium level and the higher pH. This is why I still add a little calcium in my program, although not as much. I increased the gypsum by 50 pounds per acre to help deal with the extra salt.

If I were going to add a spray program with either of these dry programs, I would apply:

Residue Treat or Detox	1-4 oz.
J & J or Agri-SC or Trans-Flo soil conditioner	2-6 oz.
RL-37	1-2 qts.
Sugar	1-6 lbs.
Vitamin B_{12}	20-40 mg.
Water	25 gal.

If my budget was very constrained, I might apply just a couple of these in the spray:

2-3 lbs. sugar
30 mg. B_{12}
25 gal. water

or:

4 oz. *Residue Treat* (IAT)	2 qts. RL-37
4 oz. soil conditioner	2-3 lbs. sugar
25 gal. water	30 mg. B_{12}
	25 gal. water

Making an analogy to grilling your dinner, you would take charcoal as your basic heat source and add lighter fluid to help get the charcoal burning. Our dry fertilizer program is the base, and the spray is the lighter fluid.

If all we could do was a spray program on the fall tillage equipment, I would consider the following:

2 gal.	ammonium thio-sulfate
2-6 gal.	liquid nitrogen
1-2 gal.	molasses
6-10 gal.	water

This will help get desirable stubble digestion and initiate the availability of more nutrients in reserve.

With the fall program, I would sow a cover crop. Oats are my preference because they help increase soil phosphate availability and freeze off so they do not get out of hand in the spring. Any cover crop is better than none. When spring or the next crop season arrives, get another Reams soil test. Inevitably, there will have been a change from the previous test.

The first thing you must understand is that crops grow from energy; therefore, you must provide sufficient energy for the upcoming crop. If you have been using large quantities of nitrogen and muriate of potash, you cannot just stop using them without replacing their energy equivalent, or you will have a failure. If you reduce the potash, you must replace it with calcium. Beyond this, you can maintain the energy flow by adding various materials according to the soil test and particularly the ergs reading. These materials include vinegar, ammonium sulfate or thio sulfate, potassium sulfate, nitrogens, phosphates, and water.

I like to put liquid calcium into all preplant or preemerge programs to stimulate calcium availability and early season growth. I add sugar or molasses to this to feed the microbial system and to stabilize the liquid calcium. If you are in an organic program, use 1 to 4 gallons of calcium lignosulfonate or several ounces of *Biomin* calcium from J. H. Biotech, calcium DL from IAT, or *Microcal* from TNA. If you are not in an organic program, you can use 1 to 4 gallons of the calcium-nitrate-based liquid calcium or several ounces of the other materials mentioned.

Avoid products made with calcium chloride. The liquid calciums other than Biomin calcium can go in with the liquid nitrogen and preemergent herbicide or can be sprayed on after planting or in the first watering, whichever is most feasible. Next, we apply the N-P-K materials, always adding a carbohydrate to stabilize them, avoiding chloride materials, straight anhydrous ammonia, and diammonium phosphate. Split applications of fertilizer are best, and we can add a biological, fish, and seaweed to the mixes. In some areas, I would add a trace element or two, although they seem more efficiently used in foliar sprays, as well as vitamin B_{12} or vitamin B complex. B_{12} helps calcium availability, and B complex feeds microbes. Whenever the soil test shows a narrow calcium-to-magnesium ratio (<7:1), I would add a soil conditioner to the preemergent or first spray, or anywhere there is known soil compaction. Some people also will add some hydrogen peroxide to their program; this may help oxygenate the soil. As with all alternative materials, caution should be exercised. Whereas a little is good, a lot could be disastrous.

During the growing season, monitor at least the ERGS, pH, sodium, and refractometer values. At least one complete soil test also should be taken to evaluate the need for corrective measures. Every irrigation should contain some nutrient so that the soil ERGS stay above 200.

If the crop is growing lushly but is setting no fruit, you need to spray or feed with something to set the fruit. Vinegar and household ammonia work well at 1 to 2 quarts and 1 to 4 pints, respectively. The household ammonia keeps the sprayer clean for the "organic" people. You can get somewhat more sophisticated and add to the vinegar and ammonia 1 to 4 pints of phosphoric acid, 1 pound of powdered fish, 2 to 6 ounces of seaweed, and 1 to 2 pounds of sugar and/or molasses mix. Another mix to consider is 5 to 10 pounds of a dry soluble like 20-20-20 plus one-half pound of epsom salts and a quart of RL-37 foliar sprayed. This may be required only once or several times, depending on soil reserves, moisture, and weather. A good manager can dictate virtually every phase of growth and fruiting of his crop if he is equipped physically and mentally to do so.

When designing a program, always remember that plants grow from energy. This energy is created by the interaction of nutrients. Remember that nitrogen will move in all directions, so it must be joined with a carbohydrate to help stabilize it. Nutrients will seek their own strata level where they are most stable relative to other nutrients *if* they are not bio-complexed. These strata extend from the air to deep in the soil. Carbon Strata No. 1, Magnesium Strata No. 2, Phosphate Strata No. 3, Potash Strata No. 4, Aluminum Strata No. 5, Zinc Strata No.6, Manganese Strata No. 7, Iron Strata No. 8, Copper Strata No. 9, Calcium Strata No. 10. These rankings were given by Carey Reams in his short courses. Nitrogen will go in all directions depending upon its form. Per this chart one can surmise that without bio-complexing nutrients in the soil nutrient activity will be segregated and calcium and carbon will be the most difficult to maintain in the root zone. Calcium will descend out of the root system within weeks, so it must be united with a carbon, which seeks to rise into the air, consequently stabilizing both of them. Phosphate will generally remain fairly stationary in the soil unless it is in a water-soluble form, which can leach out. Combine this with a sugar to help stabilize it and make it more inviting for microbial assimilation. Actually, you are managing the microbial system in the soil, which in turn manages the crop. If you choose to do so under sterile conditions, so be it. Your production quality (refractometer reading) will reflect this.

The following is a summary of guidelines, as well as several programs you can use directly or alter to suit your particular needs. Learn the concept and design your own programs.

BASIC PROGRAMS

1. Get off the undesirable fertilizers:

Muriate of potash (KCl)—No compromise—use potassium sulfate (K_2SO_4), liquid or dry soluble blends, potassium nitrate (KNO_3), organics, and compost. Get calcium into the program in order to release "fixed" potassium, but be gentle with this. You can flush too much at once and have a disaster. Use no chloride fertilizers.

Anhydrous ammonia (NH_3)—Can tolerate a partial compromise, which is to carry saddle tanks filled with at least water and a carbohydrate T'ed into the ammonia line at the knife. You can also add other materials to this solution, like 10-34-0 or liquid fish. The more water you add, the better the result.

Dolomite lime ($CaCO_3$-$MgCO_3$)—This may neutralize pH, but it results in tightened soil, restricted air spaces, and reduced nitrogen efficiency.

2. Soil test (Reams test) at least once during the growing season and preferably every week with high-value crops. This will give you the trend of nutrient availability to the crop, as well as the stability of the microbial system. Take soil and air temperatures at the same time.

3. You can probably cut nitrogen by 10% to 40% simply by adding a carbohydrate with it, laying off the dolomite/magnesiums and chlorides, and split-applying the nitrogen mix.

4. Get some calcium into the system, either liquid calcium (no chloride) at 1 to 4 gallons per acre or finely ground or pelleted high-calcium lime at 25 to 200 pounds per acre. You could apply both the liquid and the dry calcium materials. Get the material out early in the season and combine it with a carbohydrate or organic (humate, compost). If gypsum is called for, use 50 to 300 pounds of this in addition.

5. Apply your phosphate and potash fertilizer starters or sidedress materials with a carbon—sugar, molasses, humic acid, humate, fish, seaweed, compost. Remember, liquid phos-

phates must be accompanied by a sugar for most efficient use.

6. Apply a foliar spray where appropriate, according to the refractometer test. Forget the guessing; select according to refractometer increases.

7. Scout for insects and diseases, and spray only when the technician suggests. Use predators, where available. If it is too much bother to be so attentive, find another profession. As soon as one farmer says he cannot do such and such, several other farmers do it and more.

8. Keep good records and always leave a test strip that actually gets checked. You must weigh and test yields. "Windshield" estimates are useless for making management decisions.

9. Whether you are organic certified or conventional, learn the principles and build your program around those principles.

10. KISS. Keep it simple, stupid. Make simple changes that you can and will execute.

Preferred Ideal Program, Which Can Be Altered to Suit Each Individual's Needs

Preplant

- 1-3 tons compost
- 1-3 gal. liquid calcium with 3-7 lbs. sugar or ½-2 gal. molasses
- 50-500 lbs. calcium carbonate ($CaCO_3$)
- 100-300 lbs. North Carolina black rock phosphate
- 100 lbs. Idaho/Florida soft rock phosphate
- 50-150 lbs. potassium sulfate (K_2SO_4)
- 50-300 lbs. gypsum ($CaSO_4$), where appropriate
- 400-2,000 lbs. granite dust or equivalent
- 1-3 gal. Bio-Culture Tea (BCT)
- 1-3 gal. fish, seaweed, micronutrients, humic acid . . . mix
- ¼-⅓ seasonal nitrogen with carbohydrate

1× or 2× sidedress or topdress—test to determine specifics

- ¼-⅓ nitrogen with carbohydrates
- 1-5 gallons BCT/bio, micronutrient . . . mix
- Extra P & K with carbohydrate, if warranted
- Additional calcium, if warranted

Foliar—check with refractometer to determine

BCT/bio, micronutrient . . . mix
N-P-K by choice
Additives by choice

If you insist on using a liquid program, still apply the compost, calcium carbonate, granite dust, and soft rock phosphate but use hot mixes or dry soluble blends rather than the dry.

In any event, keep the materials in the aerobic zone for maximum effect. Keep the raw manure to a minimum. If you farm in the west, buy compost rather than manure. If you farm in the midwest or east, make compost or at least treat your liquid manure before applying it to the soil. Manure must be composted in the soil, if not before. This takes energy, microorganisms, calcium, and oxygen, which, for the most part, are scarce in most soils. Manures also add salt, which is another burden to the soil.

EXAMPLE PROGRAMS (COMPLIMENTS OF D. L. SKOW):

Watermelons		APPLY PREPLANT:
ERGS	170	500 lbs. soft rock phosphate
N	50	100 lbs. 15-0-14
AN	0	100 lbs. calcium nitrate
P	50	50 lbs. 11-52-0
K	120	
Ca	325	SPRAY ON SOIL:
Mg	273	5 lbs. sugar
pH	4.9	4 lbs. calcium nitrate
		3 lbs. potassium nitrate
		30 cc B_{12} (release calcium)
		3 cc injectable iron
		20 gal. water
NN	90	
AN	40	
P	110	TOPDRESSED AFTER MELON SET:
K	180	40 lbs. ammonium nitrate
Ca	80	650 lbs. 11-52-0
Mg	101	210 lbs. 15-0-14 (fill melons)
pH	5.1	70 lbs. calcium nitrate

Strawberries		
NN	25	TOPDRESSED:
AN	0	500 lbs. SRP
P	50	1500 lbs. high-calcium lime
K	100	100 lbs. sul-po-mag
Ca	570	40 lbs. ammonium nitrate
Mg	156	
pH	6.2	

SPRAYED ON SOIL:
15 cc B_{12}
8 oz. Nitromax
3 oz. soil conditioner

AFTER BERRY SET:
100 lbs. calcium nitrate (dry)
Spray the following:
4 lbs. 12-61-0
6 lbs. sugar
1-1/2 pts. RL-37
20 gal. water

SEPARATE SPRAY:
4 lbs. calcium nitrate (watch leaf burn, add sugar, 1-4 lbs. to protect)
20 gal. water

If there is low calcium or a narrow calcium-to-magnesium ratio, use vitamin B_{12} with *Nitromax, Soil Treat, Residue Treat, Z-Hume, or Trans Detox* and a little sugar.

Onions note, sift lime in the row to help control maggots.

Blueberries with ammonium sulfate in drip line, and problems growing leaves.

NN	40	
AN	0	
P	100	1. Stop ammonium sulfate in drip
K	100	2. Add more phosphate
Ca	2500	3. Try a little $CaNO_3$ in drip line
Mg	300	
pH	4.6	(too low, want at least around 5.6)

In evidence, too much fruiting (Yin) energy. In this case calcium nitrate swung the energy pendulum a little the other way. pH came up a little.

A situation I encountered in 1984 while working with Dan

Skow in Minnesota demonstrates how variable the nutrient system can be and how effective foliar sprays can be.

Oats	*6/10/84*	*6/16/84*	SPRAYED BETWEEN TESTS:
Ca	5400	14000	5 oz. 10-52-10 dry
P	100	150	soluble
K	100	130	1/2 oz. multi-crop
NN	10	10	seaweed
AN	2	2	5 gal. water in mist
pH	7.0	6.9	blower
ERGS	110	330	(6/14/84)

The postseason test that was taken showed that the calcium level had dropped back to just a couple thousand pounds. It appeared that the phosphate and potassium were being used by the oats as fast as they were being released. This all indicated that a really stable microbial system was not yet established in the soil.

As an additional comment, another oat field that I foliar fed had continued to progress over the years in this same area of Minnesota. Yields consistently had been between 130 and 150 bushels per acre. No Atrazine had been applied to the field since 1984 or thereabouts. As a result, it was assumed, backed by industry insistence, that there should be no danger of Atrazine release stunting the oats. Consequently, last year no compensation was made in this field's oat-fertility program for Atrazine. The result was a 37 bushel per acre yield. A sample of these oats was sent to A & L Laboratories for evaluation. Atrazine was isolated and determined to be the cause of the stunting. So much for the propaganda that pesticides readily dissipate. What is not considered is that pesticides combine with other substances in the soil, rendering them relatively undetectable by conventional means. But when the conditions are right, these chemicals release. This process is called detoxification.

POTATOES

Get the soft rock phosphate and calcium into the program. Phosphate, sulfate, and microorganism activity are the keys to preventing scab. At tuber set, apply 300 to 400 pounds of 15-0-14 or 4 to 5 pounds of potassium nitrate, potassium sulfate, or calcium lignosulfonate and potassium sulfate.

Another approach that has worked very well for Jay McCaman's clients growing potatoes in Michigan is the following:

Spray
1-2 pts. 8-16-0
1-2 oz. vitamin C
4 oz. hydrogen peroxide
1/2 lb. dextrose corn sugar
20 gal. water

Spray
1 lb. 14-30-15 for tuber growth
1-2 lbs. 28-14-7 for early growth of vines

Potato bruise is caused by too little calcium in the tuber. Phosphate is the key to getting calcium into the tuber. Back off the sugar on potatoes late in the season. Use molasses when an iron need is evident.

In California, a large "organic" vegetable grower, now 5 years running, raises 400 sack potatoes with 3 tons of compost and 500 pounds of an organic sidedress blend. In previous years, he has done relatively well in remineralizing the soil and can now reap the benefits.

Trees
Foliar: Preblossom, after petal fall, second week in June:
1-4 qts. 6-12-6 with fish and seaweed
½ lb. sugar
To harden wood for the fall and winter:
8-16-0 and fish/seaweed
Spring spray once a week for 2 to 6 weeks:
1-2 oz. vitamin C
30-40 mg. vitamin B_{12}
1-4 qts. molasses
1-4 qts. fish
2-6 oz. seaweed

You can add 1 to 4 ounces of cinnamon if you have leaf hopper pressure.[1]

General Suggestions:
200 lbs. ammonium sulfate (dark material from Allied Chemical)
200 lbs. soft rock phosphate
Blend and apply immediately what you can afford, or apply only 200 lbs. of the blend as a starter off the seed.

Alfalfa with a soil test of 4,000 pounds calcium (Reams test), spray per acre:
1-2 pt. white phosphoric acid
1 pt. household ammonia
1-2 qts. liquid fish
20 gal. water

Apply dry after each cutting:
100 lbs. high-calcium lime
100 lbs. soft rock phosphate

The more quickly you can feed alfalfa after it has been cut, the more quickly the crown shock is relieved. If possible, feed at cutting and at baling.

Be innovative. For the very small operation/garden, if you cannot obtain foliar fertilizers, mix 2 to 4 ounces of Coke, Pepsi, or 7-Up (no artificial sweeteners) with 1 to 3 teaspoons of Kyo-Green or Barley Green in 1 gallon of distilled water and mist plants with a Windex spray bottle.

FERTILIZER MIXING

Mixing fertilizers, either liquid or dry, can in some cases make the difference between success and failure in the field. In other cases, it makes the difference between successful application or having a mess in the fertilizer tank or truck and hours of labor to clean it out.

Mixing Liquids

There are two reasons for ordering liquid materials in a specified sequence: (A) to protect biological or sensitive materials and (B) to create a structured energetic micel in the solution. "A" is the most common reason because "B" requires the formulation of unique biological materials with no "purified" inorganics involved.

The mix-order guidelines for "A" are as follows:

1. Water, nitrogen, phosphate, chemical N-P-K's, chemical micronutrients.
2. Carbohydrates, seaweed, fish.
3. Biologicals, vitamins, delicate materials.

This sequence will ensure that the potentially harsh chemical fertilizers are sufficiently diluted by the time the organics and finally the delicate materials are added. It is also recommended that dry materials like sugar be premixed in water to prevent their plugging the mix tank. One way to maximize the energy value of the tank mix is to leave a portion of the water until last and then add the remaining water with a conductivity-meter probe in the solution, stopping the addition of water as soon as the conductivity-meter reading peaks. Some farmers have noticed that spinning the solution in a particular direction seems to alter the field response of the mix.

The mix order for situation "B" must be determined by trial and error, dowsing, or radionics.

When mixing materials, recognize that calcium and phosphate or caustics simply do not mix well. You will have a solid material in the bottom of your tank if you attempt to mix these substances unless you are very creative. Example: liquid calcium and 10-34-0.

Mixing dry materials is a little less delicate than mixing liquids, but some problems still can occur if the mix is left in the truck for any length of time. Again, calcium and phosphate are tricky, so if you mix them, spread the mix as soon as possible.

GUIDELINES TO OBSERVE

1. Plants live off of the energy released by the interaction of fertilizers and microorganisms and from the sun.
2. Nature will always follow the path of least resistance.

3. Using pumped-phase conjugation phenomena, like things can be made to "attract" or increase like things, e.g., liquid calcium drawing in more calcium.

4. Yang-energy fertilizers produce growth, and Yin-energy fertilizers produce fruiting. Growth-inducing materials include calcium, potash, chlorine, and nitrate nitrogen (Reams's viewpoint), whereas all others induce fruiting.

5. For every cause there is an effect, and vice versa.

6. The higher the balance and availability of nutrients, the faster the plant will grow in a balanced state.

7. Plant growth is limited only by energy availability/balance.

8. The greater the soil-nutrient density, the greater the crop yield, provided there is adequate moisture.

9. The greater the humus content, the greater the nutrient quantity required to maintain soil test levels.

10. Ideally, all elements go into the plant in the phosphate form with the exception of nitrogen, which can erroneously carry other nutrients in with it.

11. Nitrogen is the major electrolyte in all biological systems.

12. Nitrogen can carry potassium into the plant when there is excess nitrogen relative to phosphate, which results in minerally imbalanced, low-refractometer-reading, watery plants susceptible to a multitude of pests and conditions.

13. Phosphate is a catalyst for nutrient and energy transport, along with carbohydrate production and metabolism.

14. The higher the phosphate availability, the higher the plant's content of sugar and dissolved solids; in turn, the higher the mineral content, the higher a crop's specific gravity and the fewer its pest problems.

15. Potash determines the caliber of the stalk and leaves, the fruit size, and the number of fruit sets.

16. The phosphate-to-potash ratio (on the Reams test) should be two parts phosphate to one part potash for crops other than grass/alfalfa, which should have a 4:1 ratio.

17. Carbon is the governor of moisture.

18. Ideally, about 80% of the nutrients for biological life come

from the air and about 20% come from the soil or diet.

19. Top-quality produce will dehydrate rather than rot.

20. Magnesium is the major governor of nitrogen.

21. Whenever there is less than a 7:1 calcium-to-magnesium ratio (Reams test), there will be an increased nitrogen demand because of less efficient nitrogen use, requiring increased applications of nitrogen fertilizer.

22. Always add carbohydrate with nitrogens and phosphates.

23. Herbicides generally interfere with the natural phosphate-sugar bonding in the soil necessary for stable phosphate assimilation by plants and microbes.

24. Keep materials in the soil aerobic zone.

25. As the soil system improves, the inputs must get more "sophisticated" and finely tuned.

26. Acid/salt-based fertilizers are less effective than non-acid/salt fertilizers in making long-chain functional amino acids.

27. The condition called gummosis in plants is caused by excess nitrogen and is easily relieved with a laxative—$MgSO_4$.

28. Sun-dried humates are the best ones to use.

29. Think of the soil system as an animal (ruminant) digestive system and treat it accordingly. Some soils are very sick "baby calves." Think about how a sick baby calf must be cared for and fed to nurse it back to health, and apply these common-sense principles to your farm-management program.

30. ERGS (soil conductivity) levels must be maintained above 200 micromhos net in order to maintain refractometer values above 12 brix. This is best done by combining N-P-K's with calcium in the same fertilizer mix. If soil sodium levels are below 20 ppm, you might consider 2 to 5 pounds of table salt sprayed on per acre to raise the ergs level.

31. The calcium-humus-phosphate complex is the key to maintaining stable soil ERGS and crop quality. Without the humus component, the calcium and phosphate complex to form tricalcium phosphate rendering both the calcium and the phosphate unavailable. Ideally, all nutrients would be com-

plexed in the humus-phosphate complex more appropriately called phospho-carbonate complex, i.e., calcium-phospho-carbonate, potassium-phospho-carbonate, etc.

32. See what you look at!

THINGS TO AVOID

1. Avoid using chloride materials such as KCl and $CaCl_2$.
2. Avoid burying materials in soil anaerobic zones. Plants and microbes are poor "treasure hunters."
3. Avoid using dolomite fertilizers or additives.
4. Avoid using strong acids or caustics directly on the soil.
5. Avoid excesses in all operations.
6. Avoid any operation that moves the system in the opposite direction from that which is sustainable.

Several companies supply very effective nutrient programs that farmers, gardeners, and others can purchase. Many of these companies advertise in publications such as *Acres, U.S.A.*

There are others in various parts of the country, but these are the ones I have worked with and thus can speak from personal experience. Some of the individuals mentioned are distributors, overseeing dealers too numerous to mention, yet I acknowledge them through their distributor. My point is that this system of biological management is not some obscure, left- or right-wing movement. With the people mentioned above, I can account for well over a million acres of commercial crop land being farmed biologically, not to mention the growing numbers of golf courses, businesses, and homeowners caring for their landscapes biologically.

I have suggested a potentially ideal program. Understand what is the goal of such a program. If you cannot obtain the materials at a reasonable cost, work out a program with what you can obtain that moves you in the desired direction. Your progress might be slower, but it will be progress nevertheless.

In summary, we want to provide the stable foundation of calcium, phosphate, potash, magnesium, and nitrogen at or above the minimum levels and ratios observed using the

Reams test system. Long-term phosphate stability is provided by soft rock phosphate; a calcium foundation is provided with calcium carbonate. For the desired ratios and levels to be manifested, we must have an active biological system and active humus formation. Compost is the best material to initiate this state in the soil and it does so the quickest, but it will degrade if the mineral balance is neglected and/or destructive fertilizer materials are continued. Compost provides stabilization of nitrogens and other nutrients and provides many nutrients in bioactive form. Preferable compost would have granite dust composted into it. These are important characteristics to strive for in every program.

Finally, we must provide the mechanism for energy release, in order to grow microorganisms and ultimately the crop. This is easily monitored with a conductivity meter. Whatever you work out on your operation is best judged by the refractometer and pressures from weeds, diseases, and insects, when or if pesticides are not used. The soil tilth, compaction, drainage, erosion, color, odor, drought and flood tolerance, and structure are further indications of your progress.

—14—

NUTRIENTS AND THEIR BASIC FUNCTIONS

CALCIUM IS THE KING of elements. It is needed more than any other element by weight and volume. Calcium is the foundation of all biological systems and is the component that gives the living cell its capacitor characteristic via its place in the cell membrane. Without the proper capacitor characteristic, the cell's communication system functions poorly, resulting in cell degeneration, disease, and eventual death. Calcium is nature's best detoxifier. It is the fundamental growth inducing nutrient and the base against which other nutrients are reacted to release energy for crop and microbial growth.

The desired minimum level of calcium availability, as determined by the Reams test, is 2,000 pounds per acre. pH does not indicate the level of calcium availability. At high pH, only $Ca(OH)_2$ is soluble. Sour grass weed pressures indicate insufficient calcium availability.

Sources of calcium are as follows.

- Limestone—calcium carbonate ($CaCO_3$), the most common source of calcium.
- Aragonite—ground seashell, very-high-quality calcium carbonate.
- Calcium nitrate—15.5-0-0-19Ca; soluble.
- Calcium oxide—dehydrated lime, burnt lime; not preferred.
- Calcium hydroxide—hydrated lime, quick lime; use with caution.
- Calcium sulfate—gypsum; use a maximum of 500 pounds/acre.
- Soft rock phosphate—35% Ca.
- North Carolina black rock phosphate—40% Ca.
- Dolomite—calcium carbonate/magnesium carbonate; avoid using bonemeal—18% Ca.
- Paper mill lime—avoid or use with extreme caution.
- Sugar-beet lime—fair, depending on area of country and the soil it is going on; inferior to $CaCO_3$.
- Calcium lignosulfonate—good soluble organic source.

PHOSPHATE

Phosphate is the major catalyst in all living systems. It is necessary for photosynthesis and metabolism to occur and is the key to obtaining high crop refractometer readings.

Research in the United Kingdom has confirmed that rock phosphates outperformed acid phosphates in studies lasting longer than two years. Some agronomists contend this was due to acidic soil conditions inherent to England, saying this would not be the case in alkaline western American soils. These scientists completely miss the fundamental premise of this research, the fact that microbiological activity not pH was the impetus for increased phosphate activity. Their "chemical" mindset will not permit them to see, let alone acknowledge, that pH vs. nutrient availability becomes secondary in a biologically active soil. The key to phosphate availability is soil microorganism activity.

Soft rock or colloidal clay phosphate is preferred for long-term sustainable soil management and sustained desired crop refractometer readings. North Carolina black rock phosphate containing 4.8% carbon, 30% phosphate, and 40% calcium seems to be an exceptional phosphate fertilizer.

The desired minimum level of phosphate in the soil using the Reams test is 400 pounds. Obtaining this level through the use of acid phosphates is highly unlikely. Soft rock, being a colloidal clay material, does not tie up readily into insoluble tricalcium phosphate as do the acid phosphates. Like all clays, however, it is susceptible to clay degeneration by KCl. Active biological activity is the prerequisite to sustaining 400 pounds or more of available phosphate per acre.

When acid phosphates are used, they should be combined with sugar, which increases the effective use of phosphate as much as 100 times. As phosphates become more available and approach the 2:1 ratio with potash, there will be less leaching of nutrients, less erosion, and reduced broadleaf-weed pressure. Ideally, all nutrients will be taken into the plant in the phosphate form.

Sources of phosphate are.:

- Soft rock phosphate, good.
- Hard rock phosphate, good.
- Super phosphate—0-20-0, specialty.
- Triple super phosphate—0-46-0, no-no.
- Diammonium phosphate—18-46-0, no-no.
- Monoammonium phosphate—11-52-0, 12-61-0, okay with sugar.
- Liquid phosphoric acid—75% to 85% P_2O_5, okay diluted with sugar water.
- Soap byproduct—0-40-0, check.
- Bonemeal—6% P_2O_5, okay.
- Liquid fish—2% P_2O_5, good.
- Mycorrhizae fungi—varies with bioactivity, good.
- Oats cover crop—varies with nutrition, good.

POTASH

Potash determines the caliber (thickness) of the stalk and leaves, the size of the fruit, and the number of fruit that set. The minimum desired level of potash on the Reams test is 200 pounds per acre in a 2:1 ratio of phosphate to potash.

Excess potash can result in potash substitution for calcium, thus degrading the plant's capacitor characteristics. Potash can also be "fixed" by clays, causing a reduction in the expandability and the CEC of the clay. Black spots on the leaves and stems of alfalfa indicate potash excess or, in effect, phosphate deficiency.

Sources of potash are:

- Potassium sulfate—K_2SO_4—0-0-50.
- Muriate of potash—KCl—0-0-60/62, sometimes called red potash, white potash, or Kalium potash. This product also contains 40% chloride. Muriate of potash is absolutely the poorest choice of potash a farmer could purchase. Soils simply will fail to become biologically active, sustain 400 pounds per acre of available phosphate in a 2:1 ratio with potash, and produce to their potential as long as this product is used. KCl is one of the most effective clay-aging products (degenerators) available.
- Chilean nitrate of potash—15-0-14, mined in the mountains of Chile. Has a fairly high sodium (Na) content. Use with discretion.
- Potassium nitrate—13-0-44, a manufactured product of good solubility. Often used in dry soluble blends, this is a decent material.
- Potassium hydroxide—0-0-50, caustic potash. Be very careful with this product and use it with discretion.
- Sawdust, wood ashes—not the best because of their dehydrating effect. Use with discretion.
- Tobacco stems.
- Pecan hulls.
- Cotton seed hulls.
- Cotton burr ash.
- Sugar beet by-products.
- Rice hulls.

Every source of potash must be checked for contamination, as should any fertilizer used on the farm.

NITROGEN

Nitrogen is the major electrolyte in the soil. Every cell contains nitrogen; it is the sun in every cell. The percentage of protein is calculated by multiplying the nitrogen content by 6.25 or 6.40.

The presence of nitrogen does not guarantee the presence or manufacture of true functional protein. Protein requires nitrogen plus carbohydrate or a carboxylic acid that is reduced, having an amine group (NH_2^-) added to it, to form a simple amino acid. This amino acid then links with other amino acids to form a protein. In conventional analyses, this information is ignored when protein is evaluated by testing only for nitrogen.

Nitrogen acts as an "isotope," alternating between the nitrate form and the ammonium form. Nitrate nitrogens seem primarily to promote growth responses in plants, whereas ammoniacal nitrogens seem to promote fruiting responses. It is usually claimed that nitrate nitrogen is readily taken into the plant, whereas ammoniacal nitrogen is not. This is a limited perspective. Such observations are made in in-vitro conditions or in nonbiologically active soils typical of modern chemical farms. The intake of ammoniacal nitrogen into plants seems to be directly proportional to the density of desirable root rhizosphere microorganisms. These microorganisms fare very poorly with herbicides, fungicides, and soil insecticides. Nitrate nitrogen must be reduced to amine groups by nitrate reductase enzymes before it can become a part of true, functional proteins. Consequently, it is not the most efficient means of transporting nitrogen into the plant. It might be easiest to transport nitrogen under nonbiological mechanisms, but not the preferred.

Nitrogen is one of the most overused materials in modern agriculture. It also is one of the most polluting.

Nitrogen applications should always be accompanied by a carbohydrate such as sugar or molasses. This helps to hold the nitrogen in the root zone and provides the basic building blocks necessary for the manufacture of amino acids and finally proteins.

The minimum desired levels of nitrogens on the Reams test are 40 pounds per acre each of nitrate and ammoniacal nitrogens.

Sources of nitrogens are:

- Manure—not preferred unless composted.
- Compost—the best source of many nutrients; be quality conscious.
- Ammonia sulfate—the dark material from Allied Chemical is

best.

• Milorganite—be cautious in using this.

• Sludge—prefer to avoid using, or compost and check for heavy metals.

• Tankage—best if composted; check for contamination.

• Blood—okay; check for contamination.

• Fish products—good; best if digested to some degree or composted.

• Urea—okay but not preferred.

• Sodium nitrate—not preferred because of sodium.

• Calcium nitrate—a good product when used judiciously; can use dry, but best if in liquid.

• Ammonium nitrate—okay if judiciously used; high salt index.

• Chilean nitrate—okay except for sodium.

• Anhydrous ammonia—poor choice, very caustic. If one insists on using it, dilute with water, molasses, etc., just before injection. Eventually avoid using it.

• Aqua ammonia—anhydrous diluted in water; acceptable with sugar/molasses.

CARBON

Carbon is the element that conveys life to the system. All living systems must have carbon; it is the energy storehouse for the system. Carbon is the governor of moisture. One part biocarbon holds four parts water. The biologically active carbon (humus) content of the soil determines its sustainability, efficiency, and productivity. The greater the amount of carbon, the greater the energy reserve. Carbon buffers the soil, improves soil tilth, and improves nutrient-holding capacity.

Sources of carbon are:

• Compost—perhaps the best.

• Carbohydrates—sugar, molasses, starch.

• Crop residue—only if aerobically digested.

• Manures—only if aerobically digested.

• Humates and humic acids—must use judiciously; organic matter is not necessarily humus.

MANGANESE

Manganese is the element of life. It activates a number of enzymes, including some related to photosynthesis, and is an important component in chloroplasts. Manganese brings the electrical charge into the seed, creating the magnetic force to draw the other elements into the seed. Manganese seems to be closely correlated to iron and copper; it is very important for seed quality and germination.

Sources of manganese are:

- Manganese sulfate—very common, but overused in many places where much sulfate has already been used.
- Royal jelly—perhaps the single best source, but it is expensive.
- Biofermentation manganese—preferred.
- EDTA chelates—prefer to avoid; many better chelates are available.

IRON

Iron draws energy to the leaf by absorbing heat from the sun; it makes the leaf darker, thus absorbing more energy. It will increase the waxy sheen of the crop. Iron is necessary for the maintenance and synthesis of chlorophyll and RNA metabolism in the chloroplasts. It increases the thickness of the leaf, that will geometrically increase nutrient flow, resulting in a production increase geometrically. Forty pounds per acre is desired on the Reams test.

Sources of iron are :

- Iron sulfate—most commonly used, but a potential problem where sulfate has been overused.
- Biofermentation iron—preferred.
- Molasses—a very good source where acceptable.
- Soft rock phosphate—a good reserve source.
- EDTA chelate—not preferred; many better chelates are available.

COPPER

Copper is the key to elasticity in the plant. It is an important constituent of many proteins like ascorbic acid oxidase, cytochrome oxidase, diamine oxidase, and polyphenol oxidase. Copper is an important nutrient for many microbes, such as *Aspergillus niger*. It controls molds and often alleviates perceived zinc deficiencies. Copper interacts with iron and manganese. Sul-po-mag, applied between July 15 and September 15 up to 200 pounds per acre, seems to help in copper availability.

Sources of copper are:

- Copper sulfate—most common, but a potential problem where sulfate has been used excessively.
- Biofermentation copper—preferred.
- EDTA chelate—not preferred; many better chelates are available.

BORON

Boron is important for filling in hollow stems. It seems to have various functions, but there is little agreement among plant physiologists as to specifics. Boron can cause strawberries to taste woody. Boron deficiency causes black heart. Boron is best used where calcium also is being used. It is an effective biocide, but it must be used with caution.

Sources of boron are:

- Chicken manure or compost—preferred.
- Borax.
- Biofermentation boron—preferred.
- EDTA chelate—not preferred.

MOLYBDENUM

Molybdenum is an important component in several enzymes, particularly nitrogenase and nitrate reductase. It is necessary for proper phosphorus and ascorbic acid metabolism. Molybdenum is a catalyst for iron in the bark or epidermis, is important in the integrity of bark or plant skin, and gives a transparent look to the sheen on the bark. Nitrogenase fixes N_2 (atmospheric nitrogen) to NH_3 in the plant. Plants grown on nitrate nitrogen have an increased need for molybdenum because of the increased need for nitrate reductase, which reduces NO_3 to NH_3 in the plant. Molybdenum is most commonly used in legumes but should not be limited to them.

Source of molybdenum are:

- Molybdenum glucoheptonate—preferred.

COBALT

Cobalt is necessary for nitrogen fixation, especially in legumes' root nodules; it is important for the formation of bark and cellulose and for seed-coat formation. Cobalt works in conjunction with molybdenum and fluoride to give hardness or sturdiness. It is a constituent of vitamin B_{12} produced by actinomycetes.

Sources of cobalt are:

- Vitamin B_{12}.
- Cobalt-calcium blend from I.A.T. and T.N.A.[1]

ZINC

Zinc is an essential component of many enzymes in the dehydrogenase, proteinase, and peptidase groups. It is a minor catalyst for sul-po-mag and copper and is correlated closely with copper and active nutrient systems. Zinc helps to make acetic acid in the root to prevent rotting; it is used to control blight and allows dead twigs on trees to shed off. Perceived zinc deficiency is often only symptomatic. Research has indicated that known soil-zinc deficiencies result in symptoms of plant-zinc deficiency only about 50% of the time. Zinc is much overused and promotes the growth of many weed species.

Sources of zinc are:

- Bioactive soil

In general, it is best to avoid using the following:

- Zinc sulfate—common, but a problem where sulfate has been overused.
- Biofermentation zinc—preferred.
- EDTA chelate—not preferred; many better chelates are available.

SULFUR

Sulfur can inhibit molybdenum assimilation and reduce nitrogen fixation. It is a component of many proteins and plant oils. Sulfur is important in many metabolic processes and is required in the synthesis of certain vitamins in plants. It has insecticidal properties and a strong fruiting-energy (Yin) characteristic.

Sources of sulfur are as follows:

Sulfur should be used in the sulfate form. Elemental sulfur can readily cause fruit rot at maturity, especially where there is a shortage of available calcium.

- Ammonium sulfate.
- Calcium sulfate
- Nutrient sulfates.

MAGNESIUM

Magnesium is part of the chlorophyll molecule. It is a nitrogen regulator. Magnesium is grossly overused in modern agriculture. It causes instability of nitrogen in the soil and contributes to soil compaction if available in quantities causing a narrower than 7:1 ratio with available calcium evaluated by the Reams test. Perceived magnesium deficiencies are more often symptoms of ignored or disbelieved calcium and phosphate deficiencies. Except in rare situations, the only time magnesium is used in a progressive fertilizer program is as sul-po-mag or when there is a nitrogen excess that needs to be antidoted. It is an excellent antidote for nitrogen toxicity.

SILICON

Silicon is not generally recognized as a plant nutrient; if it were, it would not be addressed because many people believe large quantities of silicon are found naturally in the soil. This, however, is a misconception. Silicon seems to be a missing link in certain finely tuned fertilizer programs. This situation is not overly common, yet it occurs fairly often. Silicon seems to have some correlation to the carbon-calcium interaction in the plant and is generally used as a foliar additive.

Source of Silicon:

The best source of silicon seems to be the herbal extract of the horsetail plant at rates of tenths of an ounce per acre.

CHLORINE

Chlorine is a heavier-than-air, greenish-yellow gas that is highly toxic but rarely occurs freely in nature. Chlorine is generally found in a chloride ion, or in salt. It is recognized as a trace element essential to plant growth, partially for the maintenance of a healthy immune system. Plant scientists generally contend that plants obtain sufficient chlorine either from the air or from rainfall. A & L Laboratories set 3 ppm or 6 pounds per acre as the desired level of chlorine on their soil test.

As with all other trace elements, excess chlorine is very harmful to soil fertility. From a physics perspective, chlorine and common chloride salts, e.g., KCl, are diamagnetic, thus potentially inhibiting the soil's ability to absorb solar and cosmic magnetic energy. From a chemistry perspective, chloride salts aggregate clay particles, causing them to become compacted and dehydrated, thus sealing and hardening the soil. From a biology perspective, chloride compounds in excess are anti-biologic, thus suppressing the growth of desirable soil microbes.

Recent university researchers, attempting to counter our claims against the use of KCl, have contended that chloride fertilization enhances crop yield. Investigation reveals that these university researchers neglected to use a refractometer to monitor crop quality and failed to use a conductivity meter to monitor soil energy. Most American farms, as pointed out earlier, run out of energy as the season progresses. Adding chlorides to the soil will increase the conductivity, thus giving more energy to the plant. Adding any electrolyte or salt to the soil will do this and thus increase crop "yield." That does not mean the crop needs more chloride. The university mentality still neglects to factor in the result of chloride fertilization on soil biology, structure, and tilth, as well as its effect on crop refractometer values. In the few cases where a chloride salt might be warranted it would only be warranted in small quantities, 1-10 pounds per acre, "never" tens or hundreds of pounds per acre.

VITAMINS AND ENZYMES

Vitamins and enzymes are very important biological materials, which are essential for the survival of both microorganisms and plants, as well as people and animals. As more is observed in the soil biological system, it is realized that these materials were taken for granted in the past and simply are not addressed by modern agriculture. It makes sense that, if the microorganisms are suppressed, so are the metabolites they produce—vitamins and enzymes. As a result, progressive agriculturalists have incorporated these materials into their fertility programs, either as inherent components of blends or as separate mix additives.

Plants supposedly manufacture many of their own vitamins and enzymes, but further study shows that many of these vitamins and enzymes or their building blocks are obtained from microorganisms in the soil; they are not manufactured if there is too narrow a fertility spectrum. Beneficial soil fungi and bacteria cannot produce many of their own vitamins and enzymes and must rely on symbiotic microbes to supply them. If any of these organisms are inhibited or killed by pesticides, there will be deficiencies of various vitamins and enzymes for use by both microbes and plants. Sources include biofermentation products, specific enzymes, B vitamins, and vitamins C and K.

WATER

Water is a very important nutrient. Aside from oxygen, it is often the most limiting factor in soil-nutrient and microorganism function. Water quality must be considered when formulating any fertilizer program and must be compensated for in the mix.

OXYGEN

Oxygen is the most limiting factor in biological life. Humans can live for several weeks without food, several days without water, but only a few minutes without oxygen. Soil microorganisms are the same. Oxygen must be available for the system to function properly. Plants need oxygen in their metabolic cycles, specifically the Krebs cycle, just as animals do. Timely tillage is very effective at oxygenating the soil. Hydrogen peroxide can add oxygen to the soil, as can fluffing of the soil using appropriate fertilizer materials.

—15—

CULTURAL MANAGEMENT

TILL THE SOIL FOR A REASON, and know the reason you are tilling. Reasons to till the soil are to incorporate residue into the aerobic zone, to prepare a seed bed, and to aerate the soil. Tilling too much, tilling too deeply, and tilling wet soil are three major contributors to soil degeneration. Poor tillage management can completely negate a premium fertility program. Conversely, superior tillage management can greatly compensate for a poor fertility program.

Combine trips over the field. Every time you drive across the field, put something in or on the soil or feed the crop, even if it is only sugar or molasses water.

CYCLES

Several cycles are important to the farmer. Four important cycles are photosynthesis, the Krebs cycle, nitrogen, and or-

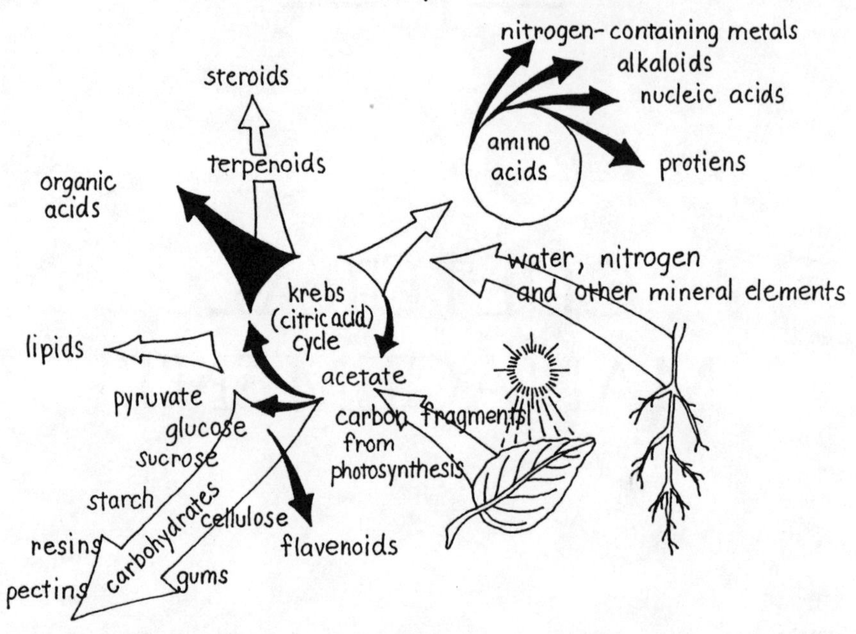

ganic matter—humus.

All of the organic matter in living things is ultimately provided through photosynthesis. Although photosynthesis is mentioned periodically in farming discussions, it is rarely understood or considered when formulating fertility programs. When farmers feed the crop and soil, they usually take photosynthesis for granted, without any consideration for its efficiency or yield. Like any functioning system, its quality is no better than the materials it has to work with. As such, notice the presence of phosphate, organic acids (ascorbic, gluconic, acetic), and carbohydrates in the various cycles. These are factors we consider when using a refractometer and when combining phosphates with sugars, selecting products, and evaluating the quality of a fertility program. Schematics of the various cycles are shown on the following pages.

An important consideration relative to nitrogen is that, before combining with organic acids to form amino acids,

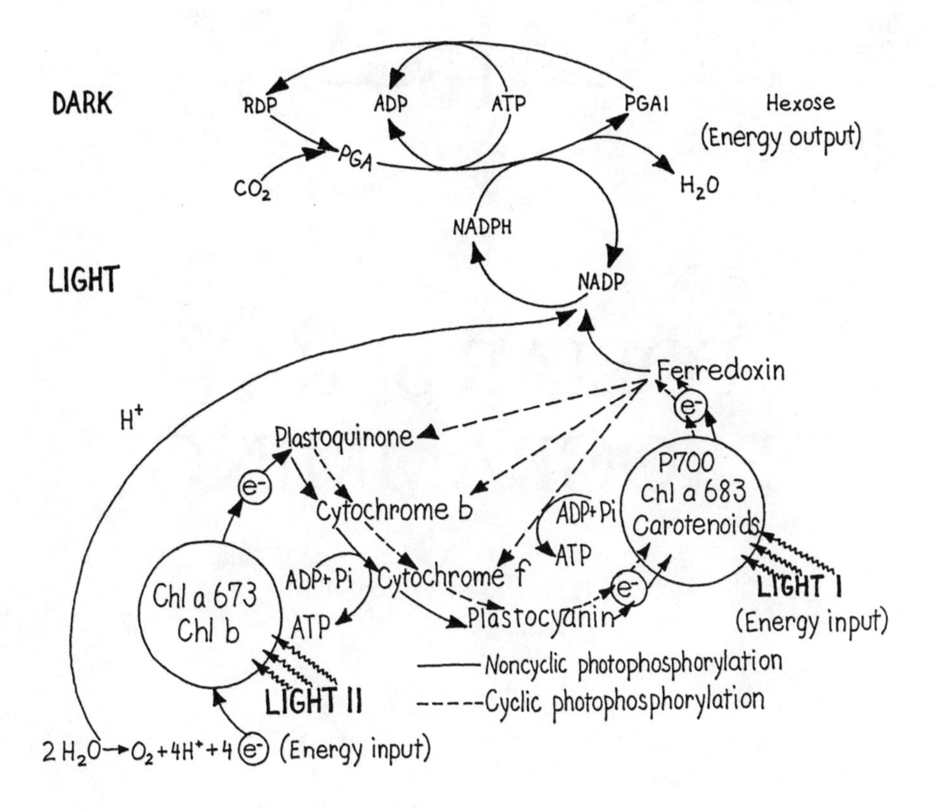

nitrate nitrogen must be reduced through a series of enzymatic reactions, e.g., nitrate reductase involving molybdenum. The reduced nitrogen groups, NH_2 and NH_3, combine with the carbon frameworks formed during the oxidation of sugars (forming organic acids) to form amino acids.

Keep in mind that materials are metabolized only if all the needed specific enzymes are present.

—16—

FOLIAR SPRAY PROGRAMMING

FOLIAR FEEDING OF CROPS is potentially one of the most efficient and economical methods of fertilization. In the 1940s, researchers at Michigan State University, in cooperation with the Atomic Energy Commission, found that foliar feeding of plants was 8 to 20 times more efficient than root feeding. Aside from the need for good soil fertility, particularly available calcium, the key to successful foliar feeding lies in the selection of an appropriate spray material and the timing of that application.

Conventional agronomists rely on analyses of plant tissues and evaluations of symptoms to determine what nutrient composition is needed for the spray. Many times this spray alters the tissue analysis and seemingly eliminates the visual symptom(s), but often it does not address causal deficiencies or improve crop quality and yield. This is puzzling to farmers and classical to the practices of conventional agronomists, who rely only on symptomatic observations and data.

Successful farmers and consultants have found a simple method of "asking the plant" whether a fertilizer spray is desirable, neutral, or undesirable. Plants do not read books, so their desires usually are quite different from the suggestions of textbook advisors. The procedure simply entails checking the crop refractometer reading before the fertilizer spray in question has been applied, and then 30 to 60 minutes after the application. Only a small test area is evaluated, thereby avoiding the possibility of spraying an entire field only to learn that the spray used was ineffective or even detrimental.

PROCEDURE

Construct two or more plastic or wire rings that encircle a 5 or 10 square-foot area. The area of a circle is 3.1417 times the radius squared. The circumference of a circle is 2 times 3.1417 times the radius. A ring encircling 5 square feet would have a diameter of 2.523 feet (30.28 inches) and a circumference of 7.93 feet (95.15 inches). A ring encircling 10 square feet would have a diameter or 3.568 feet (42.8 inches) and a circumference of 11.21 feet (134.52 inches). Because there are 43,560 square feet in an acre, 10 square feet equal 1/4356 of an acre. Using a 10 square foot ring makes calculations easier, but this size of ring is more difficult to carry around unless you make it so that it can be disassembled.

Take these rings to the field and drop them 3 to 5 feet apart. One ring will be used as a control or check area; the others will be designated as test areas. Check the refractometer reading within each ring and record these readings. Be consistent with the methodology and the parts of the plants used for these readings so each test area is sampled identically.

In a spray bottle, mix the exact fertilizer and water ratio that your sprayer is calibrated to apply. For example, if you would apply 2 quarts per acre of 6-12-6 in 20 gallons of water, you would mix 0.8 ounces of 6-12-6 in 1 quart of water. Twenty gallons equal 80 quarts, so a 1-quart mix for your spray bottle would be 1/80 of your per-acre mix. This is calculated by multiplying 2 quarts by 32 fluid ounces per quart to get 64 ounces. Divide 64 by 80 to get 0.8 ounces of 6-12-6 for 1 quart of water.

Next, you need to determine the number of squirts from the spray bottle to apply to the 10 square-foot test area in order to equal a spray rate of 20 gallons per acre. Because 10 square feet equal 1⁄4356 of an acre that would get 20 gallons of spray, you would apply 1⁄4356 of 20 gallons or approximately 0.6 ounces, which is about 2.4 tablespoons. This is calculated by multiplying 20 gallons by 128 fluid ounces per gallon (2,560 ounces), and dividing this number by 4,356. Because there are about 4 tablespoons per ounce, 0.6 ounces times 4 tablespoons per ounce equals 2.4 tablespoons.

Take a measuring cup and count the number of squirts from your 1-quart spray bottle to equal 0.6 ounces or 2.4 tablespoons. This is the number of squirts you will apply to your 10-square-foot test area to equal the amount that would be applied if you sprayed the field with your sprayer at 20 gallons per acre.

After misting this spray mix on the test area, wait 30 to 60 minutes and recheck the refractometer reading of the crop in the test area, as well as the control. If the brix reading in the test area increased by at least two full points net, the spray is desirable and would benefit the field. You can then spray the entire field with the confidence that the spray is beneficial. You can mix several different sprays and check each one. If the refractometer reading remains unchanged or drops, the spray is undesirable for that particular day for that particular crop. It is not necessarily a reflection on the product quality, but rather the plant's need at that time. The only exception to this guideline is when there might be a delayed reaction where no change in refractometer reading is observed for several hours. In this case, leave the rings in the field for 24 hours and recheck the brix reading of both the control and test areas. If an increase in the brix reading of the test area is observed, this spray would then be sprayed on the entire field. The delayed response might occur where a specific prescription has been formulated for a particular field and the spray test is a verification of its appropriateness. The delay could be a result of weather, temperature, water, and so on.

Keep in mind that, before and during a storm, the refractometer readings of the growing crop will be lower, as they will after several days of cloudy weather. The lower the nutrient reserve, the more these refractometer values will be lowered by such circumstances. Imperative to maintaining adequate crop brix readings is the continuous maintenance of the soil conductivity reading. It must remain above a net 200 ergs, or you will not be able to hold the brix above 12 when you do get it to that level.

It is recommended that you purchase an automatic temperature-compensated refractometer so that your readings will be more accurate and you will not have to calibrate the refractometer every time the temperature changes.

To reiterate, be consistent in the location of the plant where the refractometer reading is taken. The poorer the plant's health (correlated to lower refractometer readings), the more the brix readings will vary throughout the plant and throughout the day and season. Always compare the refractometer values in the test area with those in the control area. You desire a net increase, meaning an increase over the change observed in the control. If the control area had no change and your test area increased by two points, you have a net increase of two. But if the control area increased by one (which can happen as the sun rises in the morning) and your test area increased by two, you have a net increase of one.

The question often arises: How can commodities with high refractometer readings have insect and disease problems? Common examples include sweet corn at 24 brix having corn ear worms, and grapes at 18 brix having white flies, mites, and leaf hoppers. The answer is that the refractometer reading taken at the weakest part of the plant in question must be considered. The aforementioned problems indicate that the refractometer readings were taken of selected plant parts (e.g., ear, grape) only, and not of all the parts of the plant. The weakest link determines the outcome of insect and disease infestation. An ear of corn at 24 brix with corn ear worms inevitably will have leaf or stalk refractometer readings below 12. Grapes at 18 brix with insect infestation inevitably will have cane or leaf refractometer readings below 12 brix.

With apples, the opposite seems to occur. An apple with apple scab fungus will itself have a low refractometer reading (below 12); however, the leaves on the branch supporting the sick apple will have very high refractometer values (above 12 or even in the upper 20s). In any event, there is a mineral imbalance/deficiency in the crop.

Modern hybridization has produced plants that create such previously mentioned imbalances to satisfy cosmetic desires without considering mineral balance or the plant's natural ability to satisfy cosmetic desires if it is just provided with the necessary mineral to do so. Hybridization has resulted in plants that will accumulate sugars in given parts, yet not be able to metabolize, transfer, or convert them to plant parts. The result is similar to the circumstances in which animals and people are drugged or fed a diet that causes the body to accumulate fat without the necessary mineral balance for optimum health. As a result, the animal or person becomes obese. Obesity may be a desirable trait for market hogs, but not for dairy cows. Also, in some cases, improper diet causes the animal or person to be unable to metabolize sugar properly, resulting in hypoglycemia or diabetes. Similar circumstances can develop in plants when a disproportionately low or high level of sugar is present in a particular plant part.

The weakest link of the system, whether plant or animal, determines the overall strength of that system. If sugar is disproportionately accumulated in a specific part of a plant, it indicates an artificially induced condition, such as that caused by the girdling of grapes, or simply the plant's inability to metabolize such sugars properly due to a deficiency of the nutrients necessary to do so, whether they are genetically, artificially, or nutritionally induced. Because the sugar accumulated in such minerally imbalanced systems is out of balance with the rest of the plant, that imbalance is broadcast in the IR radiation emanated by the plant, thus signaling the garbage-collector insects to come and clean up. Insects get sick from healthy plants because they cannot handle the rich nutrients present in those plants. Further verification of energetic communication in nature can be found in the writings of Robert Becker, Vlail Kaznacheyev and Phil Callahan.

—17—

MANAGEMENT: MAKING IT HAPPEN

COMPETENT MANAGEMENT IS the key to any successful venture. It is perhaps even more important to a successful farming operation than to other businesses because, as Charles Mercier stated in *A Manual of the Electro-Chemical Treatment of Seeds,*[1] farming "is . . . one of the most complex and difficult [industries] to master and to pursue with success. Farming requires more intelligence than any other industrial occupation." Unfortunately, many Americans, farmers and vocational agriculture teachers included, perceive farming as a nontechnical profession.

Farm management has many connotations. Some farmers think that, because they paid income taxes last year, they managed successfully, or that they did not make a profit last year because the Russians did not buy enough American grain to relieve the imaginary American grain surplus, which drove the price of grain down and resulted in less farm income. (It should be noted that the prices of American farm commodities are suppressed deliberately through the importation of commodities.[2] However, farm inequality or disparity is not brought about by some sinister force outside the farmer's control; it is the result of mismanagement. Farm parity is achieved through good management. It is earned. Yes, many American farmers are in financial distress, and many are going out of business, yet similar percentages of other entrepreneurs also have gone out of business. Well over half of America's farmers are in very good financial condition. We just don't hear as much about them as we do about those who have been unsuccessful.

In this chapter, I shall suggest several practices to use in your farm-management program to improve your efficiency, profitability, and sustainability. The days of mining the soil which some people call farming are rapidly drawing to a close, and your ability to manage with new materials and methods will determine your success. I suggest that you view the film *Discovering the Future: The Business of Paradigms,* narrated by Joel Barker.[3]

ACCOUNTING

Many skilled financial advisors, analysts, and accountants favor particular financial spreadsheets. These are simply formats for recording and keeping track of all your monetary transactions. To some farmers this might seem trivial because they have already implemented an extensive financial record system or have corporate farms with an accounting department. Unfortunately, however, most farmers do very little financial record keeping beyond that required in filing their income tax forms. Even successful farmers can improve their record keeping in order to evaluate their status and progress more accurately. Some farmers would benefit from hiring a financial expert, at least to establish a financial analysis system. The financial-analysis system, regardless of the type of operation, needs to be functional. It must be easy to read and provide information in such a way that management decisions can be made directly from that information. It is fine to know your growing cost per bushel of rice, which can easily be subtracted from the selling price to get net profit, but this cost per bushel should be broken down into each component, e.g., water cost, rent, tillage, seed, seed treatment and handling, labor, fuel, maintenance, insurance, interest, depreciation, herbicides, insecticides, fungicides, fertilizers, soil amendments, consulting fees, percentage of total farm expenses, and percentage of total farm time. There should be an entry for harvesting cost and one for drying cost, as well as for the number of points of moisture removed, hauling cost, and marketing cost. To make an accurate judgment about your management system, each of these categories must be itemized and entered in your records. Many farmers have little or no concept of whether they are living on cash flow, profit, or depreciation. All such elements must be itemized so that they can be altered, if necessary, in a financially sensible way.

A total cost per bushel figure does not tell you that your maintenance cost per bushel is double that of five years ago, although it is the same percentage of total cost per bushel it was five years ago. Perhaps the maintenance cost is twice the percentage of total cost per bushel it was five years ago. This may indicate a need to revamp your maintenance program.

Perhaps a more important reason for thoroughly itemizing costs per unit of production is to enable you accurately to evaluate alternative agronomic programs. Alternative programs inevitably will be the standard for farming in the near future. Your success in implementing such programs will be determined by your management. A good example of how a good record keeping system can aid in evaluating alternative agronomic programs is that of David Larson of Larson Farm Management in Princeton, Illinois. Having a farm management business, his own grain farm, and a grain dealer's business, he was in the ideal position to evaluate commodity characteristics and to compare conventional chemical programs with biological agronomic programs.

Through meticulous record keeping, David Larson discovered that growing corn under conventional agronomic practices (using liberal quantities of nitrogen, potash, and agricultural chemicals) required that an average of 7% of moisture per bushel be removed at harvest. At an average drying cost of 1.5 cents per percentage point using on-farm dryers and 2.5 cents per percentage point for commercial drying, this amounted to 10.5 cents per bushel for on-farm drying, or 17.5 cents per bushel for commercial drying. Larson discovered that, after three to five years of farming with a biological agronomic program, the number of percentage points of moisture to be removed at harvest from a bushel of corn dropped to an average of 3.5 points per bushel. This amounted to a 50% savings in drying costs. Given a typical midwestern grain farm with 500 acres of corn yielding an average of 140 bushels per acre, this amounted to $3,675 more net income to the farmer. Most significant was the fact that yields did not suffer as a result of the improved agronomic management. In addition, Larson noticed that the per-bushel test weights were one to two pounds more under the biological programs, regardless of corn variety, and that the protein content was 1% to 2% higher than it was in the conventionally grown corn. The last finding is significant to livestock farmers who routinely purchase protein. Larson further noticed that the conventionally grown commodities also tended to go out of condition more readily in storage than did the biologically grown commodities, even after drying.

The preceding figures were not based solely on one farm. They were compiled from David Larson's own farm, his managed farms, and his grain-handling business—that is, from many farms, from thousands of bushels of corn, from many different varieties, and over several years of monitoring.

How many farmers know their average drying costs, their average test weights, their average storage shrinkage, and the average nutritional value of their crop?

This is just one example of how record keeping can reveal information that would otherwise be overlooked. Most farmers look only at gross yield when evaluating their operation. Thus, they are missing the real components of profit and loss.

Another example of the value of record keeping is that of a dairy farmer in Wisconsin. This farmer was feeding his cows a high rate of protein, and a feed consultant recommended that he cut the protein level by two pounds per cow per day. The farmer did so, and production declined by two pounds of milk per cow per day. The farmer was enraged and began adding protein to the ration once again. The consultant explained to the farmer that it was costing him nearly $30 per hundredweight to produce the two pounds of milk. At a selling price of $15 per hundredweight, this was very poor economics. The farmer refused to believe that the consultant knew what he was talking about.

The consultant had based his recommendation on the following calculations: soybean meal, at $250 per ton, costs 12 cents per pound. At 44% protein, that is 880 pounds of actual protein per ton of meal, at 28 cents per pound of actual protein. Hence, two pounds of actual protein were costing the farmer 56 cents. By adding two pounds of protein, the farmer was gaining two pounds of milk per cow per day. At $15 per hundredweight, one pound of milk is worth 15 cents, and two pounds are worth $.30. Therefore, the farmer was spending $.56 and receiving a return of 30 cents, for a loss of 26 cents per cow per day. This loss was not a result of international marketing manipulation but of poor management by the farmer.

The important function of any accounting system is to record every source of money and where funds are spent. Some items may seem trivial, but if you know where your money comes from and where it goes, you can budget accordingly and determine whether any changes are necessary. Technically, it might be more convenient to place some of the items discussed above, like test weight, protein cost, and the number of percentage points of moisture removed, on enterprise accounting (discussed in the following section) and transfer only the totals like "total drying costs" to your master accounting worksheet.

It is best to discuss such organizational details with your accountant and decide which method you prefer.

In any event, it is important to keep track of every aspect of your operation, including your own labor, so that you can make sensible business decisions. Many farmers survive on their cash flow, only to retire with few or no assets because, through the years, they considered only the cash flow aspect of the business, overlooking depreciation, opportunity costs, and personal labor.

I often have heard farmers comment that they could not afford to pay themselves a fair wage for their own labor, yet when asked how much it cost them to produce a bushel of grain, a hundredweight of milk, or a bale of cotton, they were hard pressed to answer. Even fewer could state what their depreciation, storage shrinkage and loss, or substandard quality costs were per unit of production. Where are the financial records to support their statement that "I can't afford to pay myself a fair wage?" Without such records, how do they know what they can or cannot afford? Quite often, those farmers who have the lowest profit at year end are the ones who have done the least record keeping and are least knowledgeable about their production costs and efficiency factors. Often it is the seemingly insignificant factors like the number of percentage points of water to be removed from a bushel of grain at harvest, specific gravity, test weight, density, and storage integrity that make a difference in the profit column. But such information will be available only if the farmer has kept good records of these figures.

ENTERPRISE ACCOUNTING

One way to project profit or loss is to use enterprise budgets; these are micro-budgets of individual enterprises on the farm. Enterprise budgets include only those factors involved in the individual enterprise. For example, a dairy farm generally warrants having an enterprise budget for hay, corn, oats, wheat, milk, and meat. You could also separate out calf and bred-heifer enterprises. With each enterprise standing alone, it is possible to evaluate possible weak points in the operation. Enterprise budgets also allow the farmer to compare management techniques before implementing new ones. This can be done by writing enterprise budgets for each management technique and comparing the results. At year end, these enterprise budgets should be revised to represent actual figures, in order to evaluate your accuracy in budget projection, to compare various management approaches such as different fertilizer programs and cultural practices, and to compare alternative agricultural methods to conventional chemical agricultural practices.

VALUE-ADDED ACCOUNTS

I recommend including another accounting component—the value-added account—in the typical accounting portfolio because, more often than not, subtle and sometimes not-so-subtle value-added characteristics are associated with superior farm-management practices. Some of these features return immediate profit to the farmer, more as public sentiment and legislation change. These include reduced dry-down requirements; increased crop density; increased crop health and nutritional value, resulting in improved animal health and nutrition and less pressure from insects and disease; reduced shrinkage of stored crops; improved handling, packaging, and shipping characteristics; increased net yield; lower cost per unit of production; improved stubble digestion; reduced soil compaction and hardpan; improved water infiltration; increased ease of tillage; heightened leaf, stem, and fruit refractometer readings; reduced pesticide residue, mold, and aflatoxin levels; increased feed conversion; and improved taste and palatability.

Some people argue that value-added characteristics are not usually considered in pricing and marketing commodities. This assumption is incorrect. Each value-added characteristic listed above affects the marketability of every agricultural commodity. Take corn, for example. Farmers are paid by the bushel, based on a standard test weight, moisture, and grade.

If your corn requires only 3.5 percentage points of moisture removed versus 7 percentage points per bushel, you will have only half the dockage for drying and shrinkage. If your corn has a higher refractometer reading, it will have greater density; less mold, resulting in lower aflatoxin levels; and higher test weight, that converts to a greater premium per bushel because bushels are calculated by weight. All of these value-added characteristics contribute to the eventual price per bushel the farmer ultimately is paid for his corn. If he is naive enough to think otherwise, he is only fooling himself.

The market trend is toward value-added pricing. In 1990, Pioneer Hybrid International contracted for one million bushels of corn in the midwest, paying premiums of $.35 and $.40 per bushel for corn meeting their value-added criteria. The corn had to be grown entirely without pesticides of any sort and also test pesticide-residue-free. The contracts appeared to be genuinely oriented to the farmer, unlike most other commodity contracts. Pioneer even agreed to pay $10 per acre contracted if they canceled the contract before April 30, 1990. At 1990 market prices, premiums of 35 and 40 cents per bushel are significant. Five hundred acres of corn with a 140-bushel-per-acre marketable yield would give a return of $24,500 to $28,000 in premiums, or $49 to $56 per acre. Although it would require excellent managers to produce corn without pesticides, is not excellence what the American economy is supposed to be all about?

The value-added accounting of one's operation not only reveals possible income-earning characteristics of the operation. It also reveals the level of consciousness exhibited by the farmer. Farmers who are concerned about their influence on their fellow man and the environment, as well as the food chain, are the leaders and teachers who are the real backbone of this country. Farmers have the greatest potential of any group to change and direct the course of society.

As George Watson pointed out in his book, *Nutrition and Your Mind*, the nutritional integrity of an individual's diet determines the integrity of that person's thinking and emotional stability. When farmers regenerate the soil, they recondition the food chain. This, in turn, improves the nutritional integrity of society, thereby enhancing people's thinking and emotional stability. No drug, psychiatric counseling, rehabilitation program, welfare check, educational system, or legislative mandate can ever have such an all-encompassing influence.

GOAL SETTING

An important part of record keeping, budgeting, and managing a business is setting goals. Many farmers have no written goals. Their goal setting consists simply of hoping to make it through this year and still be in business next year. However, to fulfill his dreams and aspirations or even to repay his loans, the farmer must establish written goals, both short and long-term, for which to plan and strive. Goal setting is an exercise in clarifying one's objectives. The clearer the objective, the more readily it can be attained and the easier it is to determine whether it has been realized. The farmer's goals will include performance objectives for each facet of the operation; they will also include some personal achievements and desires that are seemingly unrelated to the farm operation, like vacations and recreation.

Using a dairy operation as an example, possible goals might be as follows:

- Raise rolling herd average from 18,000 to 18,500 lbs. milk, 720 to 740 lbs. fat.
- Decrease services per cow from 1.5 to 1.25.
- Reduce lactation cycle from 14 to 13 months.
- Improve feed conversion from 85 lbs. to 82.6 lbs. per hundredweight of milk.
- Increase average cow longevity from 4 to 5 lactations.
- Reduce mastitis from 1 to 0 cases per month.
- Reduce retained placentas from 1 to 0 per month
- Reduce cost per hundredweight by 10%.
- Increase milk refractometer reading from 10 to 20.
- Reduce milk conductivity reading from 6000 to 5700.

General Goals

- Replace barn doors.
- Paint milk house.
- Purchase new front-end loader.
- Reduce energy bill by 5%.

Family Goals

- Take family on vacation to Disney World.
- Replace refrigerator.
- Replace draperies.

It should be obvious that some of these goals can be attained in one year, thus, the farmer will need to set new ones for the following year. Other goals will require several years to fulfill. Also, many of these goals require that the farmer formulate definite objectives for other enterprises on the farm. For example, improving the health, feed conversion, and production of the herd eventually will require that the feed quality be improved. A refractometer reading of 20 for the milk will never be achieved by feeding milk cows today's typical feed or feed rations. A value of this magnitude would require feedstuffs with at least an equivalent refractometer value. Grain would be sprouted or at least soaked for 48 hours before feeding, long-stemmed hay having at least a 20 brix refractometer reading would be liberally provided, and clean water would always be available.

Goal setting must be a deliberate process; it is an integral part of good management and a necessity if one is to achieve his desires. Goals need not be complex; in fact, it is best to keep them simple so they are understood explicitly. Set short-term (daily, weekly, monthly), intermediate (yearly), and long-term (5, 10, or even 25 year) goals. Every long-term and intermediate goal will have shorter-term objectives leading up to it. Write these objectives down and achieve them or alter them as necessary. Evaluate your goals periodically to determine whether they have been achieved or are still appropriate. Such an evaluation will serve as a critique of your management system.

Many people say that farming is a way of life. In reality, farming is a business. It is a way of life only if it is a successful business. Unfortunately, for many people, farming is a struggling way of life because their business is mediocre. In many cases, if farmers treated farming like a business, it would be sufficiently enticing for their sons and daughters to undertake as a profession.

Put the fun back into farming! Manage your farm as if you were going to be farming forever and would have to be accountable to future generations for your farming practices and land uses, as well as for the nutritional integrity and safety of the commodities you produce.

MAKING IT HAPPEN

Despite the importance of record keeping, computer analysis, planning, and goal setting to a successful farming enterprise, there is no substitute for getting out and making it happen. The key to making it happen in any business setting is never doing anything yourself that you can delegate to someone who can do the task better, more efficiently, and consequently more economically than you, the manager, can. A common difficulty for pure entrepreneurs, and most farmers fall into this category, is delegating responsibility. Most farmers have expanded their operations over the years, and their success has depended on their willingness and ability to delegate responsibility to others. In operations where this has been a problem, the farmer either becomes overextended, having only common-labor employees, or he goes out of business.

The pure entrepreneur remembers the hard work that went into building the business and is afraid to relinquish any of the responsibility. He must learn to be a business manager as well as an entrepreneur. Although he must retain the entrepreneurial spirit to keep the business vital and dynamic, he must realize that he is no longer the only one involved in the operation. This is often a problem with many father and son/daughter enterprises. The father, being the entrepreneur, often is reluctant to become the business manager and delegate responsibility to his son or daughter. Many times sons and daughters must wait until their fathers retire or die before assuming responsibility for the farm. Then they are frustrated because they have not been prepared to make major decisions concerning the business.

There are no easy solutions to this problem, other than to recognize the situation and include steps to remedy it on the farm and personal goal lists. Take some management classes. Obtain the advice of a business consultant. Both father and son/daughter should attend classes in business management, group dynamics, and personnel management. Treat the farm like a business, and you will find that your son or daughter will appreciate and run it more like a business than a bequest.

Another important aspect of business management for farmers is hiring professional consultants for specific areas of need on the farm. In certain regions of America, farmers routinely hire consultants for pest monitoring and control, improvement of cultural practices, financial analysis, and marketing. Some farmers retain animal nutritionists and/or veterinarians for their animal operations. In other parts of the country, farmers are reluctant to hire such consultants, seeing them as insults to their farming competence or as threats to their bank accounts. This reflects the entrepreneurial thought pattern. In today's business climate, farmers must realize that good consultants, like good employees, make money for the business.

Part of the responsibility of a good business manager is to choose those employees and consultants who make rather than cost the business money. One very important detail that must be tended to in making this determination is the maintenance of thorough records. If farmers do not keep good records, they cannot make unemotional business decisions regarding the employment of laborers and consultants.

Employing farm consultants has never been as important, opportune, or perplexing as it is today. With the changing public sentiment regarding chemical agriculture, farmers are being forced by both economics and legislation to change their farming practices. For decades, farmers have relied on the Cooperative Extension Service for advice about farming, with mixed results. In this era of chemical free farming, the standards are different, and so far Extension personnel have little understanding of these new regulations and practices. Private enterprises have addressed the problem of changing to chemical-free farming, have learned the new standards, and are available to help the farmer. Many competent private agricultural consultants have already established solid track records and, unlike university researchers, have had to achieve positive results in the field to stay in business.

When considering whether to hire a consultant, decide where you need guidance. If you cannot decide on that, perhaps you should consult a financial analyst, a business planner, or a farm manager to formulate a farm business plan complete with goals. In light of the current production trends, most farmers could use a consultant to help them with fertility, cultural, and pest-management practices. Professional advice is needed in these areas because the current management information that is most readily available to the public perpetuates the widespread use of chemicals. Proponents of conventional chemical programs cannot imagine farming without chemicals, nor do they comprehend that balancing soil and plant nutrition corrects the problems of soil compaction, erosion, hardpans, insect, disease, and weed infestations, and inferior commodity quality. In addition, balancing soil and plant nutrition sets the stage for achieving record yields. Consequently, the farmer must seek independent counsel in these areas.

When seeking an advisor, look for someone who recognizes the underlying importance of soil and plant nutrition. Such a person would not be your typical agronomist or soil or plant scientist, who is knowledgeable about and advocates the traditional system of soil and plant nutrition, in which it is assumed that toxic chemical intervention is necessary to grow the crop. Rather, you should seek an advisor who recognizes that biological controls and nutrition are far more effective in managing pests and that indiscriminate use of toxic chemicals causes more pest problems than originally existed.

At times, perhaps, the judicious use of toxic chemicals might be suggested, depending on the management program employed and how nutritional balancing of the soil has progressed. Economics dictate that the farmer must get a crop in order to remain in business next year; therefore, that must be the first priority. However, in producing a crop, the farmer can employ practices that lead to less chemical dependency, improved soil health, and increased healthfulness and cleanliness of the commodity. Some farmers will have the management to convert to a biological agronomic program, whereas others will use a transitional approach and perhaps never have a fully organic operation, according to the strict definition of that term. Other farmers will simply go out of business because they lack sufficient management to continue farming.

If you employed an insect and disease scout who advocated the traditional chemical approach and then hired a fertility management consultant who understood the actual correlation between nutrition and pests, you would not be able to progress as rapidly as possible because the two consultants probably would not work well together. However, if your scout is superior but traditional in his thinking, he will soon recognize the true nature of the situation. I suggest that such individuals read Phil Callahan's publications and contact competent field scouts who are employing a biological approach to pest/disease control and nutrition.[3]

A good consultant understands nutrition and its link to soil and plant health. Such a consultant can help transform your operation into the farming business of the future. The consultant should understand the information contained in this book regarding which materials to use (and those not to use), quantities of substances to use, monitoring methods, and so on. A competent consultant will make you more money than his recommendations cost you.

Farmers should expect positive results. The misinformation

spread by the media about poorer yields, more insect and disease damage, and poorer quality of today's crops is based on the failed research in farming without chemicals and the trials of uninformed farmers. The media are correct in that, under the traditional approach to nutrition, it is impossible to farm without chemicals.

That is why we use "nontraditional" nutritional practices. They work! Expect your crop to have better shelf life, improved storage characteristics, increased nutritional value, less insect and disease pressure, enhanced palatability, and better feed conversion. Expect things to work. If they seem not to work as well as you envisioned, evaluate your records to see what happened.

THE "KISS" PRINCIPLE

*K*eep *I*t *S*imple, *S*tupid is the best management strategy of all. Simple directives can be taught, duplicated, executed, and evaluated. Managers often make matters too complex; this is stressful for all concerned and requires additional management. Dan Skow points out that, whether you have an animal operation or a cropping enterprise, you must attend to certain basic factors to be successful. These include providing adequate air/ventilation and clean water, satisfactory comfort, and sufficient cleanliness. Any farmer whose operation is having problems is not adequately providing these four factors; conversely, every successful operation is satisfying these four requirements.

Sometimes managers get so involved in technical analyses that they overlook these basics. If there is insufficient air, your oxygen-breathing organisms will be stressed or sick, and the anaerobic organisms will proliferate. Bad water can cause anything from insufficient consumption or nutrient interference and tie-up to poisoning. If animals (people, cattle, azotobacter) are uncomfortable, they will not perform well; comfort allows them to carry out the task at hand. Finally, an ounce of prevention is worth a pound of cure. Cleanliness is the best mode of prevention you can implement.

Success does not happen because the manager makes things so complex that no one knows what his next move will be. Some people think management by confusion is good management. It is not, nor is management by dictatorship. These techniques may flatter the manager's ego, but they undermine teamwork. Further, the most sophisticated computer-generated feed ration or foliar spray cannot totally overcome suffocation, dehydration, repeated poisoning, or deplorable conditions.

Keep it simple, keep it fresh, keep it tolerable, and keep it clean—and it will keep you in business.

RECORD MANAGEMENT

Every time you go to the field, you should walk it and take notes on your observations and the current weather conditions. It is said that the best fertilizer a farmer can have is his footprints in the field. If you are going to farm in the 21st century, you must be serious about the stewardship of your land!

You will seldom notice changes if you do not record them specifically. For example, if you are growing grapes and you routinely check the grapes with a refractometer because it is the grapes that you sell, you will probably get the grapes to a reasonably satisfactory refractometer reading through the judicious use of fertilizer. However, until you consistently get the refractometer value of the grapevine leaves above 12, you will continue to have problems with such insects as leaf hoppers and white flies. The leaf canopy has a larger surface area than the grapes and, therefore, emits the dominant electromagnetic signal attracting insect pests, regardless of what the grapes themselves are doing.

As another example, if you never record the moisture content, test weight, and nutritional value of your grain, you will not notice that after three years on a nutritional management program the required dry down is less, the test weight is greater, and the nutritional content (e.g., protein) is greater regardless of variety. Nor will you notice that the maturity time of your crop is decreasing. Without records, you will not notice that your fields are blowing less in the wind, that tillage equipment can be pulled more easily, or that fuel consumption per acre is lower. Most important, you will not realize that your net profit per acre has improved.

On the following page is a field-observation list you might incorporate into your field notebook. You can modify the list to suit your particular operation and crop.

SUMMARY

The alternative system of agriculture works in the field, where the farmer is either made or broken. This system is not just a theory or wishful thinking. When people assert that there are volumes and volumes of replicated research substantiating the validity of the traditional system and snubbing the alternative system, I remind them that the replicated research only verifies that what was done turned out as intended. That does not verify that the experiment was scientifically valid or that the traditional system works. On the contrary, under this system of agricultural pseudoscience, we have destroyed more than 50% of the country's topsoil; contaminated a majority of the water supplies; diminished the nutritional value of all food commodities; created pesticide-resistant insects, weeds, and diseases; and injured countless people with toxic chemicals. Furthermore, we continue to do all these things.

Since World War II, pesticide use has increased tenfold, yet crop losses due to the pests targeted by this chemical approach have doubled. No one would consider this a success. The only thing that is valid in the traditional system is that it perpetuates itself and financially benefits the petrochemical industry while enslaving the planet to poisons. Where is the replicated university research proving that this system of chemical use works in the long run? It is time for the American farmer, who has been the backbone of this great nation for more than two hundred years, to reclaim the responsibility inherent in that status. He must stop raping the land and defiling the environ-

FIELD OBSERVATION LIST

General Characteristics		*Crop Characteristics*	
Wildlife Signs	____	Color	____
Birds Present	____	Height	____
Earthworm Signs	____	Plants/ft^2 or Crown	____
Topograph	____	Buds/??	____
Hardpan Depth	____	Nodules/Root	____
Hardpan Thickness	____	Nodule Color/Size	____
Soil Structure	____	Tap Root	____
Crumbly	____	RootHairs/Rhizosph	____
Block	____	Leaf Color	____
Sheety	____	Leaf Blemishes	____
Clumpy	____	Hollow Stem	____
Erosion	____	Root Bark Sluffing	____
Wind	____	Internal Node Health	____
Water	____	Pith Color/Condition	____
Soil Moisture	____	Brix—Leaf	____
Soil Conductivity	____	Brix—vine/stalk	____
Soil pH	____	Brix—Fruit	____
Soil Temperature	____	Insect Pests/Damage	____
Air Temperature	____	Disease Sym/Damage	____
Weeds Present	____	Node Spacing	____
____________	____	Fruit/Bowl Set	____
____________	____	Characteristics	____
____________	____	Lodging	____
Weed Brix Reading	____	Stalk/Vine Shape	____
Herbicided	____	Pollination %	____
_________	____	Brace Root Numbers	____
Insecticided	____	Characteristics	____
_________	____	Other__________________	
Fungicided	____	_______________________	
_________	____	_______________________	

ment, the food chain, and his fellow man and get back to farming. Only then will he find true parity.

—18—

TURF, LANDSCAPE, AND ORNAMENTAL PLANT CARE

FEW PEOPLE REALIZE THAT COMMERCIAL agriculture has an important branch aside from food and fiber production. This branch accounts for several billion dollars in business and employs thousands of people. I will point out promptly that I do not consider myself a turf or horticultural specialist, nor do I claim expertise in this field. I do, however, understand plant and soil nutrition, and, based on the experience I have had with turf and horticultural crops, I know they are controlled by the same laws of nature as are food and fiber crops. Many horticultural crops are considered weeds, in comparison to food and fiber crops. From a nutritional viewpoint, current cultivars of most horticultural crops that have been hybridized tend to react more like domesticated crops than like wild weeds.

Lawn and turf grasses require high calcium availability, in

the range of 4,000 pounds per acre or more using the Reams soil test. Ideally, they also require a 4:1 phosphate to potash ratio using the Reams test. These levels of calcium, phosphate, and potash will maintain a lush, vigorous grass growth uninvaded by sour grass weeds like quack grass, or broadleaf weeds like dandelion and pigweed. Even more important, these levels of nutrients will keep the lawn, fairway, and landscape free of diseases such as "dollarspot" and insect pests such as white grub and weevil.

Following the principles outlined in this book, successful control of weeds, insects, and diseases on turf and landscapes is being achieved at nutrient levels less than the ideals desired by various businesses across America. The point is that, even long before these ideals are achieved, weeds, insect pests, and diseases can be decreased dramatically, simply by directing soil and plant nutrition in the desired direction. Calcium availability seems to be the foundation of grass nutrition, as it is with all other crops. The rule still applies that potash can replace calcium in the system, eventually leading to less than desired results. Under a potash-and-nitrogen- based turf fertility program, you will eventually have to use pesticides and herbicides and engage in a continuous battle with seeding longevity, thatch build-up, hardpan, diseases, insects, moles, weeds, and increasing scrutiny from the Environmental Protection Agency and the Department of Natural Resources with regard to environmental contamination. The last point is no small consideration, in light of the Supreme Court's ruling that local ordinances can be enacted that greatly limit or ban the use of pesticides and herbicides beyond those regulated by federal agencies. There is also the issue of having to keep children and pets off the grass after poisonous chemicals have been applied. The Japanese are becoming concerned about using toxic chemicals on the golf course, considering subsequent human traffic.

Successful alternative turf and lawn programs are being marketed by J & J Agri-Products in Dillsburg, Pennsylvania, using NitroMax, soil conditioner, and liquid calcium; Erthrite,

Inc., in Gap, Pennsylvania, using their SK dry blend followed by liquid calcium, seaweed, and bio-copper applied in a spray; Nourish Industries in Kansas City, Missouri, applying a completely organic program with unmatched success now for four years; Freedom Formulations in Delphi, Indiana, applying selected biological mixes followed by dry soluble spray blends; Andy Lopez, in Malibu, California, applying a completely organic turf, landscape and garden program; Susequehanna Tree Surgeons in Lancaster, Pennsylvania combining the Nitron line of products with innovative management; and Venman's in Rockford, Michigan, applying a combination of dry blends followed by dry soluble spray mixes. Although each of these programs is different, their philosophies of nutrition and their results are similar.

The turf and lawn care business is different from food and fiber agriculture in an important way. One is not harvesting a crop and getting a direct return on investment. The return on investment is aesthetics and now also safety. Most people are used to the rigid chemical approach known as "weed and feed," which actually involves spraying a poison to kill weeds and perhaps insects, and applying nitrogen and potash to "green up" the grass. The two applications are self-perpetuating. The nitrogen-potash "fix" causes an imbalanced plant and soil, resulting in intervention by nature's caretakers (weeds) and "sanitation engineers" (insects and diseases), which then necessitates further applications of poison. Of course, the initial aesthetics appear desirable—green grass without weeds or pests. But the conditions are compounded each year until, finally, you either cannot grow decent grass or what you do grow is unsafe for humans and animals.

More and more golf course managers, landscape specialists, and homeowners are realizing that chemical management leads to a dead end. As a result, the demand for nontoxic management is growing. Many nontoxic turf programs initially are more expensive than chemical ones because a more complete fertilizer mix and greater quantities of it are required; also, extra labor is involved. This is mainly because chemical programs have so unbalanced the soil in the past. As with almost anything one purchases, the buyer usually gets what he pays for. It takes a little time for soil conditions to change, so the homeowner needs to be patient and realize that after one or two seasons of proper management the turf will be healthier, the aesthetic aspects will be better than before, and toxic chemicals will be unnecessary, eliminating the expense of such poisons.

The next subject of issue is on turf plants, such as shrubbery, trees, and ornamentals. Most of these plants will respond well to the same improvements that are made on turf, but there are some exceptions. Some plants are in such degenerated states of health that any change would be advantageous. Water quality is perhaps the most important factor in bringing about major improvements. Most city water is a major contributor to plant demise, largely because it contains chlorine and other undesirable components. The chlorine makes it difficult to establish a vigorous, desirable biological system in the soil. Every time one waters with chlorinated water, the system is greatly stressed. This situation must be compensated for with the fertilizer blend. Calcium is of real importance in this regard. Also, applying a buffered humic acid (one that does not precipitate at pH's below 7) can help sequester the chlorine, thus buffering its negative effects.

Many landscape and ornamental plants will respond to foliar feeding with seaweed, fish, vitamins, and dry solubles. Some, however, do not respond well to these feedings. Hybridization is the probable factor contributing to the non-response, just as hybridized corn does not respond as well as open-pollinated corn to foliar feeding. In such cases, one might have to rely primarily on soil feeding until an effective foliar material is found. Failure of foliar response often is related to soil imbalances, problems with water or material quality, timing, or simply material selection. Extensive trial and error using a refractometer might be required before a positive response is achieved. Persistence is the key. Often, the simplest item does the trick.

In any event, multiple soil feedings are preferred. Fall is the best time to get started, but regardless of starting time, a fall program is essential. In commercial situations, in which insects and disease exist, before getting the situation turned around it might be advisable as a last resort, to avoid losing the customer, to apply a pesticide, preferably a biopesticide, to buy some time. I would suggest mixing the pesticide with something like Spraytech oil from Integrated Agri-Tech so that the amount of pesticide could possibly be reduced, followed by some type of detoxification product like *Soil Treat* from I.A.T., *Detox* from T.N.A., *A-35* from Nitron, or *RL-37* from International Ag Labs, to name a few.[1] Mineralization is the key to the effectiveness of these products and subsequent success.

Where potting mixes are being used, such as in greenhouse operations, it is difficult to obtain biologically active commercial potting mixes, primarily because of sterilization practices and the use of nonbiologically active materials. Due to quick turnover rates, one is always working, for all practical purposes, with sterile or biologically juvenile growth mediums. To improve on this situation, we integrate compost into the mix. The compost is biologically active, but any mineral materials one adds, such as lime, soft-rock phosphate, rock phosphate, gypsum, rock dust, and so on, are not biologically active. The compost medium must work on them, which might take a little time. Ideally, these minerals would be composted initially so that one would end up with a complete, composted potting mix. This can be achieved, and a few commercial potting-mix suppliers come close to such a mix. The better the potting mix, the more efficient will be your liquid nutrient program. Some type of water-improvement system, such as reverse osmosis, is highly recommended for all water going to potted plants. In urban areas, such a system will easily pay for itself in improved plant growth and enhanced fertilizer efficiency.

Once the water is taken care of, the next item of importance is the dry soluble nutrient mix. Avoid using chlorinated mixes, anything with potassium chloride or calcium chloride. Use fish and seaweed as adjuncts to the N-P-K's, and include sugar and/or molasses. Most water-delivered nutrient solutions would benefit from the addition of a buffered humic acid to help buffer salt reactions on the plant roots and complex nutrients for better absorption. Typical humic-acid products are not buffered; therefore, the humic acids will precipitate at pH's below 7. The fulvic acid will remain soluble, but the humic- acid portion is the most valuable, and it will not stay soluble if it is unbuffered. The only buffered humic acid I am aware of presently is RL-37 from International Ag Labs, Fairmont, Minnesota.

Regardless of the situation, the basic principles of nutrition must be applied. Insect pests and diseases are indicators of nutritional imbalances, which must ultimately be corrected if these pest and disease conditions are to be eliminated. Although the task is not an easy one, good managers eventually will accomplish it.

HELPFUL HINTS

Two major problems common to turf plantings are white grubs (or blackheaded pasture cockchafers as the Australians call them), which are the larval stage of what we commonly call June bugs or beetles, and fungal disease.

White grubs attract various varmints, which tear up the turf to feed on the grubs. Eliminate the grubs, and the varmints will leave. Poor soil fertility and microbial activity are the cause of white-grub proliferation. Promising field research is being conducted using the predator fungus Metarhizium anisopliae as a biocontrol additive to the soil. This can be used before one's soil and turf are regenerated.[2] As for fungus control, use buffered humic acid at 10-20 ppm, copper at 2-4 ppm, and a predator bacteria or the Trichoderma growth enhancement product marketed by J. H. Biotech in Ventura, California.

A WORD ABOUT SUGAR

A question occasionally arises concerning the use of white sugar in fertilizer programs because it is said to be detrimental in human diets. It is not white sugar, per se, that is harmful to the human body, but rather the lack of minerals along with the sugar. Most people's diets are overloaded with sugar and, in this context, consuming more sugar is deleterious. Balance is the key because we do need some sugar in our diets.

It appears that, when the soil is severely degenerated, the most basic of components are needed to get the system working. Sugars other than white sugar require some degree of "digestion" because of their crudeness or the unique niche into which they fit. In other words, sugars other than glucose fit into specific metabolic pathways and require alteration to fit into courses other than these particular ones. These specific metabolic pathways may be further along in the sequence of events than what we need to do at that particular time—like kneading bread dough before all the ingredients are mixed into it.

Glucose is the most basic carbohydrate and is necessary to initiate the entire cycle of metabolic processes. If you do not have enough of this basic carbohydrate, you cannot get much to happen, regardless of what else is present. It is similar to making bread without water. Although water certainly is not the primary ingredient of bread, it is essential in getting the bread-making process started. You could add other liquids to the dry ingredients, but the flavor, texture, and consistency of the bread would be different from what you desired. Your product could even be a disaster. On the other hand, if you were to add too much water, you would have soup rather than bread, and adding still more water would only make the situation worse.

So it is with sugar relative to fertilization. Soils are in dire need of basic carbohydrates, sugar, to initiate the desired metabolic processes. Human diets, on the other hand, are so overloaded with sugar that the body can hardly metabolize it all. People's sugar craving is evidence of the body's natural quest for mineral. Sweetness and mineral content are synonymous in nature, but man has fooled himself by providing sugar without the mineral, creating an imbalance and an even greater "sweet-tooth" craving for mineral. Our bodies do need carbohydrate, but in a natural balance with mineral. Our soils have been deprived of basic carbohydrate as well as mineral and often need a little sugar (in the range of 1 to 20 pounds) to bring the food groups into a balanced enough condition that the mineral can be used.

As a general rule, the source of carbohydrate needed will vary according to the level of health or balance at which the soil exists. White sugar (sucrose—a compound of glucose and fructose) and dextrose (D-alpha glucose) seem to be the appropriate carbohydrates for the most unbalanced or sick soils. The point to remember, however, is that sugar is not the only material used. A complete prescription of nutrients, including sugar, is recommended during the cropping season.

Other common sources of fertilizer carbohydrates are molasses (cane blackstrap being preferred) and corn syrup (fructose). There are times when sucrose and dextrose (corn sugar) are the most appropriate carbohydrates and other times when molasses and corn syrup are the most appropriate. Experience and radionic evaluation help determine which to use. In any event, as Carey Reams impressed on his students, do not use more of what you already have too much of.

—19—

BASICS, OR FIRST PRINCIPLES

FIELD TRIPS IN AGRICULTURE often are more like dog and pony shows than extended classroom experiences. Typically, the circulating tour group is paraded around at a distance, being shown various test plots, tillage systems, and management practices. One very important group experience that is rarely included in traditional field tours is the evaluation of crops, weeds, and soils in a hands-on manner in the field. Keen observation of field and crop details can tell the observer more than any soil or plant-tissue analysis. This is a key point because the farmer and consultant must take note of field observations to determine whether progress is being made and whether soil and plant-tissue tests are providing accurate pictures of the situation.

THE FIELD TRIP

Upon approaching any field, notice the overall appearance of the crop. Is it droopy? Rigid? Does it smell fresh? Are there any peculiar odors like methane gas, mold, ammonia, or pesticide? Is the overall color of the crop a deep blue-green? A mild-green or pale-green color? Do the leaves have a glossy sheen or a typical absorbent appearance? Are any birds nearby? Are they singing? How does the field feel to you? In other words, what sense do you get about the field?

As you move into the field, notice what types of weeds are growing and whether they or the crops are being eaten by insects. Ask whether a herbicide was used to control weeds and what types of weeds are most bothersome. Check the refractometer reading of both the weeds and the crop; write down the value of each. As the soil improves, the weeds will decline and the crop will increase in brix readings. Pull some weeds and slice open their stems lengthwise with a knife. Look at the pith of the weed. What color is it? Is the stem hollow? The healthier the weed, the higher the brix reading, the more solid the stem and the more pearly white its pith, and the less insect damage it will have. The same, of course, holds true for the crop. The healthier the weed, the more conducive your fertilization practices are to growing weeds rather than crops unless weeds (herbs) are your crop.

Notice what type of root structure the weed has. Grasses often have shallow, dense root systems that are attempting to loosen compacted soil. Broadleaf weeds generally have long taproots that are attempting to relieve hardpans and gain access to nutrient reserves at lower depths in the soil, as well as to extend the electrical circuit of the soil/plant complex to greater depths.

CORN

Dig up a corn plant. Notice the amount of root mass present and in which direction the roots grow. Are they pretty much growing out in the top two or three inches, or are they mostly growing down? Are there many or any live roots directly below the middle of the plant? Where does the soil structure change from a loose, crumbly structure to a platey, blocky one? Most soils in America have a crumbly structure for only the top one-half to two inches of the top soil and a platey structure from there on down. This means there is only a one-half to two-inch aerobic zone; the rest is predominantly anaerobic, for all practical purposes.

In general, the roots will be concentrated primarily in the aerobic layer. You will notice that the greatest concentration of fine, fuzzy hair roots will be in the aerobic zone; this is because microbes and root hairs need oxygen. The microbes and root hairs make up the rhizosphere, which is the area of greatest nutrient exchange. Notice the color of the corn roots. Are they pearly white and soft, or are they brown and brittle? Pinch the roots slightly and pull to check whether the root bark sloughs off easily. If it does, there is a salt problem. Check for hardpan using a penetrometer, shovel, or brazing rod. This condition can be present in a sandy soil as well as in a clay soil. Notice that the roots pretty much stop at the hardpan.

Cut the corn stalk lengthwise and look at the plant's plumbing system. Is the bottom of the stalk base, the pons, brown and hard? In general, we find that it is. This is congested or plugged tissue caused by toxins in the soil. These toxins might be pesticides, or they might be metabolic by-products from anaerobic breakdown of crop residue and manures. These could be products like formaldehyde and alcohol. Notice that the corn stalk has sprouted brace roots above this area. This is the back-up or by-pass plumbing system functioning to keep the plant alive. You may find that the node where these first brace roots sprouted is also brown and hard. In this case, a second layer of "by-pass" brace roots has sprouted. Brace roots are the plant's rescue response when the plumbing system below gets plugged. If the plant did not sprout brace roots, it would die. In this case, as David Larson and others have reported, covering the brace roots, through cultivation, can increase corn yields by 5 to 15 bushels per acre. This allows the brace roots to contact the soil and, consequently, to interact with nutrients. In some areas the soil conditions are so toxic that the brace roots curl or burn off upon touching the soil surface. Compare the conditions of the plumbing systems of corn stalks between fields and note the differences in corn quality and refractometer readings.

Cut the corn stalk horizontally. The stalk should be round, not oblong or teardrop shaped. Out-of-roundness is an indication of a calcium deficiency. Look at the veins. Ideally, they should be pearly white and packed so tightly that it is difficult to count them. The pith should also be pearly white. Pull off the leaves from the stalk and notice that at each node, on opposite sides as you progress up the stalk, there is a baby or embryonic ear. This represents the plant's true potential, which will be realized only when the soil is regenerated to the point that the plant's plumbing system remains clear throughout, and the roots have an aerobic zone of 10 to 12 inches in which to grow.

Squeeze the juice from the stalk next to an ear and take a refractometer reading. If the brix level is 8 or above and maintains this reading for 24 hours a day, there will seldom be any noticeable damage to the ear silks by adult rootworm beetles. However, if this reading drops below 8, there will be progressively greater silk damage as the reading gets lower and lower. It is important to make sure that the reading is a "true" reading and not one in a dehydrated condition, which would give a false impression. This reading can be a valuable tool in management because, regardless of the beetle population, if the reading in the stalk next to the ear is 8 brix or above throughout the day and night, spraying an insecticide would be unnecessary and a waste of money. Another factor to determine concerning these beetles is the amount of pollination that has been completed. If the ear is 90% pollinated, spraying a pesticide for silk cutting beetles would also be unnecessary and a waste of money.

To determine the amount of pollination that has occurred on an ear, peel off the ear shuck, being careful to leave the silks undisturbed. Once the shuck has been removed, shake the ear to see how many silks fall off and how many remain attached to the ear. Silks that remain attached are yet to be pollinated. Those that drop off have been pollinated.

SOYBEANS

Dig up a soybean plant. Look closely for the rhizobium nodules. They look like little puff balls attached to the secondary roots. There should be many nodules. Take a knife and cut open a nodule. Its internal color should be pink. If it is some other color, the rhizobium are not fixing nitrogen in this nodule. Cut open the stalk and root of the plant as you did with the corn stalk. They should be pearly white and solid. Check the brix reading of the plant and/or leaves. Check the node spacing; nodes should be spaced every two inches or so. What insects are present, and what are they doing? Check also for hardpan, aerobic-zone depth, and so on.

ALFALFA

Dig up an alfalfa plant. Check the same items you did for the other crops. Unfortunately, if you are checking a typical alfalfa field, you will be lucky to find any rhizobium nodules; if you do, they will probably be brown or green inside rather than pink. This is due to the compacted, nonaerated, toxic conditions that often are found in conventional alfalfa fields. Fertilizing alfalfa with potassium chloride will almost guarantee poor rhizobium nodulation. The nodules need oxygen and calcium to multiply.

In addition, check the number of new sprouts coming from each crown. A dozen or more healthy sprouts should be coming along. The stems of alfalfa should be solid; this primarily accompanies higher brix readings and calcium. Notice whether there are black spots on the leaves and stems, especially the lower leaves. Such spots indicate that excess potash is being released into the plant. Notice whether there are small weeds growing up between crowns. This indicates low energy and nutrient imbalance. Keep the energy and especially calcium up, and alfalfa will remain solid and vigorous for at least eight to ten years.

SUMMARY

Although every crop has unique features, the basic observations listed in this chapter nevertheless apply to every crop. These factors are solid stems, round stems, refractometer readings ideally above 12 throughout the plant and throughout the day, pearly white stem pith and veins and roots, free-flowing plumbing systems, full root systems to 8 to 12 inches deep with extensive root hair and rhizosphere development, numerous and internally pink rhizobium nodules on legumes, glossy sheen on the leaves, an 8- to 12-inch aerobic zone in the soil, minimal or no weed pressures without herbicide applications, absence of insect pests except on weeds, absence of the previous year's undigested crop residue, absence of hardpans and soil compaction, uniformity in growth and fruiting, uniform fruit size and shape, stable soil temperature at 72 F., conductivity-meter readings between 200 and 600 ERGS during the growing season, sodium levels below 70 parts per million, presence of birds, presence of fresh scents and absence of undesirable odors, and presence of a vital feeling.

It has wisely been said that the best fertilizer a farmer can apply is his footprints in the field. Get in touch with your soils and your crops. If you are too busy to get in the field and evaluate your crops, you are either farming too many acres or misappropriating your priorities, much like the doctor who doesn't have time to listen to his patients to learn their "real" problems but hastily prescribes a medication and issues a bill.

Field monitoring is a management issue that encompasses a philosophical concern as well. The more attuned you are to your crops and soils, the better able you will be to affect them positively through your farming practices. In addition, you are managing your fellow human beings' sustenance. If you entrusted your tractor to a neighbor, how would you want him to care for it? Society has entrusted you with their food production, children's play area, recreational zone, and environment. How are you taking care of it?

The doctor who cares enough to learn about his patients and, consequently, is able to help them has a steady flow of new patients as a result. Likewise, the farmer who cares enough to learn about his crops and soils will have an ever-bountiful harvest.

PRACTICAL BIOASSAY TESTING

Traditionally, Western civilization has depended greatly on empirical laboratory testing to develop theories and draw conclusions. Most of this testing is done not only in vitro, but also with mechanical instrumentation that is totally removed from biological reality. People incorrectly assume that such testing represents real life biological responses, when in actuality it only delineates indicators of possible biological responses. The belief that a machine can serve as a biological model has stemmed from the allopathic notion that nature is flawed, irresponsible, and self-destructive. As explained elsewhere in this book, this is a linear model with its corresponding linear logic of execution in what is actually a nonlinear system. Nature, of which man is a part, is nonlinear in function and by design. Consequently, somewhere in our testing and evaluation we must implement a truly nonlinear test, called a bioassay. Put simply, in a bioassay, a living organism is used to perform the evaluation. Any farmer can easily do three such bioassays.

The first bioassay involves the use of a refractometer, growing plants, and a hand-spray bottle containing a nutrient mix. This test is explained in the chapter entitled Foliar Spray Programming, in which the use of the refractometer in determining foliar sprays is described. To reiterate, one simply checks the refractometer reading of the crop in a small test area, sprays the crop test area with the mist bottle with the predetermined number of squirts to equal actual field-application rates, waits 30 minutes to an hour, and rechecks the refractometer reading of the crop test area. If the reading rises by 2% brix or more, the nutrient spray is significantly beneficial to the crop. This mix can then be applied to the entire field with confidence that it will be of benefit. Here the plant, not some machine or textbook standard, is indicating whether a particular nutrient mix is beneficial. After all, the plant ultimately determines at harvest time whether the fertility management was suitable. In this manner, we are able to "ask" the plant what is best for it.

The next bioassay to be discussed is the earthworm bioassay. This test can be used to evaluate the quality of compost, soil, fertilizers, magnetic influences (magnets, pyramids, round towers, stone-circle medicine wheels, pulsed signals), and water, as well as the influence of pesticides, chemicals, and drugs. It can be used to evaluate the effect of cooking, microwaves, and processing on food, or any other amendment to soil or food that might be experienced. Obviously, earthworms are not people, but our digestive systems and that of the soil depend on microorganisms and enzymes. Earthworms are good surrogates for determining potential hostility to these important digestive microorganisms and enzymes. Earthworms are not parasitic like pinworms, flatworms, roundworms, or leaches; rather, they are an integral intermediate part of the desired soil digestive cycle. Therefore, the response of earthworms to various environments accurately represents desired biological compatibility. Even chemical agriculturalists consider earthworms to be indicators of desirable soil conditions. There will come a time when soil fertility will evolve beyond the point where earthworms are a necessary part of the cycle. This might take some time, but it will occur when the energy concentration of the soil is balanced beyond the need for earthworm intervention.

The actual earthworm bioassay is quite simple. Purchase four containers, generally one cup each, of earthworms. These can be crawlers, redworms, or whatever is available. Experiment with all of them. Examine the medium in which the worms are packaged. Notice its color, odor, texture, and any other definable characteristics. Take the worms out of the medium and wash them in distilled water. Be gentle and prevent drowning. Weigh them. Replace the worms in the medium and original containers. Add a measured quantity of any given fertilizer mix (one ounce of a field concentration) to one container. Add a different concentration of the same mix to a second container. Add the same amount of distilled water to the two remaining containers, which will be the controls. One ounce of liquid may be too much for optimum moisture content of the container medium. One may choose to use only one-fourth to one-half ounce of liquid [mix or distilled water] so as not to oversaturate the medium. Add one-half teaspoon of raw cornmeal to each of the four containers to feed the worms. Place the containers in an unirradiated, nonelectrified area at room temperature for 7 to 30 days. At the end of the test period, open the containers and note the appearance and behavior of the earthworms. Are they bunched together, active, lethargic? What is their odor and skin texture? Note the characteristics of the medium. Is it dry, wet, rancid, fresh? What is its texture?

Compare all of these characteristics with those noted before starting the test.

Remove the earthworms, wash them gently in distilled water, and weigh them. Calculate the weight loss or gain of each container of earthworms. If the test solution is beneficial to the earthworms, there will be no weight loss, and the percentage weight gain will not be less than that of the controls. If the product is toxic or hostile to earthworms or desirable soil microorganisms, there will be less weight gain as compared to the controls, or even weight loss. In some cases, the earthworms might even have died.

Refer to the Test Solution Formulation insert at the end of this chapter for information on mixing test solutions.

The earthworm bioassay also can be used to evaluate the soils themselves. Wash and weigh the worms as before, only this time put field soil in one of the test containers. Add one-fourth to one ounce of distilled water and one-half teaspoon of cornmeal as before. Keep two controls as before and complete the test as previously described. If the earthworms in the test container gain less weight than the controls, the soil is noticeably toxic. You can carry this test even further by putting different fertilizer solutions with the soil in additional test containers. The results might prove very revealing.

A good earthworm bioassay for every household would be to evaluate the effect that cooking has on food. For one container of earthworms, cook the cornmeal conventionally. For another container of earthworms, cook the cornmeal in a microwave oven. For another, cook both the distilled water and the cornmeal in a microwave oven. Treat the controls as before and complete the test. Upon collecting the final data, you can draw your own conclusions. You could also place a test container inside a pyramid, next to or under a model round tower, or inside a model paramagnetic stone circle or on its easterly radian and complete the experiment as you did the others.

Earthworm bioassays are very revealing. However, it is important to keep in mind that although these tests are done with a living system, that system has been removed from the earth and its magnetic field—the actual natural habitat of the earthworm.

The third bioassay to be discussed here uses microbial cultures in petri dishes. First, culture a known desirable soil microorganism in a petri dish. Soak small paper circles in various concentrations of test solutions such as fertilizers, pesticides, and well or irrigation water. Place these paper circles on top of the culture. Incubate for 72 to 96 hours and check the growth immediately around the paper circles. If the test solutions are hostile to this microorganism, there will be a clearing around the circle. This type of test is done with antibiotics on target pathogens and is called the Tollen's test. It also can be used to evaluate the effects of fertilizers and pesticides on soil microbes.

With the use of these bioassays, the farmer, consultant, or student can gain valuable insight into the reality of biology and agriculture. Many things will be revealed that are otherwise obscured or ignored in typical mechanical or chemical tests.

Collect the information and integrate it into your farming practices. Both biological and economical sustainability is the guideline to keep in mind. Bioassays will inevitably reveal information that expands on the data revealed by typical chemical and mechanical tests.

TEST-SOLUTION FORMULATION

One hundred pounds of actual N as anhydrous ammonia (122 pounds) applied per acre to the soil with one acre-inch of water (27,200 gallons × 8.33 pound/gallon = 226,576 pounds) would equate to adding 3,333 pounds of 3% household ammonia per acre or a 1.45% solution of household ammonia. To duplicate this concentration for the test, mix 1.45 ounces of household ammonia in 98.55 ounces of distilled water and apply one ounce of this to one container. If you have a gram scale, mix 1.45 grams of household ammonia in 98.55 grams of distilled water and apply one ounce (28 grams) of this to one container. For an equivalent application of 200 pounds of actual N as anhydrous ammonia (244 pounds), mix 2.9 ounces of household ammonia in 97.1 ounces of distilled water and apply 1 ounce of this to the second test container. Label each container.

If you want to evaluate a different fertilizer's effect on the biosystem, mix 2.48 grams of potassium chloride (KCl) in 97.52 ounces of distilled water and apply one ounce of this to one container of earthworms. Then mix 4.96 grams of KCl in 95.04 ounces of distilled water and apply 1 ounce of this mix to another container of earthworms. These concentrations are equal to 200 pounds and 400 pounds of KCl per acre-inch of water, respectively.

Regardless of the fertilizer or product used, a standard dilution parameter is required. Because we want to correlate test results to field application, by convention we assume a one acre-inch quantity of water for diluting whatever material application rate per acre is used.

THE MEANING OF pH

Scientists, researchers, and laymen alike long have used pH to evaluate the relative acidity and alkalinity of solutions—from water to soil, blood to urine, solvents to pesticides. In agriculture, pH has been given immense importance. Unfortunately, many educators and agronomists' misuse of this concept has led farmers to believe that pH is the major factor to consider in soil testing. This erroneous belief has resulted in severe mineral imbalances and deficiencies in soils and crops.

pH has been represented inaccurately as a measurement, in particular, of the quantity of calcium in the soil. This misrepresentation has resulted from a lack of understanding of what pH truly is.

Technically, pH is the negative log of the hydrogen ion concentration ($pH = -\log [H^+]$). The pH scale, which ranges from 0 to 14 with 7 being neutral with respect to water, is based on the ratio between acid and alkaline. This is best explained using, for an example, water (H_2O), which consists of two hydrogen atoms and one oxygen atom. Pure water naturally separates into low concentrations of two ions, which will carry a positive or negative charge. The hydrogen ion (H^+) will carry the positive charge, and the hydrogen/oxygen pair, called a hydroxyl ion (OH^-), will carry the negative charge. When these positive and negative ions are in a one-to-one ratio, the water solution is said to be neutral, having a pH of 7 because the H^+ concentration is approximately $1/10^7$. If there are more positive (hydrogen) ions than negative (hydroxyl) ions, the solution is said to be acid, having a pH less than 7. If there are more negative (hydroxyl) ions than positive (hydrogen) ions, the solution is said to be alkaline, having a pH greater than 7. The key to pH is the ratio between acid (H^+ equivalents) and alkaline (OH^- equivalents), as measured by the H^+ ions.

pH is not a measurement of the strength, volume, or quantity of an acid or base. To illustrate, let's consider a few common acids and bases. Most people are familiar with sulfuric or battery acid. Farmers in California inject sulfuric acid into their irrigation water to lower the pH. This acid is very dangerous to handle and will burn holes in one's clothing quite rapidly. If you were to take just 4.9 grams (there are about 28 grams in 1 ounce) of sulfuric acid and mix it in 995.1 grams of distilled water so the total weight was 1,000 grams, or a total volume of 1 liter (a liter equals 1.057 quarts), you would have a pH of 1. If you were to use hydrochloric acid instead of sulfuric acid (hydrochloric acid is the same acid as found in your stomach), you would need only 3.6 grams of acid in 996.4 grams of distilled water to reach a pH of 1. If you used phosphoric acid, which is commonly used in liquid fertilizers and soft drinks, you would need 9.8 grams in 990.2 grams of distilled water to reach a pH of 1. All three of these acids have nearly the same pH, yet because the strength of their acidity differs, it takes different quantities of each acid in water to achieve the same pH.

Looking at the other end of the pH scale, it would take 1.7 grams of ammonia in 998.3 grams of distilled water, 0.34 grams of sodium hydroxide (lye) in 999.66 grams of distilled water, and 0.056 grams of potassium hydroxide (common in liquid fertilizers) in 999.944 grams of distilled water to achieve a pH of 11. All three of these products have similar pH's, yet the strength of their alkalinity differs. It takes different quantities of each product in water to achieve the same pH. Thus, it can be seen that a pH of 1 or 11 does not automatically indicate what material or what quantity of that material is present.

This information helps explain why measuring the pH of a soil solution does not, in and of itself, reveal anything about the calcium (or any other nutrient, for that matter) content of the soil. Calcium carbonate (lime) can raise the soil pH, but so can magnesium carbonate, sodium, and potash. Western soils often have high pH with low calcium availability.

When a soil pH is taken, it often is compared with the cation-exchange-capacity reading to determine the liming needs in the soil. This method is unscientific, as can be seen by the previous information on pH. Many materials can alter the soil pH, and each will be found in a different quantity for the same pH. The problem with many soils centers on this point. The use of salt fertilizers along with high-salt irrigation water raises the pH of the soil solution. This creates a situation in which calcium is not applied because the pH is near neutral, 7 or higher. This, in turn, results in calcium-deficient soils and plants, leading to a chain reaction of nutrient deficiencies and soil problems. Anhydrous ammonia will initially raise soil pH sufficient to denature bio-proteins (humus complexes), suppress or kill microorganisms either directly or by altering cell membrane dielectric characteristics, and reduce nutrient availabilities. As the ammonia degrades either by oxidation to NO_3 or reduction to N_2 it acidifies the soil by adding hydrogen ions to the soil solution. If the ammonia were used up rather than degraded it would not lower soil pH. The fact that its use eventually lowers soil pH is partial testament to the fact that the ammonia is not being used up in plant growth.

pH is a result of the interaction of all nutrients, minerals, and microorganisms in the soil. It is not an indicator of the quantity or balance of these nutrients, minerals, or microbes. An example of this is the heavy application of triazine herbicides. These herbicides seem to tie up phosphates in the soil, making them unavailable. Phosphate tie-ups raise the soil pH. This phosphate tie-up phenomenon has been demonstrated in test fields where the corn in the half of the field that was sprayed with herbicide had a refractometer reading of 1 to 2 brix lower than the unsprayed side. Refractometer readings correlate directly to available phosphate necessary to manufacture sugar. Calcium is an important mineral in detoxifying such herbicides, but because the industry practice is to apply calcium according to soil pH, not enough calcium is applied to counter the herbicide.

Nutrients and compounds in the soil that are considered alkaline include calcium, magnesium, chlorine, sodium, potassium, salts, ashes, and aldehydes. Their alkalinity is "relative," however, meaning that if you add an item that is less alkaline than whatever else is present, the pH may be lowered even though you added an alkaline material. For example, adding calcium to a high-magnesium soil may actually lower the soil pH. Substances in the soil that are generally considered acids include humic acids, organic acids, phosphoric acid, ascorbic acid, and sulfuric acid. As with alkaline materials, their acidity is relative, meaning that if you add a substance that is less acidic than whatever else is present, the pH may be raised. It is important to understand these principles in order to comprehend that soil pH is actually a minor factor in evaluating a soil and a subsequent fertilizer program. pH is useful in preliminary evaluations of drinking water and, in some cases, animal urine. The pH of drinking water should be somewhere in the 6.4 to 6.8 range. The further it is out of this range, the less palatable the water will be for the animal and the less water it will consume. Some experts report that animal urine can be evaluated with a pH meter. The pH of cow urine should be around 7.4. If the pH is much higher than this, there is a possibility that the rumen is malfunctioning, allowing too much free ammonia to pass into the blood. If the pH is too low, the rumen possibly is not functioning properly because of too much acid, which may inhibit nutrient assimilation. This often results from too much acidic feed, like grains and silage, in proportion to hay. Consequently, the cow does not chew her cud enough to produce sufficient saliva to buffer the rumen pH.

Remember, pH is the ratio between acid and alkaline materials in a solution. pH does not indicate the quantity of calcium or the available quantity of any nutrient. It can be used as an aid in evaluating what effect various minerals and materials, such as salts, aluminum, potash, chlorine, magnesium, calcium, and pesticides, are having on the soil relative to acid/alkaline reactions. In any event, pH is an effect, not a cause.

Carey Reams viewed pH as a measurement of the resistance in the soil. To him, a high pH indicated that there was high resistance, causing a restriction in energy-nutrient flow—analogous to the soil being constipated. A low pH indicated that there was too little resistance, causing the energy-nutrient flow to be too fast for plants to capture—analogous to the soil

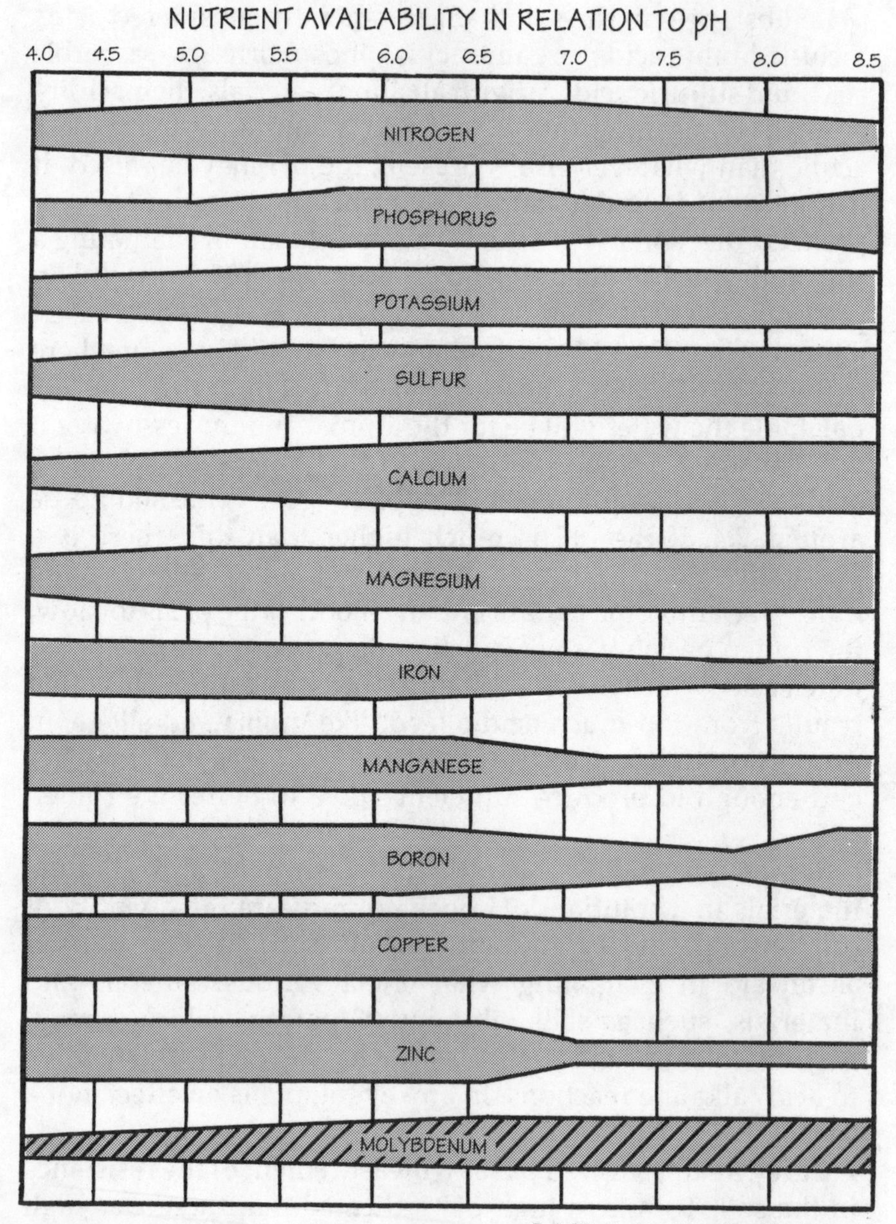

Ref.: Illinois Agronomy Handbook 1979-80

having diarrhea.

Carey Reams also used pH loosely as an indicator of whether the soil nutrient solution tended to stimulate plant growth or to stimulate fruiting. If the pH was alkaline, he considered this a favoring of plant-growth energy. Conversely, if the pH was acid, he considered this a favoring of plant-fruiting energy.

In considering the pH of soil and making statements about the availability of various nutrients (see following chart), many people will refer to the rule of thumb presented in most soil science textbooks, which states that heavy metals become toxic at low pH and tied up at high pH. This is correct and accurate in chemical and test-tube agriculture. However, when humic acids and biologically sequestered or chelated compounds enter the picture, the pH scale becomes invalid. Biological compounds interact differently than do their nonbiological counterparts. What may be true of iron availability at a pH of 8 using iron sulfate is not true of iron availability at a pH of 8 using molasses, iron humate, or iron biochelate. Once one understands biological agriculture, pH becomes a secondary consideration and concern.

THE CONDUCTIVITY METER

Soil conductivity is usually thought of as a measure of dissolved salts in the soil solution. However, the conductivity reading of a soil can give additional information.

Carey Reams used a conductivity meter to evaluate the "energy available for plant growth" at any given time in the soil. The conductivity meters we use measure in µmhos (micromhos). The new international standard term is µS [microsiemens]. Reams termed this measurement "ERGS" (energy released per gram of soil).

During the growing season, the soil ideally should have a conductivity reading between 200 and 600 ERGS. With a reading below 200 ERGS, the crop will not have sufficient energy to grow to its full potential or to maintain satisfactory refractometer readings. If the conductivity reading is above 600 ERGS, energy is possibly being wasted and is perhaps causing root damage and soil dehydration. Generally speaking, a conductivity reading above 1,200 ERGS would prevent most crops from growing successfully. Exceptions seem to exist in arid regions and greenhouse situations, to a point, yet net conductivity levels must be within this range for optimum results.

As a general rule, soil conductivity should gradually increase throughout the growing season, peaking during the time of fruit or grain filling. Ideally, it should be about 200 at planting and peak between 500 and 600 ERGS during the milk stage of grains or the green-fruit stage of fruits and vegetables.

The conductivity reading can be altered through various management practices. If the soil conductivity is less than 200, the farmer can apply a fertilizer or aerate the soil. Aeration will help raise the conductivity of wet soils by helping to dry the soil, which effectively increases the concentration of nutrients. If the soil conductivity is greater than 600 or 700, the farmer can irrigate to dilute the soil nutrients or apply a fertilizer, i.e., gypsum, hi-cal lime, sugar, humic acid, or a biological or compost product, which will buffer the excess salts that are causing the high readings.

As the farmer regenerates his soil, the conductivity reading will stabilize within the 200 to 600 ERGS range. However, periodic applications of fertilizer will be required to maintain soil balance. As the humus level of the soil increases, so will its water-holding capacity. This can dilute soil-nutrient concentrations and lower the ERGS levels. This situation simply requires rebalancing soil-nutrient levels and has the additional benefit of increasing production.

Using a conductivity meter is quite simple, as is using the other recommended meters (pH, sodium, and ORP). With all of them, take one-half glass of soil and one-half glass of distilled water, and mix gently so the soil is thoroughly wetted without breaking up the particles. Turn on the meter, lower the probe into the solution, and read the meter display. This evaluation of the soil should be done weekly throughout the growing season and especially before and after a rain, irrigation, cultivation, or fertilization. Keep accurate records and notice any differences relative to the time of day the readings are taken.

The conductivity meter also can be used to evaluate dilutions of tank mixes. This requires a meter that can measure in the hundreds of thousands of micromhos. Lower the probe into the tank mix before topping off the tank, and add water until the reading peaks. This will maximize the effectiveness of the fertilizer by giving you the ideal fertilizer dilution to maximize the energy level. Remember, it is energy that grows crops. The greater the amount of energy you can derive from a fertilizer, the greater will be its effectiveness in the field.

You can also check milk with a conductivity meter. According to some authorities, cows' milk should be about 5,700 ERGS. Above 6,800 ERGS, there are probably mastitis problems.

Distilled water should have a conductivity of less than 4 micromhos. Water used for foliar spray mixes should ideally be distilled or have a conductivity in the same range as distilled water. As a compromise, the lower the conductivity of the water, the better it will serve in the foliar mix.

In conclusion, soil conductivity, along with refractometer readings, should be checked daily in high-value crops and weekly or at least monthly in lower-value crops such as corn and soybeans.

COMPOST AND COMPOSTING

Composting is an age-old art. It has been occurring since life on earth began. Composting is the process through which organic materials and debris are digested for the purpose of building humus and humic acids in the soil. Many people think that the breakdown of organic matter is the purpose of composting; it is not. That is simply a characteristic of the process of building humus and humic acids. Building humus and humic acids is the purpose of composting. Organic matter can be broken down without producing humus and humic acids. Unfortunately, many people are unaware of this and, consequently, take the process for granted.

Several excellent books are available on composting, containing lists of preferences and procedures. A very good one to start with is *The Compost Manufacturer's Manual*, published by Ehrenfried E. Pfeiffer in 1956.

Some of the more experienced compost artists, like Fletcher Sims, Floyd Ranck, and Russell Durber in America, have developed their compost operations into commercial businesses, using high-capacity, self-propelled material turners. These people did not start 30 or 40 years ago with such machines, but rather have gone to them as a result of high-volume commercialization. The point is that good compost has not suddenly been obtained with the advent of self-propelled compost turners. Some people maintain that individual farmers, most of whom do not possess such machines, are incapable of producing good compost on the farm. It is true that these self-propelled material turners do make commercial production of good compost feasible. However, this equipment is not prerequisite to the production of good compost.

The criteria for good compost production are a good mix of organic materials, i.e., manure, straw, leaves, sawdust, food scraps, and so on, to get a 20:1 to 30:1 carbon-to-nitrogen ratio; moisture sufficient to have 50% to 60% moisture in the compost pile; sufficient pile size to get heating; and sufficient oxygen to have an ORP meter reading between 26 and 28 rH, as recommended by Sigfried Luebke of Puerbach, Austria. The compost pile must be monitored daily to maintain the moisture level and to keep the temperature of the pile between 135 and 145 degrees Fahrenheit without letting it get above 150 degrees.

It is often beneficial to add a composting inoculant to help "seed" the microbial population, especially in light of poor crop refractometer values, less-than-optimum animal nutrition and health, and the commonality of feed and injection antibiotics. All of these factors decrease the quality of the organic material, making it more difficult to establish the desired microbial system.

Good compost will have a distinctly fresh, sweet odor; a crumbly, granular consistency; a high content of active microbes, humus, and humic acid; and a soil-building, nutrient-enhancing effect. Luebke adds small debris like wood chips and twigs to improve the porosity and oxygen distribution at the interior of the pile, resulting in more uniform aerobic composting and improved compost structure; these materials are screened out at the end of the process. Good compost has no identifiable organic-matter residue, ash, or sticky, putrefied pockets. It is nontoxic to earthworms, plants, animals, and soils.

Poor compost can inhibit plant uptake and assimilation of nutrients, particularly potash. It can add aldehydes and toxic compounds to the soil. Poor compost may contain viable weed seeds and disease organisms, whereas good compost neutralizes such items as it is being made.

Ideally, compost should not be left open to frequent rainfall because the aerobic microbes in the pile will drown. However, some people make good compost out in the open.

Certain individuals insist that good compost cannot be made using a front-end loader or a skid loader for a turning machine. This might not be the easiest or surest way to turn composting piles, but some ingenious and innovative farmers are making good compost, using loaders to turn it. Good farmers/managers get the job done and make good compost; poor to mediocre farmers/managers have problems making good compost even with the ideal equipment.

A few people believe that compost should be allowed to heat to 170 degrees Fahrenheit, as it naturally will if it is unmanaged or unturned. They contend that such a temperature is necessary to kill all undesirable components in the pile, thinking that neutralization of pathogens is strictly a thermal mechanism. In reality, neutralizing pathogens is partly a thermal and partly a microbial process. Pathogenic organisms are largely eliminated by the presence of aerobic-desirable microbes and the environment they create. As for weed seeds, maintaining a temperature of 140 degrees Fahrenheit for several days is sufficient to neutralize them.

Further, if the compost pile is allowed to heat past 150 degrees Fahrenheit, the delicate humus and humic acid complexes will be degraded. Although breakdown of organic matter will be accomplished, this type of compost will not optimally build humus and humic acids in the soil or contribute greatly to the microbial regeneration of the soil. Good compost goes through digestion, not cooking.

Recommended techniques to use in evaluating composts are the earthworm bioassay test described in this book and the cress-seed germination test.

More complete information on composting can be gained from visiting a commercial operation, reading Luebke's forthcoming book on the subject, and gaining hands-on experience. Common sense, coupled with a visit to a successful farm composting operation, should be sufficient for most farmers to set up their own composting operations. It need not be an expensive undertaking or necessitate the construction of an extravagant new machine or contraption. Just duplicate what already works!

—20—

COMMON-SENSE PRINCIPLES FOR ANIMAL AGRICULTURE

ANIMAL PRODUCTION IS AN INTEGRAL part of world agriculture and an important aspect of a sustainable agricultural system. Animals need not be directly consumed for food to warrant this important place in world agriculture. Their production of milk, wool, and manure alone justifies their position. In meat-eating cultures, the additional uses of animals for protein, leather, and numerous by-products are important economic factors. As with people, the health and production efficiency of animals is directly related to their diet, comfort, and contentment. The basic provisions of fresh air, uncontaminated water, and a clean, dry place to lie down are fundamental to any sustainable operation.

The writer has emphasized numerous times in this book that the quality of our food chain is directly related to the nutritional balance of the soil. Supplementing the nutrients in animal diets is never as effective in providing the necessary nourishment as is supplying those nutrients in the feed itself through proper fertilization of the soil.

Modern livestock operators regularly add large quantities of nutrient supplements to animal rations to compensate for the poor-quality grains and forages grown by most farmers. They subsequently add antibiotics to the feeds to increase assimilation of the substandard feed and to try to ward off disease. This is termed high-tech feeding, but in reality it is a time bomb ready to explode. Most animals that are slaughtered in America are in very poor health; they are slaughtered for meat only slightly before they would die from ill health. Upon post-mortem inspection, many of these animals have been found to have toxic livers and gall bladders and highly inflamed digestive tracts, which are sure signs of feeding environments that are highly stressful and imbalanced in terms of minerals. Factory-farm advocates contend that this is the price of a burgeoning world food demand, but that it is of no consequence to the consumer. However, common sense indicates that with all systems, whether biological or inanimate, the rule still holds: Garbage in, garbage out. The irony of the situation is that this junk-food, drug-obsessed system of animal production is far from being efficient, economically sound, or ecologically sustainable, let alone humane.

Slaughter animals are raised for food and by-products. To be food, meat must provide nutrition. An animal cannot synthesize nutritionally sound flesh if it is not provided with the nutrients to do so. Illness is caused by nutritional deficiencies, and no drug on the market can correct these deficiencies. As a result, the flesh eaten by consumers will not provide the expected nutritional sustenance. Add to this deficit the drugs given to the animals, as well as the drug-resistant microbes as a by-product, and you have not only further degenerated the food value of the flesh but also added a potentially hazardous foreign material to it.

As a consequence of such unenlightened livestock management, many animal health problems that were rare or nonexistent in the past now are becoming more prevalent. Chlamydia, which was rare in unvaccinated hogs 20 years ago, is devastating today unless the animals are routinely vaccinated. Dr. Dan Skow observed that this trend over the past 20 years was directly correlated to the use of anhydrous ammonia in crop production.

In today's commercial corn, molds and their resultant mycotoxins are the rule rather than the exception. This problem can be traced directly to inferior soil management, and it causes real problems in feed assimilation, rumen function, and animal health.

The longevity of dairy cows today is, on average, only 2.5 lactations. This situation has been traced directly to nutrient-deficient feeds harvested from poorly fertilized soils, insufficient dietary roughage, and excessive dietary grain.

Despite the sophisticated computer software on feed rations, the reams of research data, and the flowery testimonials, most farmers, feed salesmen, and "animal scientists" have little or no idea how the digestive system is intended to function. Few understand the differences between the digestive systems of birds, swine (pigs), horses, and cattle. For a lesson, observe how and what animals in the wild eat.

Birds have a crop and a gizzard, which enables them to digest whole grains; they are the only animals that can do so and remain healthy. Grains, seeds, and nuts have enzyme inhibitors, which make them difficult for animals other than birds to digest. The way these inhibitors are inactivated is by cooking, which destroys the natural enzymes and vitamins; by sprouting, which enhances the nutritional value of the seed; or by soaking in water for 48 hours, which also initiates the production of sugars.

Most livestock feeders feed hogs as if they were birds and cattle as if they were swine. Such feeding causes severe metabolic stress in the animals, thereby creating favorable conditions for the various pathogenic organisms. This, of course, necessitates continuous drug intervention. What is meant by feeding swine as if they were birds and cattle as if they were swine? Many farmers feed swine with a diet of very concentrated grain that has almost no fiber. Swine should have about 7% soluble fiber in their feed rations, and protein levels between 12% and 18%, depending on the quality of the protein, fiber, and mineral components. Calcium-phosphorus ratios in the 1.2:1 range seem to work well, again depending on the quality of the overall feed ration. Too often there is mold in the feed, so amino acids and vitamins must be further reinforced by supplements.

Many cattle farmers feed very concentrated grain rations, with only minor amounts of fiber. This, of course, adds weight to beef animals and increases milk production for a short time. The physiological effect of this, however, is the gradual stoppage of proper rumen function, which results in inefficient or nonexistent nutrient digestion/absorption and metabolism; this subsequently leads to numerous health problems, including foot rot, pneumonia, retained placentas, and mastitis. These symptoms are often attributed to infectious pathogens that can be curtailed through modern drug therapy. Fundamentally, though, the problem that needs to be addressed is poor feeding management combined with even poorer feed quality resulting from mismanagement of soil fertility.

The dairy ration today is much more complex than it once was because of the current use of industry by-products, poor-quality forages, and more stressful conditions, such as water and air pollution. It used to be standard practice to formulate a dairy ration with a 2:1 calcium to phosphorus ratio, but this no longer is done. One must consider the availability of soil nutrients as well as the solubility and quantity of fiber, true protein, amino acids, soluble starch, vitamins, salts, and so on. Feed quality today simply is not what it once was. Much of the fiber in typical alfalfa is insoluble, and much of the protein is incomplete; many of the minerals are out of balance. These conditions all must be factored into the feed-ration equation and compensations made so that the animal is satisfied. Total mixed rations (TMR) are desirable in most confinement dairy operations because the rumen and stomachs can be better balanced, considering current grain and forage quality. With TMRs, one must not get soluble starches too high and avoid straight corn or corn-only grain components. A mixture of grains should be used. The addition of flax seed to the grain mix provides an excellent source of alpha omega-3 fatty acids vital for immune-system function, fat metabolism, and production of anti-inflammatory prostaglandins.

The best way to feed grain to ruminant animals is to sprout the grain, combining it with coarse forage with high refractometer readings. Because this allows the animal to chew extensively, the proper amount of saliva is produced to initiate food digestion and buffer the rumen naturally.The digestive system of ruminants is really a bioreactor, and it must be managed as such. Most "modern" feeding schemes seem viable on paper or the computer screen, but animal health and longevity statistics have shown that most of these schemes do not work in practice. Modern feeding systems are being used in an attempt to produce "test tube" meat and milk. Little or no consideration is given to the complete food value of the production, the longevity or health of the producing animal, or the ramifications of a drug- and chemical-based food supply. No drug or chemical can provide sustenance to the consumer.

The best way to learn about animal nutrition is to observe animals in the wild. Wild animals eat fresh and raw foods. Ruminants in the deer family that chew their cud eat forage almost exclusively; they are vegetarians. Yet "modern" feed-ration formulators are feeding ruminant animals animal fats and animal proteins because these items seem viable on paper. The problem is that the rumen does not perform inside the computer; it functions inside the animal. For the ruminant digestive system to function properly, the microorganisms that reside in it must be nurtured properly and remain content.

This requires proper pH balancing and appropriate foods. In actuality, one is feeding the microorganisms; this, in turn, results in the feeding of the animal. The same is true for the soil.

Let's consider how a ruminant digestive system works. Food is initially chewed and passes into the rumen or first stomach. Here the ruminant microorganisms react on the food for partial digestion. The partially digested food is regurgitated and chewed some more. It then passes into the remaining digestive compartments in sequence—the reticulum, omasum, and abomasum—where it is further digested before passing into the intestinal tract. Once the food is in the intestinal tract, the nutrients are absorbed through the villi on the intestinal walls. These villi are loaded with blood vessels. It is these villi that are affected by the feeding of antibiotics. Antibiotics make the villi membrane walls more permeable or porous so more nutrients can pass into the blood stream. However, this also makes it easier for pathogens and toxins to enter the blood stream, resulting in greater blood toxicity, liver and kidney stress, and chronic subclinical and clinical infections.

Perhaps the most prominent nutrient contributing to chronic ill health in animals is protein. In every facet of human, animal, and soil nutrition, the "experts" proclaim the benefits of protein. University agronomists are fond of using heavy applications of nitrogen on the land and crops. Dietitians, body builders, and animal nutritionists have a similar propensity to advocate large portions of protein. There is no denying that nitrogen—protein—will produce "results." What is ignored is the quality of the outcome and the price paid in longevity and quality to achieve the apparent results.

To explain this contention, we must examine what constitutes protein in the eyes of the "experts" and what protein really is and does. When a commodity is tested for protein in a laboratory, it is run through a Kjeldahl test, which expels the nitrogen contained in the sample. This nitrogen is then weighed. The weight of the nitrogen is multiplied by a factor of 6.25 or 6.4, and the product is labeled "percentage protein."

No consideration is given to the completeness, quality, or functionality of the protein, only the percentage of nitrogen that is present. Because the integrity of the protein is not considered, neither is its effect on the consumer's health. Superficial performance is the only criterion of interest. Whereas the crop might green up, grow profusely, and increase in volume, underneath the surface it is loaded with nitrates, attacked by insects, full of mold and disease, void of full nutrient balance, and has a low refractometer reading. Although cows might produce large quantities of milk, they burn out in two to three lactations, and their milk is full of pus and has a low refractometer reading. The cow develops a high blood urea nitrogen (BUN) level, which causes increased stress on the heart. The added energy that the cow expends in attempting to deal with the excess nitrogen actually results in a negative urea- fermentation potential, which is a net loss of energy from digestion. The cow develops a fatty liver, reproductive organs, and spleen because this excess nitrogen—protein—does not allow the liver to metabolize fats, starches, and sugars properly. Consequently, fats precipitate in the various organs. The most prominent symptom in cows that have been given excessive nitrogen is breeding problems.

It has been argued that ruminants can digest free nitrogen because of the ruminant bacteria. Whereas they can handle some free nitrogen, that does not mean their bodies are designed for free nitrogen. Simply observe a cow in the pasture. Having the choice, she will not eat the grass around cow dung, which is lush green. She knows instinctively that it is loaded with free nitrogen and thus is unhealthy. Where in nature do ruminants regularly consume high-protein, free-nitrogen food? They don't. For the ruminant bacteria to convert free nitrogen into true protein, they need energy, and energy comes from carbohydrates. With no feed having a refractometer reading (measurement of carbohydrate and dissolved solids) at or above 12, what is the energy source? The fiber? Fiber from forage with low refractometer readings has very little energy, so, again, what is the source of the energy?

There is none. Whereas the animal performs, just like the family car would with airplane fuel, she experiences accelerated aging and degeneration; in essence, she gets burned out. The process by which this happens is shown in the diagram on the following page.

Compounding the problem and resulting in even less NH_3 digestion by rumen bacteria, which means more NH_3 goes directly into the blood, is the grain-favored ration. Grain is chewed only minimally; thus, a negligible amount of saliva is produced. Saliva is necessary to buffer the rumen, to keep the pH in the range that is favorable for the desirable rumen bacteria. Grain acidifies the rumen, thereby lowering pH; this, in turn, decreases rumen bacterial activity, and less NH_3 is processed.

To make matters worse, some of the increased supply of acid goes directly into the blood, especially when antibiotics are fed to make the intestinal wall more permeable. The acid then migrates to the liver, causing further reduction in liver function and resulting in a chain-reaction decline in the entire digestive system. In addition, starches are not digested properly and form acids or alcohol.

Some nutritionists advocate feeding cattle alcohol as a quick energy source. That it is, but it has very detrimental effects. Alcohol suppresses rumen bacteria. It also causes calcium to precipitate and thus become unavailable. When the alcohol enters the blood, it also precipitates blood calcium (resulting in plaque build-up or hardening of the arteries); alcohol further stresses the liver, precipitating calcium and causing cirrhosis of the liver. Animals that are fed alcohol are certain to need more mineral supplementation which is convenient if you are selling both.

As mentioned earlier, when the rumen goes acid, such ailments as foot problems and mastitis appear. Mastitis supposedly is caused by numerous microorganisms such as staph, strep, and *E. coli*. However, these microbes are naturally prevalent inside and outside the body; so how do they become pathogenic and out of control? They are normally kept in check by the beneficial microbes, particularly lactobacillus, which need specific conditions in which to thrive, in order to keep the other microbes in check. When these conditions are altered too much by imbalanced nutrition or antibiotics, the lactobacillus decline and the potential pathogens become actual pathogens.

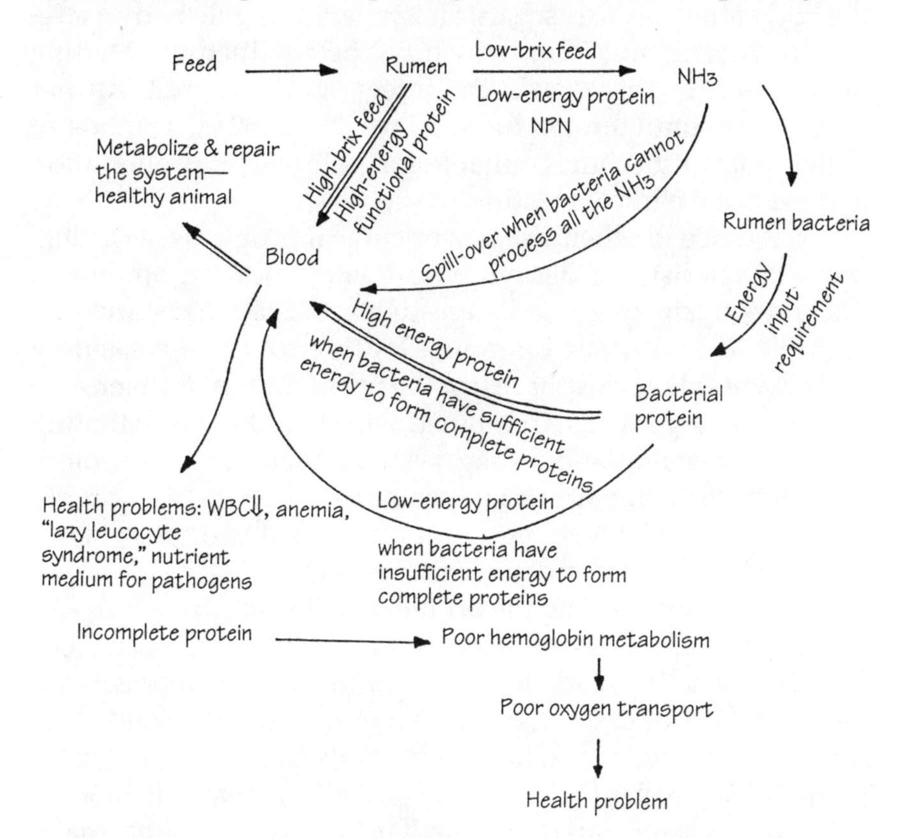

The liver robs energy from the metabolism to purge the blood nitrogens, resulting in a fatty liver. This results in other fatty organs, heart stress, reproduction problems, NPN in the meat and milk, and kidney stress.

This situation is easily demonstrated, using milk as an example. Under healthy circumstances, the refractometer reading of milk should be 20, the pH 7.4, and the conductivity 5700 micromhos. Under the conditions and type of feeding described above, insufficient usable energy goes into the cow, so there is not enough energy in the milk (correlated to a lower refractometer reading) to maintain the concentration and chemical integrity of the milk. (To clarify the significance of this, bacteria do not readily work on straight molasses because of its concentration, but when diluted it breaks down rapidly.) Next, when there is excess nitrogen in the blood (electrolytes), the body attempts to discharge it in every possible body excretion, including milk. Consequently, the conductivity reading of the milk increases and the integrity of the milk further deteriorates until the conductivity reaches 6800 micromhos, at which point it becomes suitable for pathogenic proliferation and eventual clinical mastitis.

In reference to other diseases or clinical problems, a distinctive characteristic of decaying flesh is the odor of ammonia. The previously described digestive conditions resulting in high blood and tissue ammonia levels provide increasingly more favorable atmospheres for the proliferation of anaerobic-pathogenic organisms. Combined with the lower functioning liver and metabolite-building system, tissue becomes more and more in disrepair; waste products accumulate in highly stressed areas like feet, legs, and reproductive organs; and a habitation is set up for disease organisms.

Further compounding the problem is the fact, proven by Dr. Philip Callahan, that ammonia pumps the IR, radio, and UV signals emanating from the degenerating system, intensifying the attraction of disease or breakdown organisms and flies. Also, the elevated blood ammonia levels can result in nitrates in the blood, both of which (NO_3 and NH_4) lower the blood's oxygen-carrying capacity, further reducing the animal's natural defenses and energy levels. All of these symptoms can be attributed to improper diet and poor feed quality, causing rumen malfunction. In an attempt to prevent the chain of events outlined above, nutritionists add sodium bicarbonate to the feed to buffer the rumen so that it does not become too acid. However, sodium bicarbonate is not saliva and therefore does not contain the ptyalin and other enzymes necessary for digestion of starch and other compounds. Further, $NaHCO_3$ will only partially buffer the rumen. In doing so, carbon dioxide is evolved. The more $NaHCO_3$, the more CO_2 that is evolved. If the CO_2 mixes with water, carbonic acid is formed, thereby adding more acid to the rumen. If CO_2 enters the animal's blood, which it readily will do, it can displace oxygen and cause suffocation. Short of suffocation, the elevated CO_2 in the blood causes liver stress and lowers blood pH, which in turn causes reduction in the digestive system due to reduced blood efficiency and the cycle just outlined. When properly fed, a cow should and will produce 25 to 30 gallons of saliva per day. That amount of saliva contains the equivalent of five to six pounds of the natural bicarbonates to buffer the rumen. When the "safe" level of feeding sodium bicarbonate is roughly one pound per day, it is obvious that adequate rumen buffering is not possible with sodium bicarbonate.

A number of other maladies are associated with digestion, particularly rumen malfunction. As the blood becomes loaded with more and more debris, such as NH_3, CO_2, and so on, the animal's body attempts to eliminate this waste in every possible body excretion. One place it dumps the waste is in the lungs. As in the mammary glands, this extensive dumping causes debris build-up. As debris, especially anaerobic gases, builds up in the lungs, congestion sets in and the ideal atmosphere for pneumonia pathogens is put in place and pneumonia develops. Again, the cause is a malfunctioning rumen.

Three maladies commonly occur in dairy operations. Listeria and yeast mastitis are best kept in check by Lactobacillus microbes, which are suppressed or killed by digestive upsets and antibiotics.

Candida, commonly controlled by lactobacillus even in human systems, thrives in the excessively acid conditions brought about by high-grain diets in ruminants. The problem is exacerbated by the mycotoxins produced by the Candida. Mycotoxins, like aflatoxins, are produced by molds in moldy feeds, compounding the suppression of desirable rumen bacteria. Hybrid grains, because of their narrow nutrient preferences, i.e., nitrogen and potash, are possibly more susceptible to molding.

Ketosis occurs when glycogen is not metabolized (manufactured) properly; as a result, ketone (R-CO-R) bodies are formed in excess, especially at high-stress times like calving. Ketosis is commonly treated by intravenous administration of glucose and propylene glycol. Some researchers are investigating the use of calcium peroxide for the treatment of ketosis. Again, however, the cow's inability to metabolize glycogen stems from liver malfunction, which is a result of rumen malfunction. Such malfunction at this time means the dry cow has not been properly fed.

This leads into a discussion of the animal's immune system and transference of health to its offspring. The immune system is what keeps biological systems healthy and self-sustaining. In animals it includes the lymph system, spleen, thymus, white blood cells, antibodies, and immunoglobulins. Immunity begins with digestion. A healthy digestive system will destroy a majority of the potential pathogens introduced into the animal. In addition, a healthy digestive system keeps the immune system supplied with the nutrients necessary for building healthy white blood cells, lymph, and the various metabolites needed for a fully functional glandular system.

Oxygen is an important nutrient in maintaining the immune system, as well as every other system of the animal. Pathogens generally cannot tolerate oxygenated environments. Hemoglobin is responsible for oxygen transport, and to build hemoglobin the body needs iron, copper, and protein. Unfortunately, lactating cows and animals under stress are often borderline anemic; that is, they have an iron deficiency. A primary cause of the iron deficiency is that the digestive system is under-functioning because the diet is too acidic and contains too much nitrogen. This results in an energy crisis for the liver, a deficiency in true protein, and an inability to metabolize vital components. One of these components is bone marrow, which is responsible for red blood cell and stem cell synthesis. Stem cells are precursors of immune cells. Without proper digestion, the animal is unable to assimilate the necessary nutrients to build fundamental components like hemoglobin, bone marrow, stem cells, and lymph. Without this capacity, the animal cannot sustain itself, let alone pass on optimum immunity and immunity-building constituents to its offspring.

Nutrition and digestion are the precursors of immunity, but when the immune system is suppressed the process of rebuilding must begin with detoxification. Detoxification must start with the elimination of toxic feeds. High-nitrogen feeds, heavy metals, molds and mycotoxins, drugs, and inappropriate ration balance (too much concentrate for ruminants) are all toxins to be removed from the animal. As these are purged, appropriate nutrition will help reestablish an effective digestive system, which then will provide the necessary components to rebuild the immune system and the rest of the body.

Many people think that vaccines can compensate for a defective immune system. Unfortunately, vaccines work poorly in toxic animals and may overload the white blood cell population if the animal is sick. In such cases, vaccination is of little help and may even compound the problem. Vaccines mask symptoms and keep the causes of those symptoms hidden. Ultimately, common sense and responsible management are required for sustainable livestock and crop agriculture.

Feed quality remains a keystone in animal health management, and the foundation of feed quality is soil fertility. Storage, processing, and blending are also important factors influencing feed quality. A common feed storage facility on farms is a round steel bin. Because these bins get uneven sun exposure, the south sides of the bins heat up, and condensation forms on the cooler north side, establishing an environment conductive to mold growth. Sufficient air flow through the bin can minimize mold growth, but it is not always present. The poorer the nutritional balance of the feed, the more difficult it is to prevent such mold growth. High-nitrogen fertilization seems to make the feed susceptible to mold growth and storage deconditioning. Most animal feeders attempt to eliminate the mold by diluting it (mixing moldy feed with clean feed). However, this leads to mold contamination throughout the feed.

Another practice that can lead to feed contamination is the use of mill dust. Mill dust is inexpensive, readily available, and looks good on the computer. The problem is that it is loaded with mold spores and numerous unknowns. Thus, using mill dust is not worth the risk.

Next is the issue of using round bales. Round bales are acceptable if they are stored inside, but if the bales are stored outside the outer 4 to 12 inches mold from exposure to rain and snow. This moldy portion may not appear to be a significant part of the bale or of the overall animal ration, but it does not take much moldy feed to stress the animal significantly. Dilution simply does not rectify the problem.

This same issue arises with regard to silage. Most crops grown today are so low in nutrients that they ensile very poorly; pockets of mold and spoilage result. It has been claimed that sealed silos like the "blue monuments" seen across the Midwest can eliminate such problems. Many people seem to think that, by sealing silos, good silage can be made from poor inputs. In actuality, the silage coming out of the silo will not be any better than the forage or grain that went into the silo. Silage bags seem to perform effectively and can handle direct-cut alfalfa for ensiling.

Fermentation requires energy and moisture to produce the lactic acid necessary to preserve the feed. The higher the energy (refractometer reading) of the feed, the more optimum the moisture content (55% to 60%) and hence, the more quickly fermentation occurs; this means there is less time in which spoilage can occur. The following chart, compiled by consultant Dick Donald, is a good guideline for making silage:

Feed Quality	*Comment*
4 brix	No fermentation.
4-6 brix	Poor fermentation with or without inoculant.
6-8 brix	Good fermentation with inoculant. Fair to poor fermentation without inoculant.
8-10 brix	Very good fermentation with inoculant. Good fermentation without inoculant.

10 brix	Good fermentation without inoculant. Improved feed value with inoculant.
20 brix	Optimum feed.

Silage can be improved further by adding 50 pounds of molasses per ton of silage, thus providing iron and a good source of carbohydrate for the microbes to produce lactic acid. Lactic acid is an excellent natural preservative, ruminant energy source, and antibacterial, mold antagonist. Also, adding 10 pounds of *Dynamin* per ton of wet silage helps preserve, detoxify, and nutrify the silage.

It is recommended that feeds be harvested when they are at their peak refractometer reading. In general, the refractometer reading will drop immediately before, during, and immediately after a storm. Thus, feed harvesting should be avoided at these times. Farmers who keep good records can accurately predict the weather by observing the daily refractometer readings of their crops.

Many animal farmers, like nonanimal farmers, believe that feed quality is predominantly a function of protein level and little else. What they fail to recognize is that most alfalfa with a "protein" level greater than 17% does not actually contain that much true protein. It is often urea nitrogen that contributes to the multitude of animal health problems discussed earlier, including fatty organs, breeding problems, suppressed white blood cells, and high blood nitrogen. The exception occurs when the refractometer reading is greater than 12 brix. The key to true protein formation is the carbon-to-nitrogen ratio. The carbon is necessary to carry the energy of the protein, and it comes from carbohydrates.

As mentioned before, digestion is an important aspect of animal health. A key factor in digestion is the level of digestive enzymes available. Ideally, these digestive enzymes would be inherent in the feed. However, if the feed has a low refractometer reading or is cooked, these enzymes are in short supply or nonexistent. In such cases, the ration must be supplemented with enzymes or the animal must synthesize them. The energy and nutrients required for the animal to synthesize sufficient digestive enzymes to compensate for what should naturally be present in the feed must be robbed from the energy and nutrient supplies otherwise used for normal body function, repair, and sustenance. As a result, the animal will have less than optimum production and longevity.

One of the best sources of enzymes is sprouted grains. Not only are sprouts high in nutritional value (of course, the nutritional value is only as good as the nutrient value of the seed), but once a seed is sprouted, the enzyme inhibitors in the seed are neutralized. This allows nonfowl animals to digest the seed. Although cooking can also inactivate the enzyme inhibitors of seed, it inactivates the enzymes themselves. Commercial sprouting for animal feeding is gaining in acceptance. Some French manufacturers produce and market a commercial sprouting system. Seed soaking is an acceptable alternative.

Other items that enhance digestion and nutrition are *Dynamin*, kelp, and whole oats. *Dynamin* is a unique montmorillonite clay, which NASA found aided in nitrate nitrogen and radioactivity decontamination. In addition, polypeptides (which enzymes are) line up better on clay, allowing protein formation to occur more readily than when the clay is not present.

Kelp or seaweed is a natural source of iodine, and it also is loaded with trace and rare earth elements along with vitamins and enzymes. Kelp also helps to chelate out heavy metals. The best kelp sources are those that are sun dried rather than artificially dried.

Oats are an outstanding feed. High-quality oats can replace as much as 25% of the grain in rations. They provide both soluble and insoluble fiber, giving the animal what is called "rumen lag property." This allows the rumen to catch up with the soluble proteins and starch.

Livestock management is a multifaceted business. Common sense, good observation, and compassion are the foundation of life-sustaining food production. Regardless of the management techniques the livestock manager employs, he must eventually address and improve the nutritional value of the feedstuffs as they come out of the field. The animal is no healthier than its feed, which does not surpass the nutritional integrity of the soil. A 20-brix milk requires 20-brix feed.

Some general comments and suggestions regarding livestock management are provided in the following pages.

- When making a general evaluation of dairy animals, several characteristics should be considered. The animal's coat should be shiny, smooth, and fine. If it is dull and scruffy, there is often a parasite problem and a digestive disorder. Adding diatomaceous earth (DE) to the feed ration seems to alleviate the parasite problem. DE also works well as a dusting powder for lice.
- In many herds, one will notice that animal coats have a red hue; this is indicative of copper deficiency and can usually be addressed by adding dynamate to the ration. Dynamate is the feed name for sul-po-mag, which helps make copper available. If the dynamate does not solve the problem, copper probably needs to be added to the feed. This should be done with great caution and under advisement from a competent consultant.
- Another characteristic to evaluate is the tail-head position. If it is characteristically raised in comparison to the overall smoothness of the animal's backbone, this indicates nutrient imbalance and is symptomatic of breeding problems and uterine infections. As the tail-head position raises, the pelvis tilts, and vice versa, so the reproductive organs cannot drain properly and remain healthy. One can attempt to medicate the symptoms, but no medication will correct the nutritional imbalances that have caused the problem.
- One can tell rather quickly whether a dairy cow's feed ration is satisfactory. Three ribs should be showing on her side as an indicator of good flesh composition. If more than three ribs are showing, she is receiving insufficient nutrition and is losing flesh. If fewer than three ribs are showing, she is too fat, and nutrition is going into retention rather than production. Fat-cow syndrome can result in breeding problems, ketosis, fatty organs, and, of course, poor economic efficiency.

• Perhaps the most important characteristic to evaluate when the other attributes seem to be acceptable is whether the animal appears content. Is the animal hyper or lethargic? Does she hold her ears up, or are the ears drooping? Contentment is an important attribute for getting maximum economic return from a dairy animal, or any animal for that matter.

• Hydrogen peroxide has been purported to aid animal nutrition and health. A little is good, but if one uses 20 or more ppm H_2O_2, the "good" as well as the "bad" microbes will be killed. Also, H_2O_2 in this quantity can decrease milk fat. Some research is being conducted using CaO_2 (calcium peroxide) rather than H_2O_2 because it is more stable. Lactobacillus microorganisms produce hydrogen peroxide, as well as lactic acid, B-vitamins, and other metabolites that are valuable to the nutrition of animals and soil, as well as to the inhibition of pathogenic proliferation.

• Mycotoxins, produced by molds, cause breakdown of the collagen (vitamin C) that holds cells together. As a result, the body tissue loses its tone and takes on a doughy quality.

• Using one pound of Gentian or Ultra Violet per ton of feed has been shown to kill molds in feed. The FDA has approved its use in feeds for feathered animals. Unfortunately, the politics of the situation seem to be favoring its removal from the United States market. The bottom line is: It works too well. It also is available in capsules for deworming dogs. The old teat remedy, *BlueKote*, was a Gentian Violet product. Beware of methyl violet, though; it is hazardous.

• Liquid manure can be an extremely toxic waste in some livestock operations. Animal diet greatly affects the quality of the manure, as does the amount of drugs administered and fed to the livestock. Liquid manure can contribute significantly to the salting of the soil and therefore should be composted or treated in some manner to promote digestion of the manure. Biological digestion of the manure sequesters salts out of solution into biological metabolites. In other words, the salts are transformed from undesirable compounds into desirable biological nutrients.

• When there is high somatic cell count (pus) in the milk, one should first check and correct problems with stray voltage, proper vacuum, and pulsation. Always check these while the cows are being milked, when everything is turned "on." After correcting these items, use individual towels for washing and drying the udder and teats. Then nutritionally raise the levels of anti-oxidants in the diet, using vitamins A and E, beta-carotene, selenium, zinc, manganese, copper, and cobalt. One might also consider injections of Mu-Se (available only from veterinarians) at 7-10 cc plus vitamin C at 50 cc and vitamins A, D, B_{12} at 7 cc.

• If your calves are weak or lethargic, consider an iron injection at birth. If feeding a commercial milk replacer, supplement it with amylase enzyme for digesting the starch. Preferably, during gestation, give the mother a good blood builder including copper, cobalt, folic acid, and complete proteins to prevent an anemic calf. Keep in mind that good air movement around the baby calf is necessary to curtail pathogens. Interestingly, certain viruses can travel only inches in a 20% oxygen environment, but they can travel hundreds of feet in a 19% oxygen environment.

• The practice of intensive rotational grazing is regaining popularity and sophistication. With proper management of both the pasture nutrition and the grazing rotation, maximum production from both land and cattle can be obtained, as well as optimized animal health. An excellent plant to include in the pasture mix is comfrey in 6 x 6 grids, which adds nutrition and healing substances to the pasture. Protein levels of 25% to 30% are reasonable with good soil nutrition as well as a more balanced calcium-to-phosphorus ratio (2:1) than alfalfa has.

• Herbs are very effective materials in both adding nutritional value to animal feeds and providing medicinal therapeutics to health care programs. One of the few animal feed herbologists in this country is Jerry Brunetti, with Agri-Dynamics in Martins Creek, Pennsylvania. He has practical experience and gives the following suggestions.

Herbs excellent for:

Blood, lymph, and liver. Red clover blossoms, burdock root, dandelion root and leaf, yellow dock roots.

Digestion. Gentian root, fennel, comfrey, peppermint.

Respiration. Mullein, elcampane, coltsfoot, horehound, mints.

Bowel infection. Thyme, goldenseal, comfrey leaf, slippery elm bark, nettle leaves, red raspberry leaves.

All of these herbs are commonly found growing on the farm or are available in capsules from Agri-Dynamics and Trans-National Agronomy. A good book to read on this subject is *The Complete Herbal Handbook for Farm and Stable* by Julliette de Bairacli Lerg.

• An interesting development for cattle feed is the growing of corn sprouts to 12 to 18 inches in height, yielding 25 ton dry weight of forage per acre, having an analysis of 30+% protein, 40% to 50% sugar, 25% fiber, and 70+% TDN, every 30 days for 10 crops per year. This product seems promising as a complete or nearly complete cattle feed. The only question that arises is the need for long-stemmed fiber in order to get sufficient saliva production to buffer the rumen. However, cattle do well grazing pasture not getting long-stemmed fiber, but it must be remembered that grain rations, not the lack of long-stemmed fiber, are the cause of rumen decline and excessive acidity. If you eliminate the acid-causing concentrate and provide high-energy, high-nutrition forage, the rumen will naturally be balanced. Then fiber quality becomes much more important than fiber length. For further information contact International Ag Labs, Fairmont, Minnesota.

• Some Wisconsin beekeepers report that mixing equal parts of TNA *Detox, Nitromike,* and sugar does well to keep bees alive in our current toxic environment.

• TNA *Detox* or a similar product seems to improve livestock health and feed efficiency when added to the feed ration. Other entrepreneurs report benefits from adding a good silage inoculant to the feed ration. These products simply reintroduce beneficial digestive microbes to the system. If the highly acid-forming ration is not corrected, one is still only treating symptoms.

The following are a few remedies the farmer might wish to suggest to his holistic veterinarian for possible use on the farm. No claim is made as to their efficacy. They were simply relayed to the author as possible experiments.

Chronic Mastitis

20cc Immucel from Agri-Dynamics or a comparable product.
20cc blood.

Mix blood into Immucel and inject intramuscularly every five days. This acts similarly to a hemolytic vaccine. In addition, feed the ailing female her own milk to stimulate antibody production to the antigens in the milk.

Fresh Cow

Feed the cow her own colostrum within 24 hours of calving. As it did the newborn, it will provide her with many antibodies and nutrients, including thymus excretion, which will help her regain strength.

Infected Uterus/Cervix Douche

2 quarts total mix
1/3 3% hydrogen peroxide
2/3 distilled sterile water (boil, then wait 24 hours and boil again).
1 tablespoon boric acid
1 tablespoon salt
1 tablespoon epsom salts

Douche to clean out debris and then infuse the uterus with a suitable pro-biotic/colostrum-type product like Uterex from Agri-Dynamics to help heal the infection. Be careful in using products like iodine, which can irritate the uterine lining.

Yeast Mastitis

Infuse Lactobacillus directly into the udder. Cut the excess dietary protein and acid-forming feeds, e.g., grain.

—21—

PHILIP CALLAHAN, Ph.D.

IN THIS BOOK I HAVE REFERRED to several pioneering scientists. Many of them have been misunderstood, others have been vaguely understood, and some have not been understood at all. The problem, I believe, is not these individuals or their work, but rather the assumption that peer approval determines the legitimacy of a scientist and his work.

One of the outstanding scientists of our time is Philip Callahan, Ph.D. Most people have no inkling of the work he has done and continues to do. Callahan is a natural scientist whose lifelong quest has been to understand and explain the physiology of nature. Several industrial applications have been made of his work, many of which the Department of Defense has classified for military applications.

One such application was Callahan's spinning-color-disc experiment, in which he showed that visual light pumps the scents of various targets, food or a mate, thus allowing an insect to locate its target. This finding led to the development of the AN/ALG 144 infrared countermeasures device, which is used on American helicopters to broadcast IR signal patterns that cloak the helicopters by making them appear to enemy attack missiles to be their own aircraft.

Callahan's projects range from deciphering the communications systems of dolphins to discovering why they so readily get caught in tuna fishing nets, interpreting the communications system of the AIDS virus to explain how it functions and finds the lymphocytes, and patenting a simple cure to arthritis, to researching history by participating in such activities as hiking the emigrant trail across the Sierra Nevada Mountains and building old folk medicine replicas. He is also a gifted artist, falconer, ornithologist, entomologist, and speaker.

Philip Callahan's specialty has been biological communications within and between cells, as well as between organisms. Perhaps his most profound statement is that "form and frequency go together." By this he means that every form or shape is keyed or correlated to a specific frequency or group of frequencies, and every frequency or group of frequencies matches or encompasses a specific form or shape. This is really antenna physics. I will not get into the mathematics or theory of antenna physics in this discussion but will consider the principles involved.

Most people think of antennas as metal contraptions that extend from their automobile, the roof of their house, or the mast of a ship. They acknowledge that insects have antennas but never make the connection between insect antennas and mechanical ones. Insect antennas are thought to be for sensing things, not for receiving electromagnetic waves. Sophisticated communications systems like microwaves and fiber optics are thought to be reserved for people's use and certainly to be more sophisticated than anything that exists in nature. In reality, every man-made communication system, no matter how sophisticated, was being used by nature long before man ever thought of building it.

To understand the essence of this discussion, one must comprehend what an antenna is. Once that is understood, we will be able to extrapolate this understanding into other subjects. The encyclopedia says that an antenna is a device for radiating or absorbing electromagnetic waves. Basically, an antenna is any device that intercepts, then guides, and/or focuses electromagnetic radiation/waves of any type, e.g., sound, "light," radio, X-ray, cosmic, mitogenetic.

The most obvious electromagnetic waves are radio, television, and microwave. However, the electromagnetic spectrum is much larger than these three types of radiations. And, as shown in physics, the various divisions of the electromagnetic spectrum are connected and vary only in terms of wavelength and frequency. As such, they all fall into octaves, each of which is harmonic to all the other octaves.

For example, picture the keyboard of a piano. It contains three octaves, all within the audio range of the electromagnetic spectrum. The human voice has harmonics 12 and 19 octaves away from it in the radio-wave range of the electromagnetic spectrum, which has harmonics 14 and 21 octaves away from it in the infrared range, which has harmonics 1 octave away from it in the visual range, which has harmonics 11 and 14 octaves away from it in the X-ray range of the electromagnetic spectrum. The point is that the electromagnetic spectrum is simply like one huge piano keyboard with a virtually infinite number of octaves.

Important to this discussion and particularly to antennas is the principle that antenna design is based on the desire to isolate specific regions of the electromagnetic spectrum most effectively in conjunction with the translator that is connected to the antenna. In other words, antenna design is keyed to the tuner/translator to which we connect the antenna. The antenna allows the specific frequency/wavelength range to have the maximum effect on the translator/tuner/receiver mechanism attached to it. The antenna might be compared to the outer ear and ear canal, which collect sound waves. The inner-ear components (anvil, stirrup, and so on) then translate these waves to a signal the brain can interpret.

Any antenna, regardless of its size and length, intercepts and guides electromagnetic waves in every range of the electromagnetic spectrum, although the specific frequencies and wavelengths within each range will vary between antennas. The same antenna cannot be used for every application because present technology segregates the translation of each electromagnetic-wave range into a usable signal. Different translators are required for radio, infrared, television, and optical waves. It is the coupling of the antenna with the translator that gives the desired results. That is not to say, however, that the infrared or optical antenna is not receiving and sending radio waves. It is, but the radio waves correlated to nature are low frequency and very low power. These are not readily detected by typical radio equipment; rather, a biological intermediary is necessary to sense them. Radio waves are generally effective carrier waves for biological systems, meaning they function somewhat like the train that carries the payload. The base radio wave of all biological systems is the Schuman frequency, between 7 and 8 cycles per second.

Callahan has shown that every shape functions as an antenna in some respect. The key to recognizing this is our ability to translate the intercepted and focused energy to something we acknowledge as significant, e.g., movement of a meter needle, noise, and so on. For several decades, Callahan has demonstrated that biological organisms function as infrared and ultraviolet antennas, knowing that this means biological organisms function simultaneously as antennas for every other electromagnetic energy in existence. With the aid of a simple radio-wave detector and a multiplying oscilloscope, he proved that biological organisms function as radio-wave antennas.

Let's take a moment to discuss some of the frequency interactions found in nature. Flowers come in just about every color and shape one can imagine. Generally, we think of shape as being nature's unique design to accommodate and distinguish specific species of plants. And we think of color as a visual lure for insect pollinators. Botanists usually leave it at that, but upon further evaluation and reasoning, a much more wondrous revelation unfolds. Every shape is significant and correlates to a signal or group of signals it receives and transmits.

People typically think of the flower in terms of its aesthetic qualities, but in reality it is a highly specific reflector lens. Those who understand that energy precedes matter and that a seed is a compact unit of energy (an entire plant's worth of energy) can appreciate that ovum and pollen production, with subsequent fertilization and zygote (fertilized "egg" developing into a mature seed) formation, requires a tremendous amount of energy. The reproductive parts of both plants and animals are the most highly energetic areas of the organism. It is that intangible energy that constitutes life, a neg-entropic process. This energy that is required for vitality becomes concentrated in the seed. The greater the concentration, the greater the seed's viability and vitality, provided there is sufficient biologically active nutrient present to hold the energy intact.

The flower's structure represents a unique focusing apparatus. It provides for any energy of the correct characteristic to be directed or guided to the center of the flower, where the ovum resides and the embryo develops into a seed. The differences in flower shapes enable the specific frequency pattern of each species to be focused and directed to the seed. They are very effective wave guides. This applies to a showy flower like a rose as well as to an obscure flower like a grain.

The color of a flower is also meaningful in the context of biophysics. Nature's propensity is to procreate. For plants, procreation requires pollination. This is accomplished by attracting pollinating insects, which feed on the flower nectar. As Callahan points out, the insect tracks a signal emanating from its target. This signal is the infrared and ultraviolet radiation from the scent molecules of the nectar. The better the plant can attract the pollinator, for example, the honey bee, the more successful will be the pollination.

How can nature improve honey-bee attraction? By amplifying the calling signal. Amplification can be accomplished either by increasing the direct transmission power or by "pumping" or modulating the original signal. The latter method requires less total power output, so it is reasonable to assume that nature would employ this means. This is where color comes in. The flower's color acts as a "pumping" radiation of the nectar's central signal, resulting in a net amplification of the signal tracked by the bee. As John Ott demonstrated clinically, color is an effective force that can alter almost any biological function. Remember, the color is the frequency and wavelength reflected from the flower to "pump" the nectar signal. All the other frequencies/wavelengths are absorbed by the flower, some or all of which are then guided to and focused on the ovum.

By design, nature provides for survival of the fittest/healthiest. This is accomplished very simply. The true color of the flower, as coded in the DNA (genetic material), is realized in direct proportion to the completeness of nutrient availability for the flower. In addition, nutrient completeness determines the net signal emitted by the nectar. In combination, the color signal and the nectar signal become the attractant signal that is tracked by the bee. The bee intuitively knows that it needs nutritionally sound nectar to maintain its health. Therefore, if the bee has a choice, it will track only the highest quality nectar. This explains why some crops have poor pollination, whereas others have good pollination. Unfortunately, when no nutritionally sound nectar is available, the bee must seek the best there is—that with the highest refractometer reading.

The purpose of this discussion is to show that the principles of physics, which apply on the bench and in the laboratory, have their origin in nature. True science is in harmony with nature. They are one and the same. Phil Callahan has spent a lifetime pointing this out and demonstrating that, by understanding this principle, we can solve every problem encountered in industry, agriculture, medicine, religion, energy production, and so on. His research on and explanations of round towers, pyramids, and ancient structures apply principles of physics and antenna theory to the probable intentions and uses of these remarkable works. Nature is fundamentally energetic. The ancients understood this basic principle and constructed mechanical structures/ devices to make use of this natural phenomenon. The structures were tuned antennas, and the soil and plants were the translators. Increased plant vigor and increased body energy were the "meter readings."

God created nature and endowed man with the intelligence to learn nature's design and to apply its principles to his world. However, no man-made force transcends those found in nature, e.g., earthquakes, hurricanes, tornadoes, tidal movement and waves, and the earth's rotation. Understanding these forces allows man to apply them to enhance his environment. Until man understands these forces, he will continue to pollute and rape the planet, expend its resources, contaminate his fellow man, and generally do everything the hard way.

Understanding that everything is an antenna leads to the realization that everything's function depends on energy—its interception, focusing, and translation. From this it follows that the malfunction of any system can be understood by evaluating the antenna characteristics of that system and the result of the energies involved. Callahan demonstrated this in his round-tower, insect, and dolphin experiments, as well as in other investigations. Carey Reams's teachings were based on the same principle—that nature is energetic. It can be deduced, then, that there is no such thing as an energy shortage; the only deficiency is in common sense and awareness.

Callahan's critics have questioned his work because he credits the ancients with having a scientific motivation for their structures. According to the religionists, it is absurd to think that the ancients could have had such a motivation because they perceive that these structures were pagan religious artifacts. Even more peculiar is the religionists' insistence that Callahan and other scientists like him should provide "proof" of the scientific validity of their contentions because they are demanding that such proof be given by the same reductionist scientists who vigorously insist than humans evolved from apes.

As the proverb says, "My people perish for lack of knowledge." The earth seems to be choking and dying on ignorance. It is time to get smart. I strongly recommend that you read Philip Callahan's books, particularly *Tuning In To Nature, Ancient Mysteries, Modern Visions, A Walk in the Sun,* and *Nature's Silent Music.* Only through understanding will we be able to solve the problems of our times. Callahan's writings and research provide remarkable insights into the workings of nature. All of us, who need to coexist on this planet, vote responsibly in public elections, and act intelligently in our interactions with nature, will benefit from these insights!

—22—

ELECTRONIC SCANNER TESTING

RADIONIC INSTRUMENTS, COMMONLY called electronic scanners, are tools. As with any other tool, the success one achieves with it depends on his understanding of the basic principles associated with the topic in question—in this case, agriculture. Experience has shown that students of electronic scanners fare well if they have a fundamental understanding of the teachings of Carey Reams, Dan Skow, William Albrecht, and Philip Callahan, coupled with common sense. Students who are locked into the "kill or be killed" chemical farming mind-set find it difficult to comprehend the information gained through this evaluation system because the chemical mind-set cannot understand farming without toxic chemicals. To evolve beyond this limited viewpoint, it is important to appreciate the mutual symbiosis (cooperative coexistence) present in all of nature. Weeds, diseases, and insects are symptoms of mineral and, ultimately, energy imbalances, not of pesticide deficiencies!

Anyone can learn to operate a radionic scanner; however, some people catch on more easily and quickly than others. Some people experience "paradigm paralysis" and find it difficult or impossible to learn to operate the scanner. Yet this is no different from learning to operate any other tool or instrument. Some people have a similar mental block with regard to using computers, cameras, and common household tools or appliances.

WHAT CAN BE DONE WITH THE SCANNER?

The radionic scanner is used to evaluate the potential interaction between fertilizers and the soil, between seed samples and the soil, between fertilizers and plants or seeds, and between any two items in question. Traditionally, farmers apply fertilizers and wait until harvest to evaluate whether the program was a good choice. The problem with this approach is that, if the program was not a wise choice, the farmer might not be in business the next year for another trial. In addition, each growing season is different. What did or did not work one season might or might not work the next. Every season varies and requires its own unique program if the farmer is involved in regenerative agriculture. If the farmer is simply involved in "mining" the soil along with spraying all the rescue pesticides inherent in this traditional management technique, every season is the same, and the problems of weeds, insects, diseases, soil compaction, and erosion continually worsen.

By using the scanner to evaluate the energy patterns of the soil or plants and potential fertilizers, the farmer or consultant can determine whether there is harmony or disharmony between the two before applying fertilizer. If there is harmony, the program is desirable; conversely, if there is disharmony, the program is undesirable. The product(s) in question might be excellent materials, but inappropriate for use in this situation or at this time. The electronic scanner allows the farmer or consultant progressively to build the soil to a more balanced state, resulting in a reduction in weeds, insects, and disease;

improved crop quality and production; and, ultimately, better nutrition for consumers.

For example, using the scanner allows the farmer or consultant to determine whether sucrose, dextrose, or molasses would be appropriate as a carbohydrate source for the soil microbiology or whether a carbohydrate source is even necessary at that particular time. The scanner can aid in determining whether a particular vitamin, e.g., B_{12}, or a given biological product, e.g., algae or enzymes, would be of benefit. The scanner can aid in selecting the most compatible seed for each individual field and in determining whether a commercial seed treatment would be helpful or whether a custom seed treatment is needed. Perhaps the most valuable aspect of the scanner is its use in solving crop and soil problems. Many such problems cannot be addressed through traditional soil or plant-tissue tests.

USING THE SCANNER FOR PROBLEM SOLVING

A soil test might reveal a deficiency of some nutrient, such as potash or iron, but the test does not tell how to solve that deficiency. The traditional procedure would be to apply the nutrient in which the soil is deficient. The problem with this approach is that the deficiency might not be due to an actual lack of that particular nutrient. Rather, the nutrient in question might be inhibited by something, e.g., a pesticide, or perhaps its availability depends on the availability of another nutrient, e.g., calcium, which is more frequently lacking. Calcium deficiencies often do not show up on soil or tissue-test reports because many people judge the availability of soil calcium by soil pH and tissue calcium levels according to an arbitrary standard, irrespective of refractometer readings. An adequate quantity of calcium might be present, but it might be deficient qualitatively.

Correction of the calcium deficiency might require applying such materials as vitamin B_{12} and white sugar to the soil. This, in turn, could correct the potash deficiency. The scanner allows the operator to make these determinations and solve the "real" problem because he is evaluating the energy patterns of and interactions between the items in question. This allows him to evaluate accurately the status of soil and plant health.

The various types of soil-test devices (CEC, LaMotte, scanner) are all tools to help the farmer or consultant determine the potential status of the soil. The electronic scanner can be a valuable device in identifying a solution to the problem or laying out the course to achieve one's goal. It is important to remember that there are no magic products or solutions, nor will good nutritional programming make up for poor tillage and cultural practices.

HISTORICAL EVOLUTION OF THE SCANNER

The term radionics evolved from the work of Curtis P. Upton in Harrisburg, Pennsylvania, in the 1940s. Along with several other engineers, including General Henry M. Gross, Upton formed a successful agricultural consulting company that was effective in controlling agricultural insect pests throughout America. Upton was meticulous in compiling his research data and did much to solidify the use of radionics in agriculture in this country.

Several patents have been issued on radionic instruments around the world. Perhaps the most noted patent was issued to T. Galen Hieronymus, who received U.S. patent number 2,482,773 in 1949, as well as Canadian and English patents.

The first radionic-type instrument was called a reflexophone; its improved version was termed an oscilloclast. Both instruments were invented by Albert Abrams, a world-renowned pathologist who practiced at Stanford University Medical School in the late 1800s and early 1900s. Abrams's original instruments integrated the medical diagnostic procedure of percussion with a surrogate patient. Succeeding scientists integrated a reaction plate in place of the surrogate patient and diagnostic percussion. This improvement simplified the operation of the instrument, but it did not eliminate the need for a skilled operator. Critics of the use of radionic instruments cite dependence on an operator as a major drawback of these instruments. What the critics do not recognize is that every man-made instrument, from automobiles to computers, is operator dependent.

OPERATOR TRAINING

Training for operators of radionic instruments usually begins with a four-to-six-day instructional period. The first two to three days often are devoted to developing a basic understanding of bioenergetics, instrument function and operation, and fundamental skills. The second two to three days are devoted to agricultural applications such as product evaluation, fertilization programming, problem solving, and feed analysis, with emphasis on basic biological principles. In addition, the various choices of available instruments and their sources are addressed. After completing the basic course, one is usually in a position to decide whether purchasing an instrument would be appropriate. The Radionic Association in England has developed a fairly extensive training program resulting in licensing of professional practioners. Instructors must be competent in both instruction and operation of radionic instruments. Great strides have been made in reducing the incidence of selling instruments without proper instruction. Unfortunately America cannot boast of such a program. The U.S.P.A. in July 1991 began taking steps to "clean up" the American system and be a source of professional radionics. It is suggested that one get instruction from a competent instructor before purchasing and operating a radionic instrument.

SOURCES OF INFORMATION ON RADIONICS

Additional sources of information on radionics:

U.S. Psychotronics Association
2141 Agatite
Chicago, Illinois 60625
(312) 275-7055

Global Sciences Association
3273 E. 119th Place
Thornton, Colorado 80233
(303) 452-9300

Health Research
Box 70
Mokelumne Hill, California 95245
(No telephone number is available; write for an extensive catalog).

The Radionic Association
Baerlein House
Goose Green
Deddington, Oxford 0X5 4SZ
England

The following books are recommended:

Callahan, Philip S. *Nature's Silent Music*. Kansas City, Missouri: *Acres, U.S.A.*, 1992.

Russell, Edward. *Report on Radionics—Science of the Future*. Essex, England: C. W. Daniel, Ltd.

Radionics instrument sources:

Peter Kelly
Dimensional Sciences
P.O. Box 167
Lakemont, Georgia 30552
(404) 782-2524

Sandy Asbill
Agricultural Energizers
Route 1, Box 1843
Tiger Georgia, 30576
(404) 782-2287

Lutie Larsen
Little Farm Research
993 W. 1800 N.
Pleasant Grove, UT 84062
(801) 789-5130

Robert Fridenstine
New Horizons Trust
53166 St. Route 681
Reedsville, Ohio 45772
(800) 726-3808
(614) 378-6155

Mrs. T. G. Hieronymus
Advanced Sciences
P.O. Box 3326
Eatonton, GA 31024
(404) 485-3565

AFTERWORD

I HOPE THAT SERIOUS AGRICULTURALISTS will use this book as a ready reference in their quest for improvement. Although some of the content might seem a bit theoretical, I hope each reader has found at least some beneficial information. Our future is truly in our own hands!

In summary, keep in mind the following major points:

1. Collect data such as soil test results and crop refractometer readings as often as practical. Test the soil at least twice a year—at least once during the growing season. Learn how to interpret the data from these tests, especially what various natural indicators such as weeds, diseases, and insect pests mean and their correlations to soil and plant test numbers. Get out into the field and observe what is happening. Whenever you go into the field, take a conductivity reading of the soil and a refractometer reading of the crop. In the fall or at the end of the growing season, the difference between crop die-down and natural dry-down of grain/seed crops is the soil conductivity and crop refractometer readings. If the conductivity reading drops below 200 micromhos, the crop is running out of energy and will die down if this starving condition remains. Ideally, the refractometer values should be kept above 12. Be careful not to use too much nitrogen to maintain ergs; this will cause an imbalance and call in the insects, especially aphids.

2. Avoid detrimental practices such as using chlorinated fertilizers (chlorine/chloride is diamagnetic) and toxic materials, applying too much of what you already have, tilling wet soil, and overtilling. Once you have decided you want to

improve the soil, it makes sense to curtail practices that cause soil degeneration.

3. Treat the soil as a living digestive system that is in need of balanced nutrition and is susceptible to destruction by poisons, imbalances, and physical destruction. If it is treated correctly, the soil can be your greatest asset. If treated incorrectly, it can be your worst enemy. War is neither fair nor inexpensive. It eventually wears down all of the combatants.

4. Learn basic nutrition so that you understand the values of minerals, vitamins, enzymes, carbohydrates, oxygen, water, and so on. You will then be equipped to evaluate your crop and make viable dietary decisions with regard to its success. If you decide to use rock-dust minerals, as Hamaker suggested, check the rock-dust with a magnet. Ideally, it should be paramagnetic (attracted to the magnet). If it is not, it had better provide some nutrient complex or beneficial characteristic to your soil. Be sure it is free of contaminants. Remember that oxygen is a strong paramagnetic nutrient and that adequate oxygenation of your soil enhances the soil's paramagnetism.

5. Combine sugar and/or molasses with your phosphates and nitrogens to improve their efficiency. Add a carbon source such as compost or humates to powdered lime to keep the lime in the root zone of sandy soils.

6. Regardless of whether you follow an organic or a biological procedure, your success will be reflected in the refractometer reading of the commodity and its freedom from insects, diseases, and weeds. A wormy organic apple is substandard, pesticides or no pesticides.

7. Develop a plan of action and execute the plan. Management is the key to success.

8. Be optimistic. Keep your eyes and ears open. The simplest solutions are often right before our eyes.

9. Read *Soil Microorganisms and Higher Plants* and patent number 2,908,113 for October 13, 1959, as well as the references listed at the end of this book. Note in your study of microbial growth mediums that carbohydrates, e.g., sugar, are almost

always included. The microbe is your most important crop. Grow it well!

There is a grass-roots movement of people who are practicing the principles outlined in this book. Although it currently accounts for only a small percentage of the entire American agricultural economy, this group is well established, profitable, and growing. Large, leading family operations have begun to join this movement, primarily for business reasons. Business economics and legitimate commodity quality will be the norm for future agriculture. World population continues to expand and urbanization continues to remove land from food production. Since 1967, more than 25 million acres of cropland have been converted to urban sprawl, according to the American Farmland Trust. Therefore, production efficiency must improve. Increased urbanization exerts pressure for pesticide and fertilizer control, necessitating nontoxic management. At the same time, this process expands the potential business clientele of nontoxic turf and landscape professionals.

Biological agriculture is the new paradigm, a rekindling and modernization of ancient wisdom. It opposes the current doctrine, but truth often does. Just because chemical agriculture has been practiced since World War II does not mean it is appropriate. For hundreds of years, bloodletting was a prevalent treatment for the common cold. Does that mean it was correct? I think not!

Biological agriculture has stood the test of time and currently endures the scrutiny of nature. Neither of these claims can be made for chemical agriculture. But like every other recurring truth, biological agriculture causes considerable agitation among its opponents. These people need to be reeducated and redirected. Changes come hard for some people and easier for others.

Food, however, is only half of the basic raw material; energy is the other half. Food and energy are the raw materials of prosperity, life, and leisure or the levers of oppression and war.

A "biological" energy is also coming of age, paralleling that of biological agriculture. It entails tapping the zero-point energy of the ether, converting its endless supply of potential to usable commercial, industrial, residential, and agricultural energy. I am speaking primarily of two devices that soon will be sold commercially. They are the Popp noble gas engine and the Hyde generator.

The noble gas engine operates by pulsing an electromagnetic field across a mixture of noble gases, causing an expansion of these gases and a net production of usable electricity as a by-product. There is no pollution and no intake or exhaust, as typically expected. Its operating temperature is 130 Fahrenheit. A 400-horsepower engine weighs 400 pounds and will operate for 4,000 hours before the noble gas capsule needs recharging, at an estimated cost of $500. The gas is not consumed but merely leaks out over time. Combine the noble gas engine with the Hyde generator, which produces greater than unity above a given rpm, and you have a four-plus-megawatt electrical generator at a fuel cost of $500 per 4,000 hours. The noble gas engine has patent number 4,428,193 for January 31, 1984, and originally came out of the Nazi war machine. The Hyde generator has patent number 4,897,592 for January 30, 1990, and was invented by a General Motors engineer.

The curious reader will find patent number 2,006,676 for July 2, 1935, an intriguing revelation regarding the practical, safe conversion of water to hydrogen and oxygen for automotive fueling.

Further, we have the technology to produce the biomass and subsequent ethanol for less than $.70 per gallon. Government, news media, and puppet scientists claim that these inventions won't work, that we must have oil and more of it to supply our energy needs. That is a blatant lie. These people need oil for their continued profits, at our expense. Only through public awareness and subsequent demand will these "white" energy sources become the societal norm.

Integrate biological agriculture with its nontoxic turf and landscape management, and combine nutritionally sound, abundant, profitable food and fiber with abundant, clean, economical energy, and we have the ingredients for a future we can anticipate with excitement. Dare you be a pioneer in this quest?

PEER REVIEW AND POLITICS

It is ironic that would-be scientists insist on seeing new discoveries and work printed in peer-review literature because they really have no understanding what they are asking. Pioneers have no peers and certainly no peer publications to publish their work. When Bruno suggested that the earth revolved around the sun, he was put to death by his peers. Galileo was threatened with torture by his peers for suggesting the same thing. Simmelweis's peers ran him out of his homeland for suggesting that physicians wash their equipment and hands between patients. Nikola Tesla was laughed at by his peers, including Thomas Edison, for suggesting that alternating-current electricity ought to be the electricity of the day. Although Tesla patented more than 1,000 inventions, his works in "free energy," resonance, and biophysics are still ostracized in the peer literature. Albert Abrams was considered a genius until he demonstrated a cure for cancer and other diseases thought to be incurable; then his peers labeled him a madman. Wilhelm Reich was jailed by his peers for his work in orgone energy and cancer therapy.

Peer review is actually political review, designed to determine whether the work alienates the monopoly. Are nonastronauts peers of astronauts? Are nonpresidents peers of presidents? Are nonpioneers peers of pioneers? I say, No. Pioneers have no peers except other pioneers. The emphasis on peer review should be secondary to results in the field. It is in the field that farmers, gardeners, and landscape "doctors" are either made or broken.

Statistics are another flag commonly waved by many classroom agriculturalists. There are volumes and volumes of statistics that supposedly validate modern chemical agricultural practices, yet the system is still failing. Statistics have the inherent flaw that they represent only what the researcher wants them to portray; this information is often skewed from reality. If I surveyed all the alfalfa fields in America, I would probably find that 99 out of 100 had hollow stemmed alfalfa. From those statistics, I would conclude that the hollow-stemmed alfalfa was normal and the solid-stemmed alfalfa abnormal. In reality, hollow-stemmed alfalfa might be common, but it is undesirable/abnormal compared to optimum alfalfa. Solid-stemmed alfalfa is uncommon, but it is normal for healthy alfalfa to have solid stems. In addition, the refractometer values of the hollow-stemmed alfalfa will be significantly lower than those of the solid-stemmed alfalfa, so according to our statistical data, alfalfa should have low refractometer values. This we know is incorrect because alfalfa should have refractometer values above 12.

According to statistics, weeds, diseases, and insect pests infest crops regardless of the nutritional balance (according to conventional testing established by statistical research) of the soil and crop. This information is used to justify the continuous call for pesticide use in agriculture and the lie that Americans would starve if pesticides were not used. The reality in the field is that pests are directly correlated to a nutritional-balance threshold, below which these pests eradicate the crop and above which they leave the crop alone. Simply because the majority of the agricultural "scientists" (data collectors) in this country are unable to achieve or surpass this threshold does not invalidate the threshold. If you personally are unable to run a four-minute mile, does it invalidate the fact that it is possible for a person to run a mile in four minutes? If you are unable to make music with a piano, does it invalidate the fact that music can be made with a piano?

Agricultural authorities would like us to believe that because they have been unable to achieve nutritional thresholds in soils and crops at or above which no pest pressures occur, where yields are at record levels, and quality is unsurpassed, it simply cannot be done. Research data verifying the achievements of many "real-world" agriculturalists are needed, not to benefit the researcher or the customer because they already acknowledge the validity of the new paradigm, but to assist those who are unable to conduct such research themselves. Farmers, homeowners, and small business owners are purchasing biological products and services because they work in the field, not because there are volumes of research data sanctioning them.

There are volumes of research verifying the position of biological agriculturalists in the works of Callahan, Steiner, Albrecht, Northern, Senn, the Soviets, and others, yet it is ignored by the Land Grant University agriculturalists. Neither agriculture nor society needs the inhibition of progress so that the old guard can reinvent the wheel. Saving face is an ego trip we can ill afford, and unless agricultural institutions shed that arrogance, admit their misguided feats, and participate in viable agricultural science, they are obsolete, deterrents to progress, and an unnecessary burden on the public pocketbook. The fundamental question they need to address is: Are you going to continue to teach a lie, or are you going to participate in the solution?

POLITICS

Conventional agriculture claims to be scientific. Then why does conventional agriculture . . .

1. Ignore the works of Callahan, Becker, Popp, and Kaznacheyev in biophysics, who repeatedly have proved that all living systems are fundamentally energetic?

2. Ignore basic principles of chemistry concerning the interaction of compounds, the meaning of pH, the use and value of humic acids, and the formulation and manufacture of fertilizers?

3. Ignore biology and refuse to acknowledge that proper nutritional management solves the very problems conventional agriculture attempts to circumvent by means of genetic engineering, e.g., insect-resistant crop varieties?

4. Ignore basic geology relative to the interaction of soil particles, minerals, and humus and their correlation to soil

tilth, compaction, and hardpans?

5. Ignore basic ecology in their often-indiscriminate applications of toxic poisons, overuse of leachable fertilizers, and apathy about soil erosion and environmental integrity?

6. Ignore the volumes of research documents in microbiology, proving and reproving the biological characteristic of the soil and the necessity of its maintenance for sound farming?

7. Ignore the basic business-management principles of maintaining sustainability, keeping records on quality, and maximizing self-sufficiency on the farm?

8. Ignore the fundamental common-sense precept, which is to follow the path of least resistance and acknowledge nature as the scientific model?

9. Ignore British research showing that nonacidized, rock phosphates are far superior to high-analysis acid phosphates in long-term farming systems.

10. Ignore Soviet research showing that natural beneficial soil microorganisms can completely control soil-borne disease and pest organisms if they are provided the proper nutrition and conditions to do so.

11. Ignore research by T. L. Senn at Clemson University on the value and use of seaweed as a fertilizer and on the characteristics and uses of humic acids in conjunction with fertilizers.

12. Ignore the extensive use of humic acids by European farmers, for at least 15 years, to enhance the efficiency and reduce the leachability of chemical fertilizers.

13. Sanction and perpetuate the obscuring and demoting of William Albrecht's landmark work in soil science, as well as his forced early retirement, in order to secure substantial financial grants from a major chemical company for research having a predetermined outcome contrary to Albrecht's documented work.

Conventional agriculture claims scientific integrity. However . . .

1. Since World War II, American farmers have increased their use of agricultural pesticides tenfold—to about one billion pounds (500,000 tons) per year, yet crop loss due to

agricultural pests has doubled.

2. Soil erosion is occurring at 20 times the rate of natural replenishment, even faster than during the Dust Bowl, which occurred before the chemical Green Revolution.

3. More than 50% of our ground waters, lakes, and streams have been contaminated, some beyond use, with agricultural poisons and fertilizers.

4. Pesticide-resistant weeds, diseases, and insects abound and are increasing in number. The farm population is declining and aging. Agriculturalists' awareness and understanding of farming sustainably, profitably, and without the use of toxic chemicals is scanty in most and nonexistent in many areas of the United States.

Are these traits of good science, sound farm business management, and common sense? Absolutely, unequivocally No! These are traits of an agricultural system held captive by special-interest groups and petrochemical exploiters. It is an agricultural system held at arm's length from true science, farm business management, and common sense, by a "religious dogma" readily exposed for what it really is by true science, sound business management, and common sense. I dare say that there is not one university agricultural department in this country that can raise any crop consistently over 12 brix at its weakest point or that has any clue as to the nutritional management necessary to do so. Yet there are farmers all across this country with little or no college education who routinely achieve such results.

The motto of conventional agriculture seems to be analogous to what the old Sicilian Mafia accountant said when asked what one plus one equaled: "What do you want it to be?" Thanks to true scientists like Philip Callahan, T. L. Senn, William Albrecht, and many others functioning primarily incognito within the conventional system, the answer to "What does one plus one equal?" is returning: "Exactly what nature intended it to be!"

REFERENCES

Andersen, Arden B. *Applied Body Electronics.* Andersen Publishers: Stanton, Michigan; 1986. "Managing Electromagnetic Technology to Aid in Soil Regeneration." In *Management of Technology II: Proceedings of the Second International Conference on Management of Technology* (p. 1269). Industrial Engineering and Management Press: Norcross, GA; 1990.

Bearden, Thomas E. *Gravitobiology: A New Biophysics.* ADAS: Huntsville, Alabama; 1989.

Blake, Rev. John L., D.D. *The Farmer's Every-Day Book.* Derby, Miller & Co.: Auburn, North Dakota; 1851.

Callahan, Philip S. *Ancient Mysteries, Modern Vision. Acres, U.S.A.:* Kansas City, Missouri; 1984. *Tuning In To Nature.* Devin-Adair: Old Greenwich, Connecticut; 1975.

Coles, Robin. *Pasture Cockchafers—On the Way Out? Acres Australia* (Adelaide, South Australia), Issue 3 (1990): 43.

Devyatkov, N.D., Ed., *Applications of Low-Intensity Millimeter Wave Radiation in Biology and Medicine,* IRE Akad. Nauk. SSSR, Moscow, 1985.

Drilling Fluid Engineering Manual. Dresser Industries, Inc.: Houston; 1977.

Felleman, Hazel. *The Best Loved Poems of the American People.* Doubleday: Garden City, New York; 1936.

Hamerman, Warren J. "The Musicality of Living Processes." *21st Century Science & Technology* (March/April 1989): 32+.

Krasil'nikov, N. A. *Soil Microorganisms and Higher Plants.* Academy of Sciences of the USSR: Moscow; 1958.

Leet, Don L., Sheldon Judson. *Physical Geology.* Prentice-Hall, Inc.: New York; 1954.

Lillge, Wolfgang. "New Technologies Hold Clue to Curing Cancer." *21st Century Science & Technology* (July-August 1988): 34+.

McCaman, Jay. *Weeds: Why?* Author: Sand Lake, Michigan; 1989.

Mercier, Charles, M.D. *A Manual of the Electro-chemical Treatment of Seeds.* University of London Press: London; 1919.

Nieper, Hans A. *Revolution in Technology, Medicine, and Society.* Druchhaus Neue Stalling: Oldenburg, F.R.G.; 1983.

Palmer, Lane. "Seven Wonders of American Agriculture." *The Farm Journal* (1965): 38+.

"The Polar Solar Phenomenon." *Compressed Air* (February 1990): 8+.

Reich, Wilhelm. *The Cancer Biopathy.* The Orgone Institute: New York; 1948.

Siegel, Bernie S. *Love, Medicine & Miracles.* Harper & Row: New York; 1986.

Skow, Daniel L. *Mainline Farming for Century 21. Acres, U.S.A.*: Kansas City, Missouri; 1991.

Sources of Mud Problems. N.L. Industries, Inc.: Houston; 1985.

Borehole Instability. N.L. Industries, Inc.: Houston; 1985.

Walters, Charles Jr. "The Last Word." *Acres, U.S.A.* (July 1991): 42. *Weeds, Control Without Poisons,* Acres U.S.A.: Kansas City, Missouri; 1991.

Watson, George. *Nutrition and Your Mind.* Harper & Row: New York; 1972.

Yule, John-David. *Concise Encyclopedia of the Sciences.* Van Nostrand Reinhold: New York; 1982.

SOURCES

PAGE *NOTE*

CHAPTER 1. PARADIGMS OF AGRICULTURE

3 1. Callahan, personal conversation, 1990.

3 2. Bearden, 1989.

4 3. Lillage, 1988.

4 4. "The Polar Solar Phenomenon," 1990.

4 5. Nieper, 1983.

6 6. Andersen, 1990.

10 7. Felleman, 1936.

10 8. *The Secret of Life of Your Cells*, Robert B. Stone, Ph.D.

CHAPTER 2. CHEMISTRY

28 1. John-David Yule, *Concise Encyclopedia of the Sciences*.

CHAPTER 5. ENERGY: THE BASIS OF LIFE

62 1. "Pottenger's Cats," by Francis M. Pottenger, M.D. An *Acres U.S.A.* summary by George E. Meinig, Desk Reference page 5260, November 1983.

62 2. Personal communication. This report has not been released because of feared litigation.

67 3. Deryatkov, 1985.

68 4. Report for the Bio-information Institute, October 1991.

CHAPTER 6. PLANT FUNCTION

73 1. Article by Edward L. Breazeale, et al in *Soil Science*, Vol. 71.

76 2. Books by Thomas Bearden, available from Tesla Book Company, P.O. Box 1649, Greenville, TX 75401, or Health Research, Box 70, Mokelumne Hill, CA 95245.

78 3. E.T. Whittaker, "On an Expression of the Electromagnetic Field Due to Electrons by Means of Two Scaler Potential Functions," *Proc. Lond. Math.* Soc., Series 2, Vol. 1, 1904, pp. 367-372, as well as writings by Thomas Bearden.

78 4. Fritz Albert Popp, "Photon Storage in Biological Systems," in F.A. Popp et al., *Electromagnetic Bio-Information: Proceedings of the Symposium*, Baltimore, Maryland: Urban & Schwarzenberg, 1979, pp. 123-49, as well as writings by Phil Callahan.

81 5. *Soil Microorganisms and Higher Plants*, available from U.S. Department of Commerce, Springfield, Virginia, publication No. TT-60-21126.

81 6. Krasil'nikov, 1958, p. 412.

CHAPTER 7. CONCEPTS IN MICROBIOLOGY

87 1. Krasil'nikov, N.A., *Soil Microorganisms and Higher Plants*, National Technical Information Service, Publication No. TT-60-21126, United States Department of Commerce, Springfield, Virginia 22151. The abbreviated citations in these notes are fleshed out as 640 bibliographical entries in Krasil'nikov's work. These cannot be reprinted here for want of space. Well-known scientists will be recognized. Less well-known names can be searched out in the Krasil'nikov reference material.

88 2. Ibid, page 3-4.
88 3. Ibid, page 265.
89 4. Ibid, page 295.
89 5. Ibid, page 311.
89 6. Ibid, page 281.
89 7. Ibid, page 264.
90 8. Ibid, page 264. Krasil'nikov cites his own work in 1940, and also states that an American specialist, Professor Clark (1949) "considers that micro-organisms living in the rhizosphere perform the same work as the intestines of animals." Also cited is "Academician Lysenko (1955)," who "thinks that the microflora of the root zone act as the digestive organ of plants."
90 9. Ibid, page 281.
90 10. Ibid, page 281. Krasil'nikov cites Weinstein (1954) and Gerretsen 1948 to support this paragraph.
91 11. Ibid, page 273.
91 12. Ibid, page 12. Krasil'nikov cites Dittmer for the 1937-1938 studies, and Savvinov and Pankova for the fescue plants study.
92 13. Ibid, page 293.
92 14. Ibid, page 294.
93 15. Ibid, page 294. Researchers cited in instance were Katznelson (1946) and Shtina, 1953, 1954.
93 16. Ibid, page 284.
93 17. Ibid, page 284.
94 18. Ibid, page 290.
94 19. Ibid, page 4.
94 20. Ibid, page 327.
94 21. Ibid, page 329.
94 22. Ibid, page 152. Krasil'nikov cited for the 1946-1950 finding, and Chalodny: for the "vitamins given off by air" revelation, circa 1955.
94 23. Ibid, page 305. Cited for this paragraph are Lockhead and his associates.
94 24. Ibid, page 238. The 1942 work was accomplished by Thompson.
94 25. Ibid, page 237. Krasil'nikov cites Ondratschek as the 1940 worker.
95 26. Ibid, page 235.
95 27. Ibid, page 221. The researchers cited by Krasil'nikov were Virtanen and Hausen.
95 28. Ibid, page 241.
96 29. Ibid, page 210. Cited scientists are Gollerbach and Polyanskii, circa 1951.
96 30. Ibid, page 237. Cited by Krasil'nikov for this work in 1944 were Burcholder, McWeigh and Mayer.
96 31. Ibid, page 237.
96 32. Ibid, page 323. The 1941, 1943 researchers were Nickerson and Thimann. The 1965 work was done by Burnett.
96 33. Ibid, page 296.
97 34. Ibid, page 259. The two workers were Kotiler and Garkovenko.
97 35. Ibid, page 256.
97 36. Ibid, page 232.
97 37. Ibid, page 253. In their research, Petrov (1912), Shulov (1913), Pryanishnikov (1952), Byalosuknya (1917), Hutchinson and Miller (1911), Klein and Kisser (1925), Virtanen and Laine (1937, 1946), Tanaka (1931), Miller (1947), Sander and Burkholder (1948), and Virtanen and Lincol (1936), all confirmed that plants absorb and benefit from amino acids.
97 38. Ibid, page 251. The B_1 researcher was Shavlovski, in 1954.
98 39. Ibid, page 251. The researcher was Kuhn.
98 40. Ibid, page 246. Kogl and Haagen-Smit were cited by Krasil'nikov.
98 41. Ibid, page 246. T. and H. Bonner are the citations for this 1948 work.

98 42. Ibid, page 248. The vitamin C, B_1, PP, H, pantothenic acid and B_2 studies were conducted by Matveev and Oveharov in 1940. Krasil'nikov's own studies were conducted in 1958.
99 43. Ibid, page 250. Krasil'nikov's citations are Stephenson (1951), Schopfer (1943), and Zeding (1955).
99 44. Ibid, page 277. Krasil'nikov made these determinations in 1958.
99 45. Ibid, page 276.
99 46. Ibid, page 256. The two 1903 researchers were Laurent and Laurent. The scientists who concentrated on glucose, saccharose, lactose and levulose were Pryanishnikov (1952), Shulov (1913), Byalosuknya (1917) and Krasil'nikov (1958).
99 47. Ibid, page 255. The 1955 work was attributed to Shavlovskii. The work on yeasts as effective metabolizers of phosphate was done by Akhromeiko and Shestakova in 1954.
100 48. Ibid, page 233. The biotin and thiamine work was done by Meshkov in 1952.
100 49. Ibid, page 231. The researchers cited by Krasil'nikov were Schmidt and Starkey, 1951.
100 50. Ibid, page 161. Work cited here was done in 1951 by Hoffman.
100 51. Ibid, page 160. The scientist was kuprevich.
100 52. Ibid, page 159. This work was done by Kejima in 1947.
101 53. Ibid, page 271. Cited to support this paragraph were Sabinin (1940), Winter and Rumker (1952), Harley 1952) and Virtanen and Tornianen (1940).
101 54. Ibid, page 162.
103 55. Ibid, page 263.
103 56. Ibid, page 263.
103 57. Ibid, page 260. Cited with this quotation, Peterburgskii (1954), and Krasil'nikov (1958).
104 58. Ibid, page 260. Coon was grown under sterile conditions by Ratner and Kozlov (1954). Gerretsen (1948) made the observation on phosphates.
104 59. Ibid, page 208. Ressel summarized Rothamstead data, and in 1933 came to the conclusion in the last paragraph.
106 60. Ibid, page 365. The 1924 attempts were made by Porter. Investigators who discerned protection for cucumbers and peas from Rhizoctonia, and wheat from fusariosis are Allen and Hoenseler (1935), Bisbu, James and Timonin (1933), Millard and Taylor (1927). Other biocontrol studies were conducted by Gregory, Allen, Riker and Peterson (1952), Rehn (1953), Gorrard and Lockhead (1938), Ark and Hunt (1941), Johnson (1935), Eaton and Rigler (1946), Allen and Haenseler (1935), Reeney and Garibaldi (1948), Kenknight (1941), and others.
106 61. Ibid, page 357. The 31 cultures were isolated by Petrusheva. The 33 species of phytopathogenic fungi mentioned were observed by Leben and Keitt. The wilt on cotton plant observation was made by Kublanovskaya. The other observations on actinomycete antagonists were made by Stevenson (1954) and Stessel et al (1953).
107 62. Ibid, page 358. Novogrudsk isolated the species. Stessel, Leben and Keitt isolated the 170 fungi.
107 63. Ibid, page 358. Anwar and Gregory et al were cited by Krasil'nikov.
107 64. Ibid, page 359.
107 65. Ibid, page 367. The cited results were corroborated by Kuzina (1951) and Kublianovskaya (1953).
108 66. Ibid, page 367. Weindling (1946), Garret (1946), and Winter and Rumker (1950), gave confirming data.
108 67. Ibid, page 368. The tests were made by Seiketon.
108 68. Ibid, page 370. Wood and Tevit (1955) were cited by Krasil'nikov.
108 69. Ibid, page 370.

109 70. Ibid, page 318. The researchers were Samtsevich et al, 1952.
109 71. Ibid, page 318. This study was made by Afrikyan.
109 72. Ibid, page 319. The scientist who worked out the percentages was Popova.
109 73. Ibid, page 319.
109 74. Ibid, page 371. The workers were Wagner (1915), Cholodnyi (1939), Yachevskii (1935), Vavilov (1919) Naumov (1940), Carbone and Arnaudi (1937). The last observation was made by Kramarenko (1949), and Gorlenko (1950).
109 75. Ibid, page 371.
110 76. Ibid, page 374.
110 77. Ibid, page 335.
110 78. Ibid, page 399.
110 79. Ibid, page 408. Morgentaller (1918) was cited.
110 80. Ibid, page 411.
110 81. Ibid, page 330.
110 82. Ibid, page 331.
111 83. Ibid, page 332-333. The chlorosis observation was made by Von Euler (1947) and Hagborn (1956). The flax studies were made by Berezova and Sudakova.
111 84. Ibid, page 338.
111 85. Ibid, page 335.
111 86. Ibid, page 335.
111 87. Ibid, page 339.
111 88. Ibid, pages 341-342.
112 89. Ibid, page 302. Citations for this paragraph are Khudyakov (1953), Novogrudskii (1936), Raznitsina (1942), Kuzina (1951 and 1955). The tea plantation observation was made by Daraseliya (1949). The root-rot disease findings were made by Hildenbrand and West (1941).
112 90. Ibid, page 306.
113 91. Ibid, page 272. The "special substances" were found by Brown and Edwards (1944).
113 92. Ibid, page 357.
113 93. Ibid, page 357. Viruses and tumors relate to the work of Waksman (1953), Kashkin (1952), Kurylowicz and Slopek (1955). The 11 cultures were found by Lockhead and Lauderkin (1949). The *Phythium graminicola* observations were made by Meredith and Seminink. The Chalaria quercina description was made by Stallings (1954).
113 94. Ibid, page 358. The *Pseudomonas phaseoli* description was made by Bisby (1919). The Fusarium types were isolated in 1935 by Khudiyakov, and other investigators promptly confirmed, namely Raznitsyna (1942), Berezova (1932) Korenyako (1939), and Kublanovskaya (1953).
114 95. Ibid, page 359. The cucumber finding was made by Gorlenko (1955).
115 96. Ibid, page 343.
115 97. Ibid, page 364.

CHAPTER 8. MICROBIOLOGY IN THE FIELD

117 1. Krasil'nikov, N.A., Soil Microorganisms and Higher Plants, National Technical Information Service, Publication No. TT-60-21126, United States Department of Commerce, Springfield, Virginia 22151, pp. 286, 287.
117 2. Ibid, page 109.
117 3. Ibid, page 82.
118 4. Ibid, page 77. The "new variants" thesis was stated by Kasikov (1950 and Kudryagvtsev (1954).
118 5. Ibid, page 78.
118 6. Ibid, page 35.

119 7. Ibid, page 299.
119 8. Ibid, page 299.
119 9. Ibid, page 301.
119 10. Ibid, page 301.
120 11. Ibid, page 302.
120 12. Ibid, page 308.
120 13. Ibid, page 118.
120 14. Ibid, page 373.
120 15. Ibid, page 383.
121 16. Ibid, page 227.
121 17. Ibid, page 225.
122 18. Ibid, page 224.
122 19. Ibid, page 223.
122 20. Ibid, page 221.
123 21. Ibid, page 220.
123 22. Ibid, pages 220-221.
123 23. Ibid, page 162.
123 24. Ibid, page 132.
124 25. Ibid, page 133.
124 26. Ibid, page 148.
124 27. Ibid, page 146.
124 28. Ibid, page 153. Krasil'nikov cited Dursanov (1954) and Samokhvalov (1952) for this work.
125 29. Ibid, pages 347, 348. Sushkina (1949) was cited as the authority for the opening statement.
125 30. Ibid, page 345. References are to the Northern Hemisphere.
125 31. Ibid, page 155.
126 32. Ibid, page 289. The worker was Linfold (1942).
126 33. Ibid, page 356.
126 34. Ibid, page 412.
127 35. Ibid, page 412.
128 36. *Vitamin B_{12} as a Growth Factor for Soil Bacteria*, *Nature*, June 23, 1951, page 1034.
128 37. Krasil'nikov, op cit, page 275.
128 38. Ibid, page 272.
128 39. Ibid, page 209.
129 40. Ibid, page 385.
129 41. Ibid, page 264.
129 42. Ibid, page 316.
130 43. *The Old Man and the Secret*, *Best of Business Quarterly*, Winter 1990, page 76.
130 44. Krasil'nikov, op cit, page 244. Cited were Murry (1948), Rakitin (1953).
130 45. Ibid, page 349.
131 46. Ibid, page 343. The "against the rules" suggestion came from Ishcherekov (1910), Whitney and Cameron (1914), Greig-Smith (1913, 1918), and others. Citations having to do with monocultures are Resses (1933), Pryanishnikov, (1928), Kossovich (1905), Hutchinson and Thaysen (1918).
131 47. Ibid, page 234.
131 48. Ibid, page 283.
132 49. Ibid, page 282.
132 50. Ibid, page 312.
133 51. Ibid, page 315.
133 52. Ibid, page 318.

CHAPTER 9. CLAY CHEMISTRY

139 1. *Physical Geography*, pages 74-75.

139 2. *Sources of Mud Problems*, page 6.
139 3. *Physical Geography*, page 56, *Sources of Mud Problems*, page 9.
139 4. *Physical Geography*, page 75.
144 5. *Sources of Mud Problems*, page 5.
144 6. *Borehole Instability*, page 18.
145 7. *Drilling Fluids Engineering Manual*, pages 4-12.
145 8. Ibid.
148 9. Ibid, pages 7, 8.

CHAPTER 10. CAREY REAMS' METHODS OF TESTING AND EVALUATION

150 1. Much of the material covered here was made a matter of audio record in numerous Reams short courses. Well worth reading is *Mainline Farming for Century 21*, by Dan Skow, D.V.M. and Charles Walters Jr.

CHAPTER 11. THE TERMINOLOGY OF CAREY REAMS

176 1. Krasil'nikov, page 250.
180 2. Generally available from ADAS, P.O. Box 1472, Huntsville, Alabama 35807.
180 3. In Popp et al, editors, Electromagnetic Bio-Information, Baltimore: Urban & Schwarzenberg, 1979, pages 123-149.
180 4. Whittaker, E.T., On An Expression of the Electromagnetic Field Due to Electrons by Means of Two Scaler Potential Functions, Proceedings of the London Mathematical Society, Series 2, Volume 1 (1904), pages 367-372; and *On the Partial Differential Equations of Mathematical Physics*, *Mathematical Annual*, Volume 57 (1903), pages 333-355.
180 5. Bearden, *Gravitobiology*.
185 6. Ibid, page 56.
186 7. Ibid, page 15.
186 8. Ibid, page 16.
186 9. Ibid, page 28.
186 10. Ibid, page 11.
186 11. See Andersen, Arden, *Life and Energy in Agriculture*; Reich, Wiehelm, *Cancer Biopathy*; and Bird, Christopher, *The Galileo of the Microscope*.

CHAPTER 12. WEEDS: CARETAKERS OF THE SOIL

190 1. Generally available from *Acres U.S.A.*, P.O. Box 9547, Kansas City, Missouri 64133.
191 2. Ibid.

CHAPTER 13. FERTILITY PROGRAMMING

224 1. For information on fruit trees, contact Marie Rasch, 280 Dickenson, Conklin, Michigan 49403.

CHAPTER 14. NUTRIENTS AND THEIR BASIC FUNCTIONS

238 1. TransNational Agronomy, Phil Wheeler, Suite 101, 470 Market St. SW, Grand Rapids, MI 49503. International Ag Labs, P.O. Box 788, Fairmont, MN 56031.

CHAPTER 17. MANAGEMENT: MAKING IT HAPPEN

252 1. Published in London by the University of London Press in 1919.
253 2. *Acres U.S.A.*, July 1991, page 2.

253 3. The film can be rented from Chart House Learning Corporation, 221 River Ridge Circle, Burnsville, Minnesota 55337, or from TransNational Agronomy, 470 Market Street, Suite 101, Grand Rapids, MI 49503.

CHAPTER 18. TURF, LANDSCAPE, AND ORNAMENTAL PLANT CARE

275 1. TransNational Agronomy, 470 Market Street, Suite 101, Grand Rapids, MI 49503, and International Ag Labs, P.O. Box 788, Fairmont, MN 56031.

276 2. Coles, Robin, *Pasture Cockchafers—On the Way Out?*, Acres Australia, number 3, page 43.

Appendix 1

REFRACTOMETER CHART

The chart on refractive indexes of crop juices was originally compiled by Carey Reams. Many people have asked about the readings listed in the "Excellent" column that are below 12. I believe and have found that these values would indeed represent excellent specimens of said crops, particularly in today's market, but I still contend that sub-12 readings are less than desired. On page 80 of my book *The Anatomy of Life and Energy in Agriculture,* I list the refractive index of various crops at or above which no disease or insect-pest infestation will occur.

REFRACTIVE INDEX OF CROP JUICES CALIBRATED IN % SUCROSE OR ° BRIX

Refractometers are easy to use even for inexperienced personnel. To make a determination, place two or three drops of the liquid sample on the prism surface, close cover and point toward any convenient light source. Focus the eyepiece by turning to the right or left. Read percent sucrose or solids content on the graduated scale.

	Poor	Average	Good	Excellent
FRUITS				
Apples	6	10	14	18
Avocados	4	6	8	10
Bananas	8	10	12	16
Cantaloupe	8	10	12	14
Cherries	6	8	14	16
Coconut	8	10	12	14

Grapes	8	12	16	20
Grapefruit	6	10	14	18
Honeydew	8	10	12	14
Kumquat	4	6	8	10
Lemons	4	6	8	12
Limes	4	6	10	14
Oranges	6	10	16	20
Papayas	6	10	18	22
Peaches	6	10	14	18
Pears	6	10	12	14
Pineapple	12	14	20	22
Raisins	60	70	75	80
Raspberry	6	8	12	14
Strawberry	6	10	14	16
Tomatoes	4	6	8	12
Watermelon	8	12	14	16
GRASSES				
Alfalfa	4	8	16	22
Grains	6	10	14	18
Sorghum	6	10	22	30
VEGETABLES				
Asparagus	2	4	6	8
Beets	6	8	10	12
Bell pepper	4	6	8	12
Broccoli	6	8	10	12
Cabbage	6	8	10	12
Carrots	4	6	12	18
Cauliflower	4	6	8	10
Celery	4	6	10	12
Corn stalks	4	8	14	20
Corn, young	6	10	18	24
Cow peas	4	6	10	12
Endive	4	6	8	10
English pea	8	10	12	14

Escarole	4	6	8	10
Field peas	4	6	10	12
Green beans	4	6	8	10
Hot peppers	4	6	8	10
Kohlrabi	6	8	10	12
Lettuce	4	6	8	10
Onions	4	6	8	10
Parsley	4	6	8	10
Peanuts	4	6	8	10
Potatoes, Irish	3	5	7	8
Potatoes, red	3	5	7	8
Potatoes, sweet	6	8	10	14
Romaine	4	6	8	10
Rutabagas	4	6	10	12
Squash	6	8	12	14
Sweet corn	6	10	18	24
Turnips	4	6	8	10

For reference, pure water has a reading of "0." Within a given species of plant, the crop with the higher refractive index will have a higher sugar content, higher mineral content, higher protein content and a greater specific gravity or density. This adds up to a sweeter tasting, more minerally nutritious food with a lower nitrate and water content and better storage characteristics. It will produce more alcohol from fermented sugars and be more resistant to insects, thus resulting in a decreased insecticide usage. Crops with a higher sugar content will have a lower freezing point and therefore be less prone to frost damage. Soil fertility needs may also be ascertained from this reading.

Appendix 2

BIOENERGETICS PROTECTS PLANTS AGAINST THE WEATHER

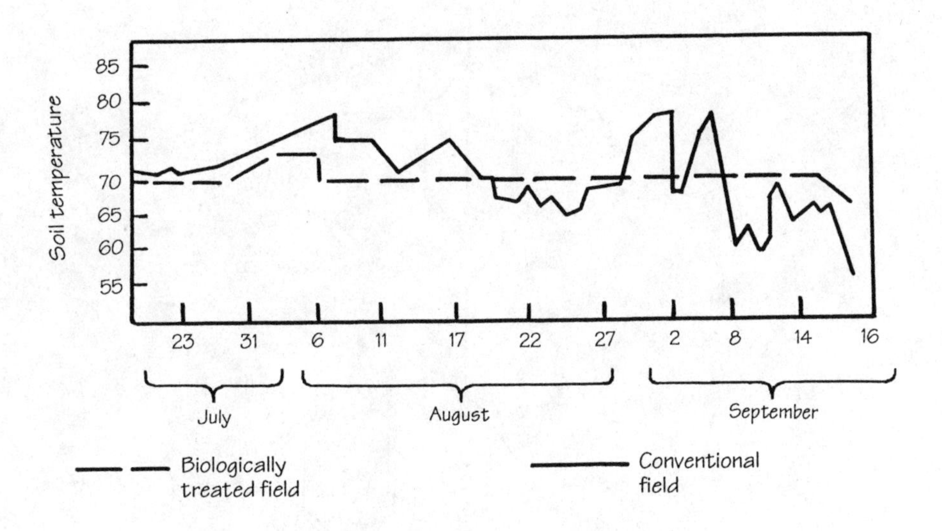

Keeping plants healthy with a nutritionally balanced program can maintain the soil at a near-steady 70° F., regardless of variations in air temperature. Compared here are the soil temperatures of a biologically treated field and a conventionally treated field in Bureau County, Illinois, in July, August, and September 1984. Typically, temperatures can vary by more than 30 degrees during the summer growing season in the midwest.

Source: Larson Farm Management

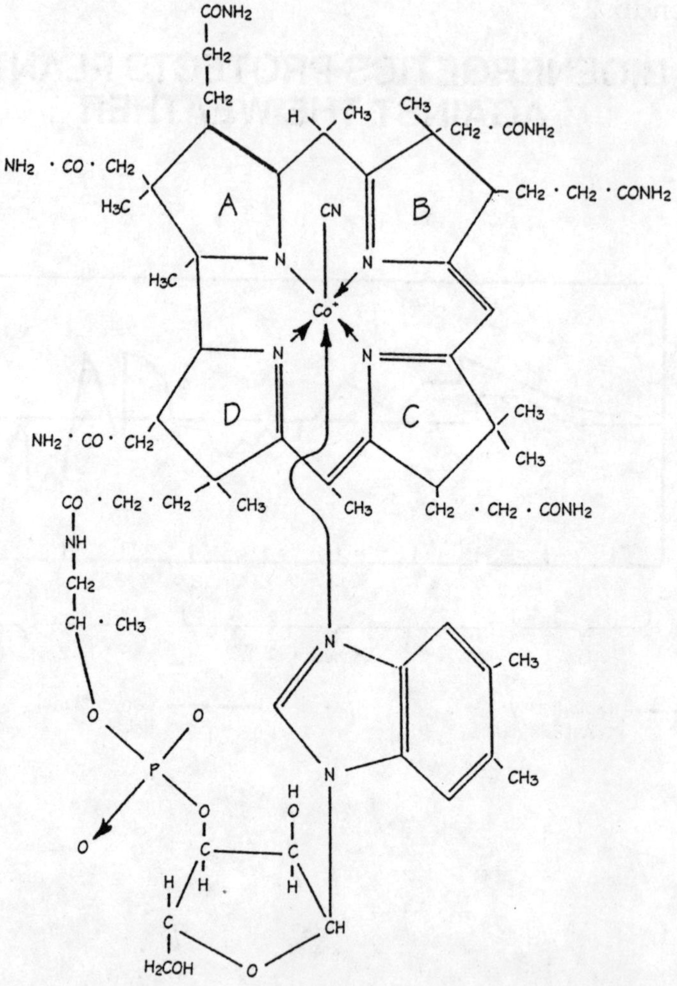

Many microorganisms are unable to synthesize vitamin B_{12} and require it for growth, thus providing the basis for a microbiological assay for the vitamin. Shemin and his associates have shown that Actinomyces, which synthesize vitamin B_{12}, can make the porphyrin portion of the vitamin molecule from aminolevulinic acid in a manner analogous to the utilization of this acid for the synthesis of the porphyrin of hemoglobin. The additional methyl groups present in the vitamin are derived from methionine.

Appendix 3

SCALE ADJUSTMENT OF REFRACTOMETER

1

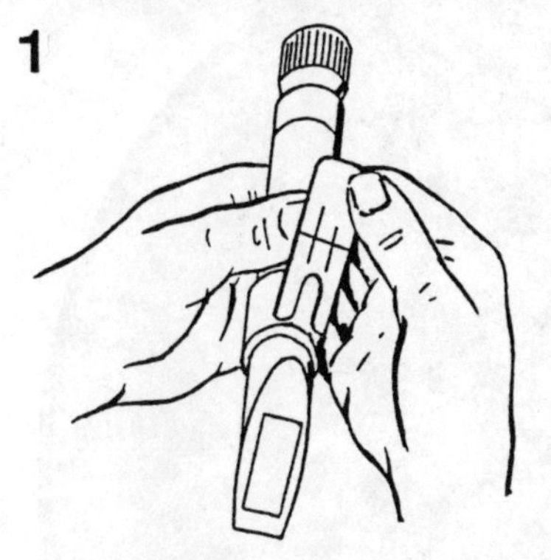

Open the daylight plate.

2

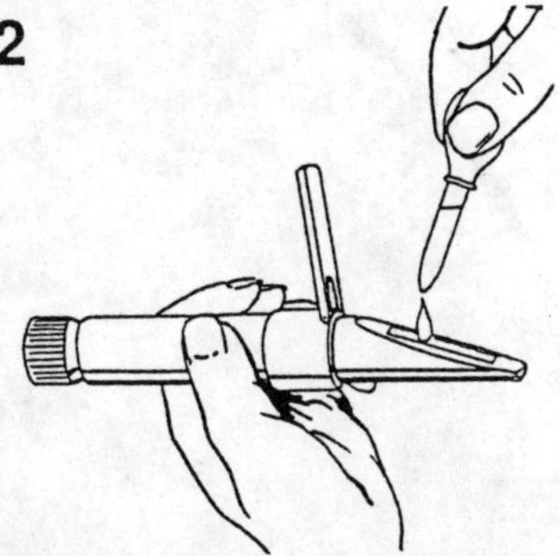

Put one or two drops of distilled water on the prism.

3

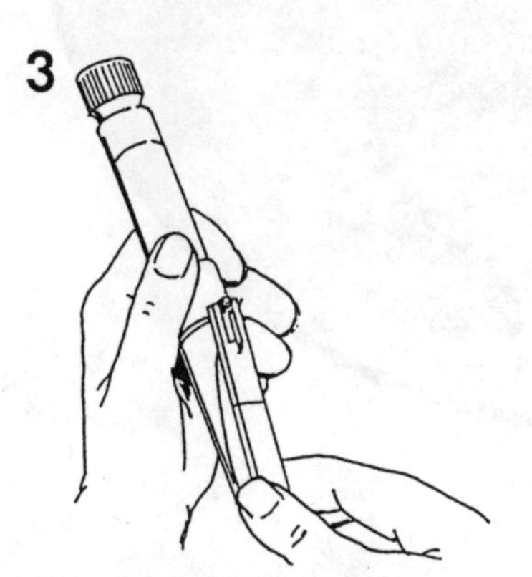

Close the daylight plate gently.

4

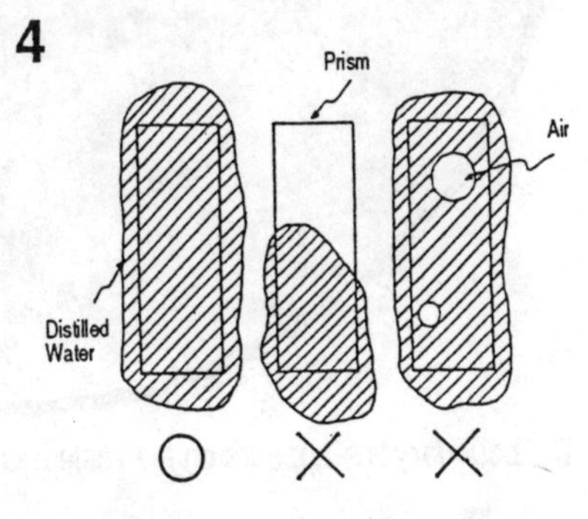

The distilled water must spread all over the prism surface.

5

Look at the scale through the eyepiece.

6

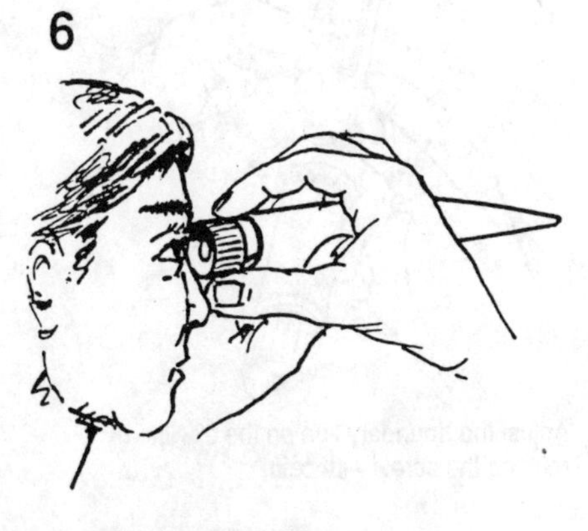

Turn the eyepiece to focus the scale.

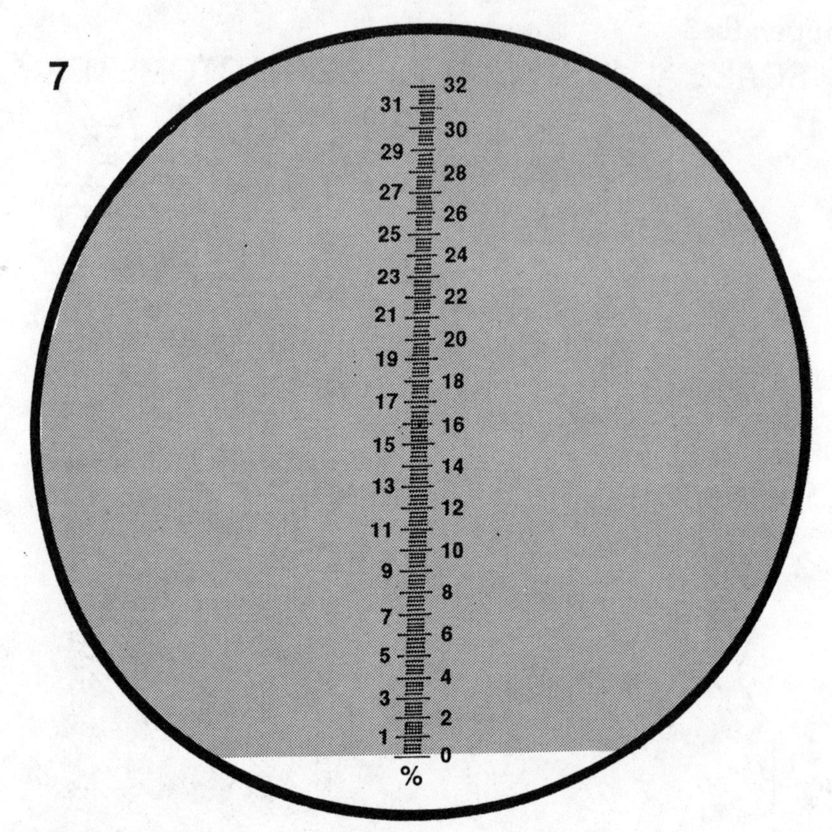

The boundary line appears on the scale near 0%.

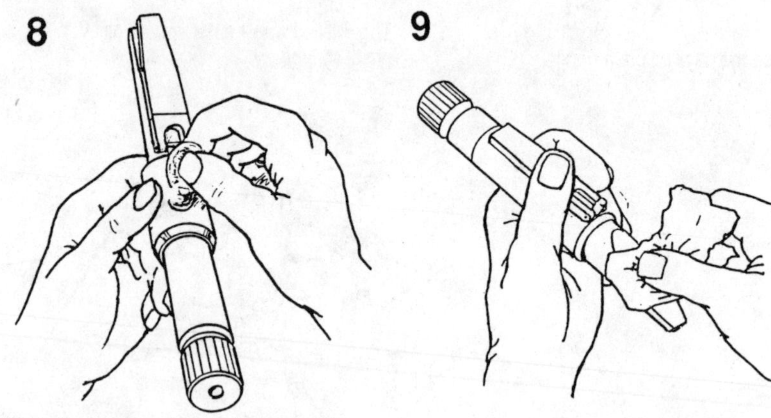

Adjust the boundary line on the 0% line by rotating the screw with coin.

Wipe off the distilled water perfectly.

HOW TO MEASURE DISSOLVED SOLIDS

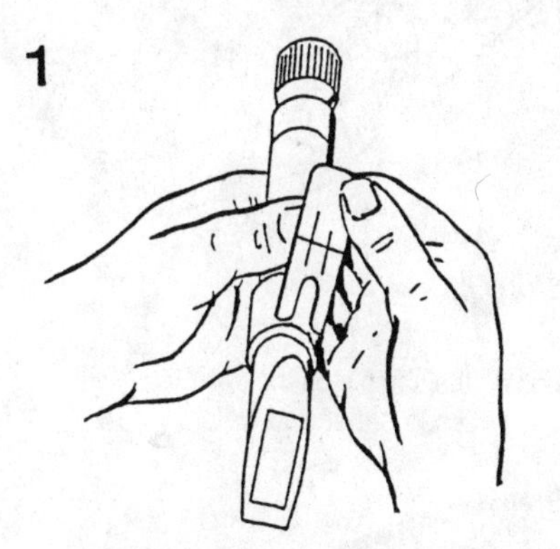

Open the daylight plate.

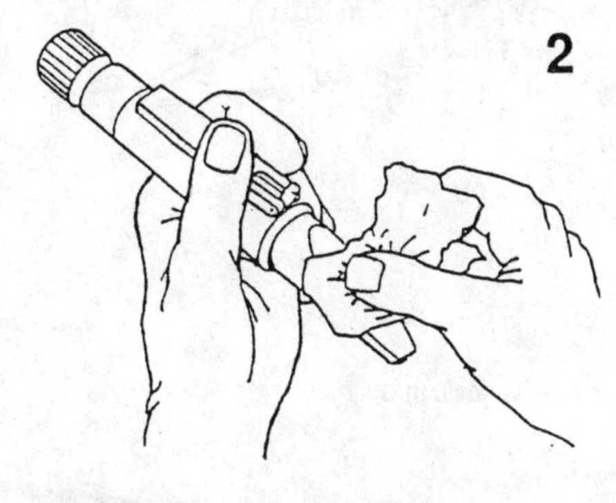

Wipe clean the prism surface and the daylight plate.

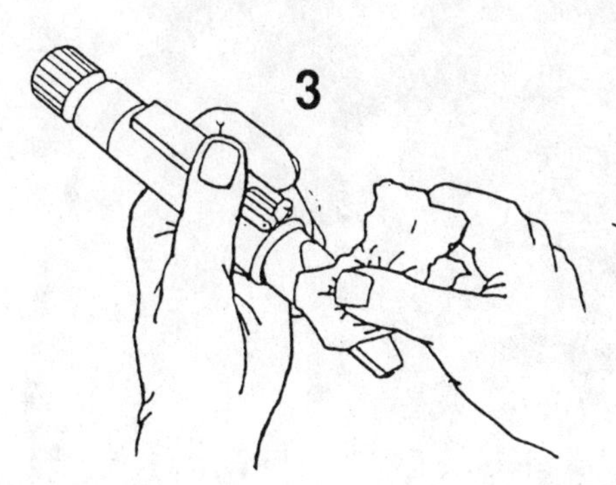

Still more, wipe off perfectly the moisture.

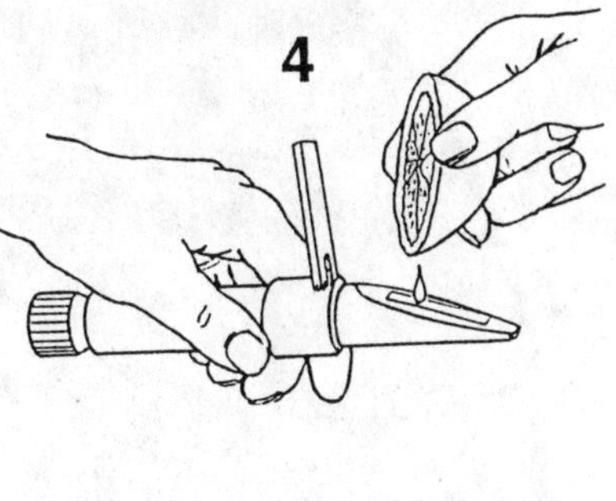

Put one or two drops of sample on the prism.

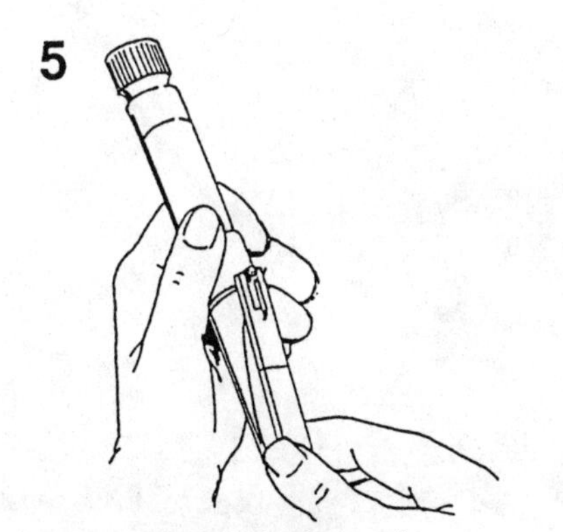

Close the daylight plate gently.

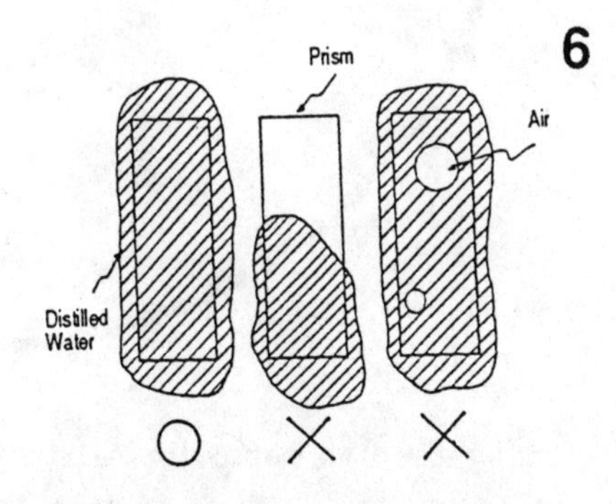

The sample must spread all over the prism surface.

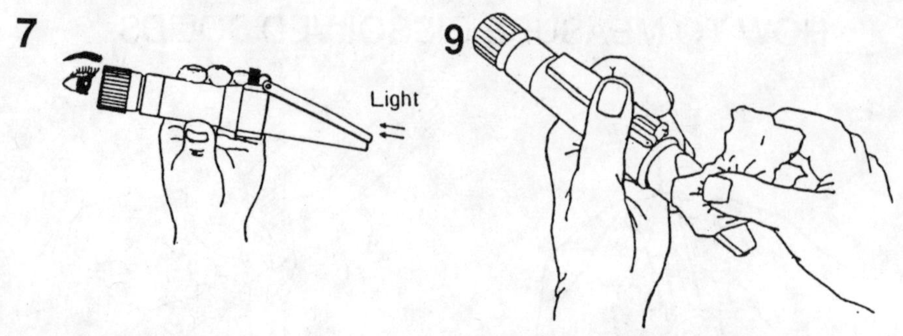

Look at the scale through the eyepiece.

Wipe off the sample on the prism with a *tissue paper and water.

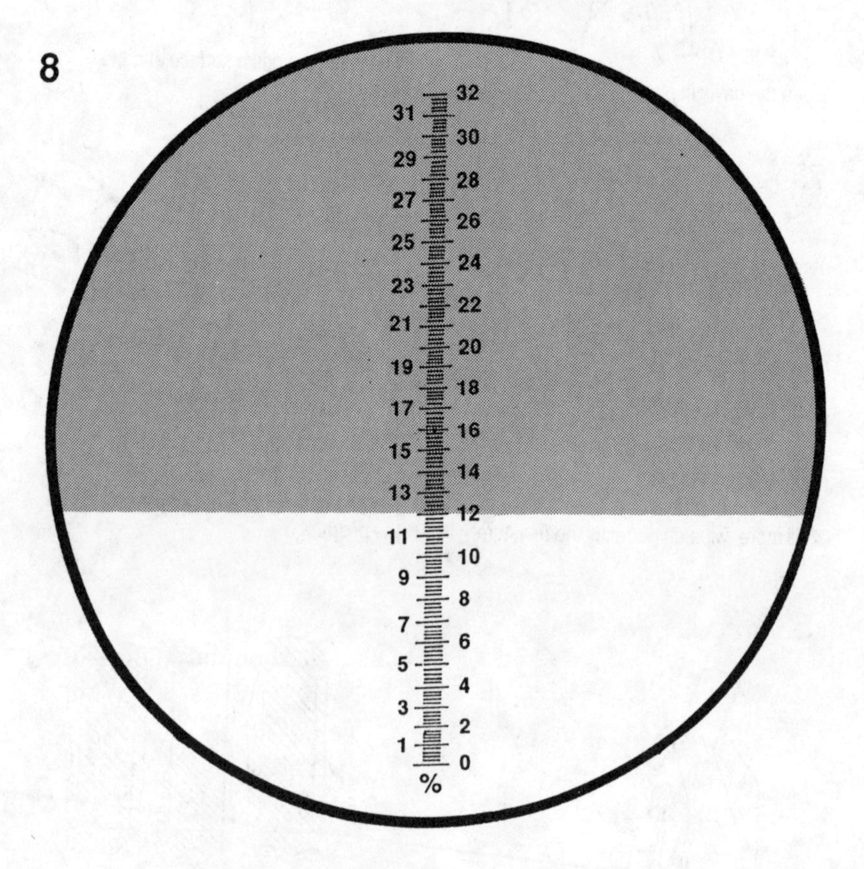

Read the scale where the boundary line indicates.

*Use a soft tissue paper.

INDEX

21st Century Science & Technology, 55

A&L Labs, 158-159, 222
Abrams, Albert, 3, 65, 172, 338
accounting, 253-260
acid-base reaction, 73
acids, 23, on skin, 47
Acres U.S.A., 62
actinomycetes, 108, 111, 113, 117-118, 120
adenosine triphosphate (ATP), 27, 28
aerobic process, 124
Ag Labs, 275
aging process, reversing the, 149
Agri-SC, 214
agriculture, conventional, since WWII, 343-344
air, nutrition from, 194
Albrecht, William, xi, xv-xviii, 174, 330
alcohol, and rumen bacteria, 308
aldehydes, 46
alfalfa, 283
algae, 46, 86-115
alkali extracts, 101
Alkaline solutions, 7
Alpha Environmental, Inc, 130
American College of Orgonomy, 133
amino acids, 24, 101, 103
ammonium, ion, 17, nitrate, 17
Anatomy of Life and Energy in Agriculture, 191
Ancient Mysteries, Modern Visions, 329
anhydrous ammonia, 21, 46, 206, 218
animal agriculture, 301-321
anions, 21, 174-175, 183, 186
antennas, 6, 80, 323-324, plant, 181
antibiotics, 110
antimicrobial properties, 109
apples, 250
Applied Body Electronics, 186
Aschoff, Dieter, 65
atmospheric systems, 82, 83
atomic weight, 14-15
atoms, 12, 13, 15, 16
Azomite, 213
Azotobacter, 108, 119, 125, 133

Babylon, xviii
Backster, Cleve, 67
bacteria, 44, 107, 117-118
Barker, Joel, 192, 253
Baroody, Rose, viii
Baroody, Ted, viii
bases, 23, on skin, 47
Bastion, Henry, 172
Bearden, Thomas, 6, 13, 60, 76, 174, 176-180, 181, 184, 185-188
Becker, Robert, 66, 72, 251
Beckman Research Institute, 55
Berg, Nancy, viii
bioassay testing, 285
biochemists, 31
bioenergetics, 357-358
biogeny of soil, 88
biological antennas, 80
biological cells, 4
biological communication, 4
biological energy systems, 67
biological insect control, 170
biological theory of ionization (RBTI), 171
biology, 43-59
Biomedical Health Foundation, xi
Biomin, 216
bions, 65, 117, 186
biosphere maximum radiation, 6
Biotech, J.H., 216, 275
Bird, Christopher, 119

birds, 303
black rock phosphate, 213
black night shade, 198
black bindweed, 197
blackfoot, 106
Blake, John, xv
BlueKote, 318
Borehole Instability, 141
boron, 237, starvation, 111
breath test, 64-65
Breazeale's theory, 73
bristle-grass, green, 199
brix, 53, 54
brix-nectar connection, 53
broadleaf, 191-192
Brown Revolution, xvi
Brunetti, Jerry, viii, 319
Buchelew, Floyd, 53
Butz, Earl, xi

calcium, 6-7, 16, 155, 218, 230 carbonate, 22, hydroxide, 7, sulfate, 210, testing, 164
calcium DL, 216
calcium-to-magnesium ratio, 157, 159
Callahan, Philip, xvii, 3, 32, 40-41, 55, 63, 67-72, 75, 78-79, 83, 85, 154, 171, 191, 251, 265, 310, 322-329, 330
calves, 319
Cancer Biopathy, The, 119, 176
cancerous tumors, 66
candida, 311
carbohydrates, 23
carbon, 14-15, 23, 174, 235, dioxide, 124
carboxylic acids, 24
Carlson, Dan 182
Carthage, xviii
castor bean, 51
Cation Exchange Capacity (CEC), 134-135
cations, 21, 174, 177, 183
cell, 43-46, 67, 118, 177, 323, electrical charge, 76, 77, eucaryotic, 45-46, procaryotic, 45-46
cell culture, 66
cellulose, 28
charges, repel, or attract?, 185-186
chemistry, 11-30
chlorine, 16, 17, 240
chlorite clay, 135, 139
chloroplasts, 55
chromosomes, 48
clay, aging of, 139, degenerated, 142, 143, effect of salt on, 148, evolution of, 143, 144, non-swelling, 135, normal, 142, swelling, 135, water absorbing, 136
clay chemistry, 134-149
clay particles, association of, 138
clay-platelet hydration, 146
cobalt, 238
coccoid, 120
coenzyme, 121
common purslane, 198
common mullen, 198
compact disc mechanism, 40, 41
compost, 229, 297-300
Compost Manufacturer's Manual, 297
compounds, 12
conductivity, 76
conductivity meter, 295-297
Congressional Record, xi
contamination, selective, 78
conventional system, 2, 3
Cooley, Sue, viii
Cooper Medical Clinic, 65
Cooperative Extension Services, xii, 169, 170
copper, 192, 236
corn, 51, 280-283
corporate agriculture, xvii
cotton plant wilt, 107
crystal formation information, 66
cycles, 243-245
Cygon, 128

dairy, 317

dairy cows, 303, 317
dandelion, 199
Darras, Dr., 65
Dechamp, Antoine, 172
Detox, 320
Devernejoul, Dr., 65
diamagnetic soils, 83
diatomaceous earth, 93
diatoms, 93
dicotyledons, 53, 56, 57
dielectric coating, 78, 79
dielectric constant, 67
Discovering the Future: The Business of Paradigms, 192, 253
DNA, 4, 27, 44, 55, 77-78, 82, 327
dolomite lime, 218
dry solubles, 208
Durber, Russell, 298
Dynamin, 213, 316

earthworm bioassay, 286-288
electrical pollution, 66
electrical wiring, 33, 34
Electro-Chemical Treatment of Seeds, A Manual of the, xv
electroencephalogram (EEG), 33
Electromagnetic Blood Test, 65
electromagnetic coupling, 180
Electromagnetic energy, externalized, 186
electromagnetic spectrum, 36-37, 78
electromagnetism, 35
electronic antennas, 80
electroretinogram (ERG), 32
ELF radio waves, 68
energetics, 3-4
energy, 60-71, 154, 174-175
energy field, 64
entropy, 9-10
enzymes, 241, in grain, 316
ERGS, 151, 153-154, 160, 227, 287
Erthrite, Inc, 272
eucaryotic cells, 47
exponent, 36

extrarhizal feeding, 127

Factor R, 121
Farm Journal, xviii, 87
Farmer's Every-Day Book, xv
feed quality, 313-315
Fertile Crescent, xviii
fertility program, 204-229
fertilizer, 29, 30, basic programs, 218, for specifics, 190, mixing, 224-225
Feynman, Richard, 76, 179
field bindweed, 199
field trip tips, 280
flour, 63
flower, as focusing aparatus, 325
foliar feeding, 126, 127
food, energy of, 63
Foord, Alarin G, 62
Fordham, Bill, 197
foxtail, giant, 199
Freedom Formulations, 273
frequency, 82, 83, 181
fresh cow, 321
fungi, 46-47, 86-115, 92, 106-107, 118, 122

Galileo of the Microscope, The, 119
Garvey Center, 63-64
genes, 48, 49
genetics, 48-50
gentian violet, 318
glycine, 24
glycogen, 28
goal setting, 260-262
gold, 15
grasses, 192
Gravitobiology: A New Biophysics, 179, 185-188
Green Revolution, xi, xii, xvi, xvii
Gross, Henry M, 333

Hamerman, Warren, 4, 55
harmonic phenomena, 4
Heisenberg, 10

helium, 17
herbs, 319-320
Hieronymus, T. Galen, 186, 333
hogs, 303
humic acids, 147, 148
humic substances, 123, 124
humus, 46, 121
Hunt, Valerie, 186
Huntington Memorial Hospital, 62
hybridization, 50
Hyde generator, 339
hydrated lime, 22
hydrogen, 16-17, peroxide, 318
hydrophilic, 22
hypertonic environment, 46
hypochlorous acid, 46

IAT, 216
illite clay, 135, 140
immune system, animal, 312
infrared, 4, 251
inhibitors, 111
insects, as inspectors, 78, nature of, 2, signal trackers, 325
Institute of Microbiology, Academy of Science, Moscow, 87
intercellular communications, 54
International Ag Labs, 53, 58
IR radiation, 251
iron, 192, 236

J & J Agri-Products, 214, 272
johnson grass, 200

kaolinite clay, 135
Kastychev, P A, 123
Kaznacheyev, Vlail, 3, 83, 179, 251
keratin, 47
ketosis, 312
kidney bean, 51
kinesiology, 64
Kirlean photography, 65
Knapp, Dieter, 65
Kotelev, 102
Kouchahoff study, 63
Krasil'nikov, N.A., 6, 86-115, 87, 116-133
Krebs cycle, 28, 243-244
Kuhn, Thomas, 192

labware, cleaning, 166
Lactobacillus, 127
Lakhovsky, Georges, 3, 172
lamb's quarters, 200
LaMotte testing kit, 151
Larson, David, 53, 58, 254-255
Larson Farm Management, 53, 58
leaf, cross section, 56, structure, 53-56
leukocytes, 63
Levanson, Norman, 179
life, 174
light, 74, 76
linear, 7, 8, 9
linear physics, 2
lipase, 99
liquid crystal antenna, 69
Logue, Gene, viii
Lopez, Andy, 273
Love, Medicine, and Miracles, 186
lucerne, 119, 120
lysine, 24

magnesium, 6, 22, 239, testing, 164
magnetic field, 73
magnetic susceptibility, 5, 6
Mainline Farming for Century 21, 191
maltase, 118
manganese, 192, 235
Manual of Drilling Fluids Technology, 135, 136, 137, 138, 140, 145
Manual of the Electro-Chemical Treatment of Seeds, 252
manure, 121
manure tea, 130
Martin, James, 130
Masterful Persuasion, 71
mastitis, 320
McCaman, Jay, 193, 223
Mercier, Charles, xv, 252

Michigan State University, 246
microbes, training, 118
microbial antagonists, 108
microbial biocoenoses, 89
microbial societies, 88, 91
microbiology, 86-115, 116-133
Microcal, 216
microorganisms, 47-48, 86-115, 116-133
microwaves, 66-67
milfoil, 200
milkweed, 200
millhouse units, 178-179
mineral fertilizers, 111
mitogenetic radiation, 6
mitosis, 4
mixing fertilizers, 224-225
molecules, 12, 13
molybdenum, 237
monocots, 53, 56-57
Morgan extract, 161
Morgan procedure, 151
multiple inputs, 8
muriate of potash, xii, 17
mushrooms, 58, 59
mycobacteria, 117, 120, 122
mycococci, 117
mycolytic bacteria, 112
mycorrhiza, 92
mycotoxins, 318
Mycrobial Studies, 113

N-P-K, 96-97, 128, 130-131, 204, 211, 216, 227
Naessens, Gaston, 119, 186
nature, 2-3, energetic symbiont, 173
Nature's Silent Music, 329, 335
neurotransmission, 4
Nieper, Hans, 76
nitrate ion, 17, nitrogen, 152
nitrogen, 16, 23-24, 100-101, 152, 218, 223-235, testing, 163-164
nitrogen cycle, 243
Nitromax, 221, 272
Nitromike, 320
Nitron, 273, 275
non-linear, 6, 8-9
non-sporeforming bacteria, 113
Noordenstrom, 66
Northern, Charles, xi, xii, xv, 171
Nourish Industries, 273
nucleic acids, 27
nutgrass, 200
nutrient flows, 74
nutrients, 82, 230-242
nutrition, 49, 313
Nutrition and Your Mind, 170, 259

Oersted, Hans Christian, 73
open-pollinated corn, 49, 50
optical wave guides, 76
organic compounds, 99
ornamental plant care, 271
orthophosphates, 145
oscilloscope, 67
osmotic mechanism, 147
oxygen, 16, 242, 312

paradigm, 1, old, 3, 9-10
paramagnetism, 83
pathogenic microbes, 117
pathogens, 110
Peace Corps, xviii
pedology, 88
periodic table of elements, 14
pesticides, resistance to, 50
petrochemistry, xvi
pH, 6, 7, 17, 18, 150-151, 153-154, 290-295
phosphate, 129, 212, 218, 231, ion, 25
phosphate-to-potash ratio, 152, 160, 191-192
phospho-carbonate complex, 22
phosphoric acid, 17
phosphorus testing, 165
phosphorus-32, 102
Photon Storage in Biological Systems, 180
Photonic Ionic Cloth Radio

Amplifier Maser, 67
photons, 4, 40, 78, 180
photosynthesis, 4, 6
physical properties, 12
Physical Geology, 139
physics, 2, 31-42
pig study, 62
Pike, Bob, 161
Pike Lab Supplies, Inc. 166
Pioneer Hybrid International, 259
pitch of C, 183
plant structures, 50-52
plant function, 72-85
plant growth, non-linear, 75
plant tissue testing, 155-157
poison ivy, 201
polar solar phenomenon, 4
politics, 340-344
pollens, 98
pollination, 53
polygraph, 67
polyphosphates, 145
Popp, Fritz-Albert, 4, 40, 72, 83, 180, 186
Popp noble gas engine, 339
potash, 232
potassium, 218, hydroxide, 23, testing, 165
potatoes, 222-223
Pottenger, Francis M, 61
Pottenger study, 61, 62
Pottenger's Cats, 61
proactinomycetes, 117
protein, 24, 306, factor, xi, synthesis, 4
putrefactive bacteria, 113

quackgrass, 193, 201

radio waves, 38
Radionic Association, in England, 335
radionics, 9, 65, 69, sources, 334-335
ragweed, 192, 201
Ranck, Floyd, 298
rations, feed, 190
RBTI, 187
Reams Soil Test, 69, 70, 72, 239
Reams testing method, 150-168, 212, 218
Reams, Carey, xv-xvii, 20, 75, 83, 85, 90, 169-188, 191, 328, 330
Reams' terminology, 169-188
record management, 267-268
red root pigweed, 201
reductionism, 2, 3, mindset, 173
refractometer, 191, 354-356, 359-362
Reich, Wilhelm, 65, 84, 118, 172, 176, 186
Reich Blood Test, 65
Report on Radionics-Science of the Future, 335
respiration, 4
rhizosphere microflora, 89, 90, 91, 93, 131
RNA, 27
Robertson, Hersel, viii
rock phosphates, 208
root, 56, excretion, 91-92, enzymatic activity of, 99, 100, secretions, 100, 101, zone, 89, 91
root-nodule bacteria, 119
Rothamstead, 104
ruminants, 305

salt, effect on clay, 148
scaler waves, 78
scanner testing, 330-335
scientific notation, 36
Secret Life of Plants, The, 72
seed, 183, structures, 50, 51, 52, treatment of, *A Manual of the Electro-Chemical Treatment of Seeds*, xv
selective contamination, 78
shepherd's purse, 202
Siegel, Bernie, 186
silicon, 239
Sims, Fletcher, 298
sine wave, 183-184

Skow, Dan, viii, 53, 58, 69, 90, 98, 128, 156, 161, 192, 220, 303
sodium, 154, chloride, 17, , ion, 17
soil, 2, environment, 105, extract/filtrate, 161-163, sampling 166, testing 166
Soil Microorganisms and Higher Plants, 6, 87, 116-133, 193, 337
Soil Treat, 221
Sources of Mud Problems, 139, 146, 147
soybeans, 283
spin, 179
sporeforming bacteria, 92
sporiferous bacteria, 109
Spray Tech Oil, 206, 209
Stanford University Medical School, 65, 333
static, 79
Steiner, Rudolf, xvi, 84
stem, 56, structure, 53-56
Structure of Scientific Revolution, The, 192
subtle energies, 67
succulents, 192
sugars, 24-27, 29, 46, 276-278
sul-po-mag, 57, 317
sulfate test, 152
sulfur, 238
Susequehanna Tree Surgeons, 273
system, the ag, 195-196

tachyon, 75
Tesla, Nikola, 172, 188
test foliar spraying, 247-248
test solution formulation, 289
testing, calcium, 164, magnesium, 164, nitrogen, 163-164, phosphorus, 165, plant tissue, 155-157, potassium, 165, Reams method, 150-168
tetrahedral, structure of, 136
topsoil, lost, xii
total mixed rations (TMR), 305
toxic bug killers, 92
toxins, 111
training microbes, 118
Trans Dextox, 221
Trans-Flo, 214
TransNational Agronomy, viii, 207, 216, 273, 320
trees, as capacitors, 84
tuned circuit, 83
tuner, plant, 181
tuning, 67
Tuning In To Nature, 32, 191, 329
turf management, 271-278

ultraviolet, 4
University of Chicago, 66
University of Florida, 62
University of Southern California, 62
Upton, Curtis, P, 333
urea, 17, 27
uridine triphosphate (UTP), 28
Urt, H.T., 71
USDA, 169, 170, 334
uterus, infected, 323

vaccines, 313
velvetleaf, 191, 202
Venman's, 273
vetch, wild, 203
vision, 4
vitamin content, of plants, 130
vitamins, 98-100, 241, 319, for soil, 210
Vogel, Marcel, 39
Voll instrument, 65
Von Neuman, John Eric, 179
Voronin, M S, 113

Walk in the Sun, A, 329
Waller, 73
Walters, Charles, viii, 189, 191
Ward, Ron, viii
water, 241
Watson, George, xvi, 170, 259
wave guides, 79

wavelengths, 35
weeds, 2, 189-203, as helpers, 78
Weeds, Control without Poisons, 189
Weeds! Why?, 193
western nutgrass, 202
wheat, xv
Whittaker, E.T., 172, 179, 186, 188
wholistic, 3
Wilbanks, 66
wild carrot, 202
wild dewberry, 203
wild oats, 203

X-ray, 67

yeasts, 118
Yin/Yang energies, 175-180, 182

Z-Hume, 221
zinc, 238